Geometry Civilized

Arîstippus Philosophus Socraticus, naufragio cum ejectus ad Rhodiensiu
litus animadvertisset Geometrica schemata descripta, exclamavisse ad
comites ita dicitur, Bene speremus, Hominum enim vestigia video.
Vitruv. Architect. lib. 6. Praef.

Fig. 0.0.1 Geometry as evidence of civilization. The inscription reads: 'Aristippus the Socratic philosopher being shipwrecked in Rhodes noticed some diagrams drawn on the beach and said to his companions, "We can hope for the best for I see the signs of men."' Euclid, *Opera omnia* (1703), frontispiece.

Geometry Civilized

History, Culture, and Technique

J.L. Heilbron

CLARENDON PRESS · OXFORD
1998

Oxford University Press, Great Clarendon Street, Oxford OX2 6DP

Oxford   New York
Athens   Auckland   Bangkok   Bogota   Bombay
Buenos Aires   Calcutta   Cape Town   Dar es Salaam
Delhi   Florence   Hong Kong   Istanbul   Karachi
Kuala Lumpur   Madras   Madrid   Melbourne
Mexico City   Nairobi   Paris   Singapore
Taipei   Tokyo   Toronto   Warsaw
and associated companies in
Berlin   Ibadan

Oxford is a trade mark of Oxford University Press

Published in the United States
by Oxford University Press, Inc., New York

A catalogue record for this book is available from the British Library

Library of Congress Cataloging in Publication Data
Heilbron, J.L.
Geometry civilized : history, culture, and technique/J.L. Heilbron.
Includes bibliographical references and index.
1.   Geometry, Plane.   I. Title.
QA455.H413   1988   516'.05–DC21   97-28432
ISBN 0 19 850078 5 (Hbk)

Typeset by EXPO Holdings, Malaysia

Printed in Great Britain by
The Bath Press, Avon

Preface

Our frontispiece, showing the shipwrecked philosopher Aristippus, represents geometrical proof as the sign of civilization. According to this conceit, persons of culture will recognize in the mathematical diagrams in the sand the features that distinguish them from their fellow citizens. Deciphered, the diagrams indicate that civilized people require proof of what is obvious to everyone else (a straight line is the shortest distance between two points, Euclid I.20); that they deduce a lot from a little (the sum of the angles in a triangle is a straight angle, Euclid I.32); and that they base aesthetics on arithmetic (the divine proportion is a ratio of areas, Euclid II.11).[1] To be sure, this doctrine confuses culture with quantitative thought, the whole with the part; but it conveys at least one truth, that, for many centuries, geometry was part of high culture as well as an instrument of practical utility.

The source of the story about Aristippus, the Roman architect Vitruvius, made the philosopher's diagrams indicators not only of culture but also of wisdom. Accordingly, Aristippus decided to remain with his fellow philosophers in Rhodes. He bade his companions carry this message back home: 'Children ought to be provided with property and resources of a kind that could swim with them even out of a shipwreck'. Hence it was, still according to Vitruvius, that the Athenians required grown children to support their elderly parents only if the parents had educated them properly. 'All the gifts which fortune bestows she can easily take away; but education, when combined with intelligence, never fails, but abides steadily on to the very end of life'.[2] For many centuries, a good education included the study of Euclidean geometry. Most of the so-called geometry books now used in schools are not agents of culture, wisdom, or even education, however, but opportunities to exercise students in the use of pocket calculators.

Our emblematic frontispiece was commissioned by the Oxford University Press for the magnificent edition of Euclid's *Opera omnia* that it issued in 1703. The same engraving, redrawn with different diagrams on the shore, appears in the Press's equally sumptuous

---

1. These propositions are discussed below at §1.1 (Figure 1.1.2), §2.3 (Figure 2.3.1), and §5.3 (Figure 5.3.8).
2. Vitruvius, *Architecture* (1914), 167–8.

editions of the *Conics* of Apollonius and the surviving works of Archimedes. The shipwrecked philosophers relieved by the sight of mathematical drawings in the sand may therefore be taken as a symbol not only of the place Euclid held in general culture, but also of the care the Oxford University Press exercises on the geometry books it publishes.[3] I am much obliged to the Mathematics Editor, Elizabeth Johnston, to her assistant Julia Tompson, and to the Press's archivist Peter Foden, for their adherence to this grand tradition.

Several of the unusual illustrations from early modern sources were found with the generous help of Tony Simcock, Librarian of the Oxford Museum for the History of Science. It is a pleasure to thank him and the Keeper of the Museum, Jim Bennett, for making available the books from which the illustrations were made. The remainder of my indebtedness for the figures is acknowledged in their captions.

This book began life as part of a project supported by the United States National Science Foundation to introduce history into the teaching of science, and science into the teaching of history, in American high schools. I thank the Foundation and my colleagues in the project, Daniel J. Kevles and Sheldon Rothblatt, for their forbearance as the book outgrew the limits assigned to me. I also thank my former colleagues in the administration of the University of California, Berkeley, especially the chancellor, Chang-Lin Tien, for ignoring my lapses from the business before us when a geometrical problem seized my mind, and Diana Wear, who showed her usual artistry in preparing the final manuscript. Most of all, I thank those dearest to me, who could not ignore my obsession, and yet indulged me in it.

May 1997   J.L.H.

3. A recent collection of papers published by the Mathematical Association of America and intended to integrate historical research with mathematics teaching also, appropriately, has Aristippus and his band on its cover. Calinger, *Vita* (1996); Rickey, in ibid., 251, 255–6.

Contents

The colour plate section can be found between pp. 152-153

1 An old story

We owe geometry to the tax collector. According to the old Greek historian Herodotus, the Egyptian king Sesostris divided all the land in Egypt equally among its inhabitants in return for an annual rent. But every year the flood of the Nile washed away parts of the plots. 'The country is converted into a sea, and nothing appears but the cities, which look like the islands in the Aegean'. Those whose land had disappeared into the waters naturally objected to paying the rent on what they had lost. The good king Sesostris thereupon decreed that anyone so affected might appear before him. Herodotus: 'Upon which, the king sent persons to examine, and determine by measurement the exact extent of the loss; and thenceforth only such a rent was demanded of him as was proportionate to the reduced size of his land. From this practice, I think, geometry first came to be known in Egypt, whence it passed into Greece.'[1]

What happened next we have from a medieval scholar, Richard Kilwardby, whose fanciful account in his widely read *Origin of the sciences* added another element to the sociology of geometry. The Nile surveyor, says Kilwardby, had time to consider the relationship of lines and angles when the flood that provided his livelihood subsided. 'This new man was still a native of the soil and not yet hungry for land ... by nature eager to know, he perceived the measurable dimension in continuous bodies ... and through assiduous study discovered the basis of determining not only lengths and breadths, but also heights, both accessible and inaccessible.'[2] Hence developed a serviceable practical science of land measurement, which passed to the Romans as 'agrimensura', or the art of measuring fields, and thence to the middle ages. Textbooks written by Roman *agrimensores* provided most of the little information about geometry available in Europe during the so-called dark ages.[3]

'Geometry' expresses in Greek what the Greeks received from the Egyptians. It retained its root meaning of land measurement or surveying at least into the late middle ages.[4] That expresses in Greek what the Greeks received from the Egyptians. But geometry also

---

1. Herodotus, ii, 97, 109, tr. Rawlinson (1858), *2*, 161, 179; How and Willis, *Commentary* (1912), *1*, 221–2.
2. Kilwardby, *De ortu* (1976), 29.
3. Curtze, *Abh. Gesch. Math., 8* (1898), 29–68; Cantor, *Die römischen Agrimensoren* (1875).
4. Smith, *History* (1958), *2*, 273, dates the first modern usage to 1482.

Fig. 1.0.1 'Geometry sets the distances and limits of the earth, distinguishes the parts, and surveys the mountains and rivers.' Thus our subject as depicted by a Flemish artist, Marten de Vos, around 1600. Note the towers in her headress, to suggest application to military architecture; and the snake and toad, perhaps to indicate the general opinion of mankind toward mathematics. From Shea, *Storia delle scienze*, 2 (1992), 283.

Terrarum ſpatia, et metas GEOMETRIA *ponit*   6
Diſtinguitq; plagas, monteſq; ac flumina luſtrat.

signifies a preparatory science, a mental discipline, an abstract discourse, and the exemplar of rigorous reasoning. Geometry in this second guise is the invention of the Greeks. If geometry had the everyday practical certainty of death and taxes in Egypt, in Greece it passed for unassailable truth. That is because the Greeks, or rather Euclid, who codified the subject around 2300 years ago, excluded from it the measurement of things and transformed it into a game

Fig. 1.0.2 The area of a triangle, according to the Rhind Papyrus (ca. 2000 BC) from Chace *et al.*, *The Rhind papyrus* (1927), problem 51, vol. 2, plate 73.

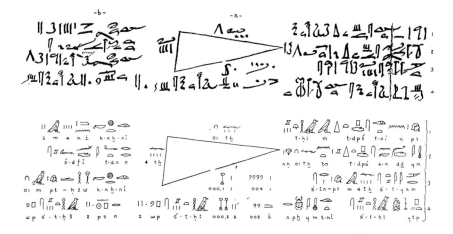

with its own unchallengeable rules, definitions, and presuppositions or axioms, supporting a structure of abstract propositions and rigorous proofs. The result was a monument to human thought and, too often, a torment to high-school students.

Students have difficulty imbibing the spirit of proofs because their need and purpose are not obvious. The proof of this assertion follows from the fact that all the classical civilizations—Egypt, Babylon, India, China—had a practical geometry, but none treated geometry as a deductive science. The papyri, clay tablets, and other written material that have come down to us invariably state problems in numbers and solve them by recipes. Nowhere do they give the reason for anything. For example, the Rhind papyrus, which dates from about 2000 BC, some 17 centuries before Euclid, directs anyone wishing to know the area of a triangle measuring 10 units on one side and 4 units on the base to multiply 10 by 4 and divide the result by 2 (Fig. 1.0.2).[5] (The recipe is only correct if by 'side' the Egyptians meant 'altitude', a point on which historians disagree.) Similar recipes, more or less correct, were followed and learned for another millennium and a half—about as long a time as separates us from the fall of the Roman Empire—before the Greeks devised a logical system that enabled them to demonstrate, on very general assumptions, that the area of a triangle *must* always equal half the product of its altitude by its base. It appears that the Greeks, and they alone, invented this system, since the notion of proof does not appear in the geometry of other civilizations before they made contact with Greek thought. Students should not become impatient if they do not immediately understand the point of geometrical argument. Entire civilizations missed the point altogether.

5. Gillings, *Mathematics of the pharaohs* (1982), 138.

1.1. Euclid and his modern rivals

The creator of *Alice in Wonderland*, Lewis Carroll, was an accomplished geometer and teacher of geometry. He made a thorough study of the dozen or so elementary treatises in geometry vying for public favour in Britain around 1880 and found them all inferior to Euclid's ancient *Elements*. Earlier he had had to cross out many passages in a geometry book he had bought for his sister lest she be corrupted by modern explanations.[6] Carroll's primary criterion was logical rigor, whereas most of the modern rivals to Euclid he examined stressed ease of access. In attempting to smooth the path for beginners, the moderns had dropped some theorems, rearranged the order of the remainder, rephrased here, papered over there, and, to say the worst, perpetrated many a *petitio principii*, or circular argument, assuming the truth of a later proposition to demonstrate an earlier one on which the later in turn depended. Carroll also had a practical argument for retaining Euclid as a whole against the perpetrators of illogicalities. If none of the rivals were plainly better than Euclid, school teachers should stick to the tried and true.[7] Whatever their faults, Euclid's *Elements* provided a standard ordering and numbering of propositions for reference and examination. Carroll's bottom line, from his edition of the *Elements*: 'Euclid's treatise is, at present [1883], not only unequalled, but unapproached.'[8]

In our democratic age, geometries abound that know nothing of Euclid. An example, taken at random from a large collection of similar books, offers a cornucopia of illustrations of basic concepts such as lines, angles, and polygons, a little instruction in logical inference and elementary algebra, but not a single proposition enunciated or proved in the manner of Euclid. He is introduced for the first time only at the very end of this very long text as the systematizer of 'the geometry you have been studying.'[9] This withdrawal from Euclid not only sacrifices rigor and eschews standardization, it also helps achieve a low level of exposition and expectation.

The present text follows Euclid's to the extent that it covers, more or less in order, the material in Books I–IV, and some of that in Book VI, of the *Elements*; but it does not give every proposition, nor prove all it does include in Euclid's manner. However, it adheres to an order that avoids circular reasoning and offers enough formal proofs, in the traditional double column format (which, incidentally, has

6. Wakeling, *Jabberwocky*, 8:3 (1979), 53.
7. Carroll, *Euclid and his modern rivals* (1885), ix–xi, 1–15.
8. Dodgson, *Elements* (1883), vi.
9. Wells, Dalton, and Brunner, *Using geometry* (1981), 442 (quote).

Fig. 1.1.1 The dual nature of geometry as symbolized by the Jesuit Mario Bettini: the prince provides the money with which the Society of Jesus educates people in mathematics for aesthetic and practical purposes. The picture is a pun on the title of Bettini's book, *Aerarium philosophiae mathematicae* (1648), that is, *The treasury of mathematics.*

been traditional for only 250 years), to make clear the full-dress Euclidean manner. It may be a recommendation, and should be a reassurance, that the order observed in Book I corresponds more or less to the logical sequence of propositions worked out by Lewis Carroll.

The text contains much more than Euclid, however.

Geometry as science

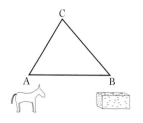

Fig. 1.1.2 A donkey's geometry, re Euclid I.20.

The followers of the Greek philosopher Epicurus, who esteemed feeling over reasoning, had no patience for the arguments of Euclid. His science is ridiculous, they said, pointing to a proposition half way through the first book of the *Elements*, in which Euclid labours to show that no side of a triangle can be longer than the sum of the other two sides. 'It is evident even to an ass.' For a hungry ass, standing at A (Fig. 1.1.2) will go directly to a bale of hay at B, without passing through any point C outside the straight line AB. The beast's geometrical intuition tells him that AB must be shorter than AC + CB. The charge that Euclid stops to prove propositions evident even to an ass has echoed through the ages. One of the echoers was the seventeenth-century philosopher and mathematician Blaise Pascal, who accused his fellow geometers of six perennial faults, among them 'proving things that have no need of proof'. He took as his example Euclid's compulsion to demonstrate that two sides of a triangle taken together exceed the third.[10] To this objection the Greek philosophical geometer Proclus, who wrote a lengthy commentary on the *Elements* early in the fifth century, had replied that a proposition evident to the senses 'is still not clear for scientific thought'.[11]

Proclus' line. For Proclus, scientific thought meant logical reasoning from axioms and postulates assumed to be true to propositions and theorems that, if the assumptions be admitted, must be true. But this was not the end of the business. Sciences differ among themselves as to subject matter. Astronomy is the science of objects seen in the heavens, physics that of things on this earth. Of what things or objects is geometry the science? Proclus thought it a problem that the idealizations that make up the matter of geometry—points without dimensions, lines without breadth, planes without depth, perfect circles—do not exist, even roughly, in the visible world. Where then did Euclid, and the geometers who preceded him, obtain these idealizations? As a good Platonist, Proclus had a ready answer: the geometrical idealizations are ideas innate in the human mind, present before all or any experience, and available when 'awakened'. Awakening can occur by looking at a triangle drawn in the sand; the idea of a perfect triangle does not arise by abstraction from figures in sand, however, but by recollection of the concepts that Nature, or the Mind of the Universe, placed in the human mind or soul.[12]

10. As given in Arnauld, *Logic* (1851), 338–9, a text originally dating from the mid-seventeenth century.

11. Proclus, *Commentary* (1970), 251. The proposition in question is Euclid I.20, proved *infra*, APS 3.1.3. ('APS' is defined *infra*, p. 59.)

12. Proclus, *Commentary* (1970), 11, 14, 33–40, 62.

Fig. 1.1.3 Gentlemen out surveying the course of a conduit to bring the water on the left to the town on the right through the hill in the center. Alberti, *Ten books of architecture* (1755), 222.

The notion that mathematical ideas have a special, supersensory status gave the study of geometry an exalted place in Platonizing philosophies. For, as Proclus has it, reasoning about geometrical objects prepares the mind for the harder task of recovering in itself ideas for which there exist in the visible world not even the rough analogue of triangles in sand. In this way, our imperfect minds can ascend to a knowledge of the Good and the gods. The highest purpose of mathematical sciences is 'intellectual insight and the consummation of wisdom'. The study of geometry needs no justification in practical utility. 'It is worthy of earnest endeavour both for its own sake and for the sake of intellectual life.' The poor Epicureans, sunk in their senses, will never acquire wisdom.[13]

The root meaning of 'mathematics' in Greek is 'learning'. Geometry is a synonym for learning, a way of learning, a test of learning. Therefore cultivate it; it will help to make you a better friend to yourself and to others. Proclus again: 'It arouses our innate knowledge, awakens our intellect, purges our understanding, brings to light the concepts that belong essentially to us, takes away the forgetfulness and ignorance that we have from birth, [and] sets us free from the bonds of unreason.'[14]

Proclus was the last important ancient writer on geometry. He died at about the time that the Northern tribes—whom the Greeks called 'barbarians'—overran the Roman Empire. The barbarians proved their barbarity by neglecting geometry. The West forgot Greek, ceased to copy and study the *Elements*, and soon lost everything of Euclid but a few elementary propositions detached from their proofs. Fortunately, the Arabs took a strong interest in geometry and preserved Euclid. When, during the twelfth century, European scholars began to make useful contacts with their better educated Islamic counterparts, the *Elements* stood ready for study, in Arabic. A few Westerners ambitious for learning mastered the tongue of Islam and translated Euclid into Latin. Proclus' commentary on Euclid followed a similar path. The original Greek texts of Euclid and Proclus, which had been preserved in Byzantium, came West during the fifteenth century, at the beginning of that heightened devotion to antique models known as the Renaissance. The same period saw the invention of printing. The *Elements* was one of the first books printed. It came out in many Latin editions, and also in Greek; and then, in 1570, in English. Most of these early editions had prefaces interpreting geometry in the sublime style of Proclus.

The preface to the first English Euclid was the work of John Dee, a magician as well as a mathematician, and an advisor on astrological questions to Queen Elizabeth I. Echoing Proclus, Dee insisted that the subject of the *Elements* not be called 'geometry', which smacked of its earthy origins, but 'megathology', or *Megathoscopiae elementa*, the study of magnitudes devoid of matter. By 'Megathological Contemplations', he wrote, we can train our minds to 'foresake and abandon the grosse and corruptible of our outward sence', whereupon we can 'conceive, discourse, and conclude of things

13. *Ibid.*, 18, 24.
14. *Ibid.*, 38.

Fig. 1.1.4 The famous crop circle that appeared in the Devil's Punchbowl (in England) in 1987. The circles were made by pranksters who trod down the growing grain. Before they claimed their performance, admirers of circles sought the cause in odd weather or extraterrestrial aircraft. Delgado and Andrews, *Circular evidence* (1989), 69.

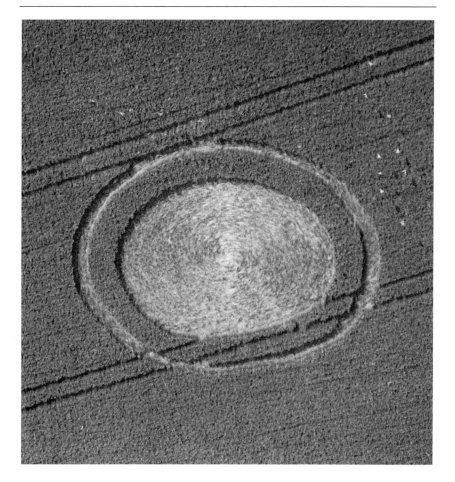

Intellectual, Spirituall, aeternall, and such as concerne our Blisse everlasting'.[15] The study of geometry may give a clue to salvation. Dee's preface was reprinted in 1660, in another English Euclid, whose editors could imagine nothing 'more ample or satisfactory'.[16]

Proclus' line had a well-placed advocate among the Cambridge dons who drummed geometry into undergraduates a generation before the author of *Alice in Wonderland* (who was an Oxford don) took on Euclid's modern rivals. This latter-day Proclus was William Whewell, the most influential man at Cambridge around 1850. He held that the idea of space, upon which all geometry depends, comes not from experience, but, rather, from the perceiving mind; the idea of space is prior to experience, indeed, it is a precondition of sensory perception, and the origin of concepts like dimensionless points,

15. Dee, in Billingsley, *The elements* (1570), sig. a. iijr; 'Megathoscopiae elementa' is Dee's addition to the title of his copy of *Euclidis elementorum libri XV, graecè & latinè* (Paris: Cavellat, 1557), now in the British Library.
16. Leeke and Serle, *Euclid's Elements* (1661), sig. A. Ir.

Fig. 1.1.5 Diagram of crop circles appearing in wheat fields in England between 1980 and 1989. The hoaxers had a taste for geometry. Noyes, *Crop circle enigma* (1990), 21.

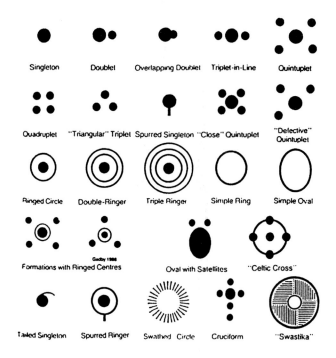

breadthless lines, and depthless planes. Hence, the Greeks could cultivate geometry 'in a form peculiarly free from admixture of extraneous elements'. Like Proclus, Whewell recommended his doctrine of the aboriginal innateness of geometrical concepts with the argument that abstraction from sensations could never lead to rigorously true propositions: he gave as an example Euclid I.20, the truth known to every ass, that the sum of any two sides of a triangle must exceed the third.[17]

Excellent as Euclidean geometry is, it does not exhaust the reach of the necessary truths originating from the necessarily true idea of space that we have as our birthright. Whewell foresaw the creation of a more general geometry that, by unfolding these truths, would form the basis of 'Liberal Education of the highest level'. This style of education—'an intellectual discipline, such as the world had never yet seen, and such as it has not even learned to hope for'—remains where Whewell left it, that is, nowhere.[18]

Parting from Euclid. The British were peculiarly slavish to Proclus' characterization of Euclid's proofs, 'all of them impeccable, exact,

17. Whewell, *History of scientific ideas* (1858), *1*, 88–9 (quote), 91–7.
18. Resp., Whewell, *Of a liberal education* (1845), 76–7, quoted by Becher, *HSPS, 11* (1980), 32; and Whewell, *Doctrine of limits* (1838), xii, in Becher, *ibid.*, 30.

and appropriate to science'.[19] Here is a representative declaration of this slavery, from a text of 1763: 'We [Britons] have generally liked Euclid's Elements best, and know them too well to be at the trouble of compiling new Elements of Geometry (as many of the French, and other foreigners have done) ... for the method and order of Euclid's Elements of Geometry can neither be mended nor altered, but for the worse.'[20] As late as 1850, the British still took their geometry in the ancient manner, according to Euclid's text as established by Robert Simson in 1756. The editor of a popular version of the first six books, published in 1855, could admit Euclid's many obscurities and prolixities and yet insist upon sticking to the *Elements*. Why? Because all Britons with a taste for science had always done so. 'The Books of Euclid, and their propositions, are as familiar to the minds of all who have been engaged in scientific pursuits, as the letters of the alphabet.'[21] As we know, Lewis Carroll maintained a similar view after the middle of the nineteenth century. He was not alone.[22]

The British could maintain their slavery so long because the University of Cambridge, the highest tribunal for mathematics in the country, insisted upon using purely Euclidean methods in teaching

Fig. 1.1.6 Inspired geometry. Interlacing grids based on a ten-pointed star. From Wilson, *Islamic designs* (1988), no. 47. (See also Plates V–VI.)

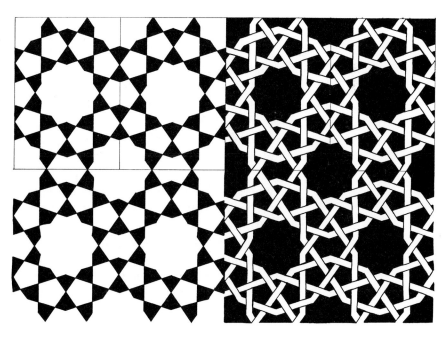

19. Proclus, *Commentary* (1970), 57.
20. Stone, *Euclid* (1763), viii.
21. Cantor, *Vorl. (1907), 3*, 489; Jones *MT, 37* (1944), 3–4; Bouton, *School sci. math., 11* (1911), 347; Playfair, *Elements* (1864), 'Preface', Lardner, *First six books* (1855), x–xii (quote).
22. E.g., Gillet, *Euclidean geometry* (1896), iii. Harrison, *Nature, 67* (1903), 578, names four then new geometry texts 'not sufficiently free of the Euclidean tradition to make them suitable for boys at school'.

geometry. Since up to the middle of the nineteenth century geometry was a university subject, Cambridge dons like Whewell had a sizeable student body, including many future teachers of mathematics, in whom to instil their ideas. Indeed, 'instil' is but a weak characterization of their method. For about a century, from around 1750 to around 1850, there was but one route to academic glory for Cambridge undergraduates: they had to excel in an examination called the Mathematical Tripos, held over three or four days in an unheated room in January. Survivors of this ordeal were classed in order of merit: those at the top of the list, known rightly as Wranglers, had been drilled so thoroughly in ancient geometry that (to quote the Senior Wrangler of 1806) they could 'repeat the first book of Euclid word by word and letter by letter.'[23]

Questions for the Tripos of 1802, ranked by order of difficulty, have survived. Students seeded as likely Wranglers were asked to find the sides of a triangle given its angles and the radius of its inscribed circle; students ranked in the next class, Senior Optimes, faced the problem of showing that, given two circles touching externally and any pair of parallel diameters, the lines connecting opposite ends of the diameters pass through the point of tangency; and probable Junior Optimes had to prove that the opposite angles of a quadrilateral inscribed in a circle are supplementary. Figures 1.1.7–1.1.9 show what is required in these problems. None is difficult. All are solved later in this book.[24]

Whewell was the last powerful voice at Cambridge for maintaining geometry pure of algebra and other pollutants. He carried the day with two weighty arguments. He observed, first, that geometry exercises the memory and the imagination more thoroughly than algebra: once a problem has been reduced to an algebraic equation, the student can manipulate the symbols without further thought, whereas a geometrical argument requires concentration and comprehension throughout. Secondly, he rejected the mixing of geometry with algebra. That would dilute the innate idea of space with elements foreign to it. Euclid calculated with ratios of lengths of lines, or ratios of areas, whereas the algebraist promiscuously multiplies heterogeneous magnitudes, divides lengths by areas, and confuses figures with numbers.[25]

No doubt geometry mixed with algebra suffers in rigor and relaxes demand on the imagination. Yet the argument for mixing algebra (including trigonometry) with geometry is also very strong: it

23. Ball, *History* (1889), 183, 192; the quotation, from Sir Frederick Pollock, *ibid.*, 112–13.

24. Ibid., 209–15; infra, §5.2 at Figs 5.2.27, 5.2.33, and APS 5.2.2.

25. Cf. Todhunter, *Elements* (1884), x.

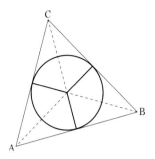

Fig. 1.1.7 Exam problem at the University of Cambridge, 1802. Given the radius of the inscribed circle and the angles A, B, and C, the examinee was to find the lengths of AB, BC, and CA.

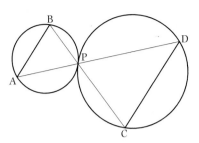

Fig. 1.1.8 An easier problem from the 1802 exam: Given two circles touching at P and any parallel diameters AB, CD, to prove that APD and BPC are straight lines.

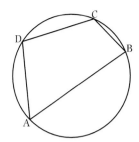

Fig. 1.1.9 The easiest problem of 1802: Given any quadrilateral ABCD inscribed in a circle, to prove that the sum of the opposite angles (A and C, B and D) is a straight angle.

permits the exploration of a wide range of material, some of great interest and importance, and it provides a sense of security that a geometrical problem might yield to analysis even when one's innate sense of spatial relationships falters. Pedagogues who accept both arguments have a problem. What degree of mixing will still allow the mental discipline that all advocates of Euclid extol as the highest purpose of the study of geometry? How, and how much, can the *Elements* be diluted or recast without destroying them? These questions became acute in the second half of the nineteenth century with the widening of educational opportunity and the gradual removal of instruction in geometry from the universities to the secondary schools.[26]

Inspired by the widening and the removal, British reformers banded together in 1871 in an Association for the Improvement of Geometrical Teaching, which tried mightily to produce a textbook more modern in spirit than Euclid's. They failed. They could not evade the university examination system, which still required mastery of the first books of the *Elements* for a bachelor's degree. At the same time, the British Association of Mathematical Teachers came into existence to support the attempts of Euclid's modern rivals ridiculed by the reactionary author of *Alice in Wonderland*.[27] At the very end of the century, the reformers at last began to prevail under the leadership of a physicist, John Perry, who emphasized practical measurement and the use of algebra, trigonometry, and squared paper. Perhaps relevantly, Perry had developed his teaching technique by practising on the Japanese, who invited him to Tokyo as one of their earliest instructors in Western science and mathematics.

26. Bartol, *Elements* (1893), iii.
27. Cf. Eperson, *Math. gaz.*, *17* (May 1933), 95.

Fig. 1.1.10 Some native American geometry. A Navajo blanket of the nineteenth century. From Wilson, *North American Indian designs* (1984), no. 30.

By 1911, Euclid was no longer required for examination in Britain; any good proof or solution to the problems would suffice.[28]

Geometry instruction in the United States began under the influence of the British, as befitted its colonial status. Euclid unvarnished was followed faithfully at Harvard in the eighteenth century, and Simson's and Playfair's traditional texts were widely used well into the eighteenth century. But from about 1820, American textbook writers began to follow the French, who had abandoned Euclid as the basis of instruction before the storming of the Bastille. The French in turn followed the lead of Jesuit pedagogues, who, as the schoolmasters of early-modern Catholic Europe, had extensive experience in drumming geometry into non-mathematical heads. A representative example is André Tacquet, the author of a very successful textbook composed toward the middle of the seventeenth century. Although he omitted propositions he deemed useless and introduced algebra into proofs, he did not think that these liberties ruined geometry for the higher exercises of the mind. 'Go on', so he addressed the count to whom he dedicated his book, 'go on, love, and esteem this science, which I hold to be worthy of its renown; pre-eminent for its certainty, because it is a single structure; appealing to the finest intellects, because true; appropriate even to clerics, because good ... and, thus, conducive to praising and loving the one, the true, and the good, that is, the incorporeal God'. Thus assured that right

28. Bouton, *School sci. math., 11* (1911), 350–2; Beman and Smith, *New geometry* (1903), iii–iv.

thinking might be compatible with algebra, European purveyors of the science of Euclid moved ever further from their source.[29]

The most successful of the French versions was published in 1794, at the height of the Terror, and went through 20 editions in 40 years. It was the work of a first-rate mathematician, A.M. Legendre, who mixed geometry with algebra and trigonometry, introduced practical examples, dropped Euclid's tedious treatment of incommensurables, omitted proofs of the obvious, and did it all clearly, cleanly, and elegantly. Unfortunately, later editions of his book reduced or eliminated the use of trigonometry and practical examples.[30] Algebra remained prominent, however ('we should be wrong not to make use of it'); it appears to particular advantage in derivations of the lengths of the diagonals of a cyclic quadrilateral and Hero's formula for the area of a triangle. Readers will be able to make their own judgments later, when both derivations are given algebraically and geometrically.[31]

British influence remained too strong in the United States to permit immediate acceptance of the full range of sacrilege in Legendre's approach.[32] A compromise ensued in the form of a text that reached its 42nd edition in 1890, *Elements of geometry and trigonometry, from the works of A.M. Legendre, adapted to the course of mathematical instruction in the United States*.[33] It was superseded as a standard by G.A. Wentworth's *Text-book of geometry*, a clear, economical, good-natured paraphrase of parts of Euclid's first six books, in five parts: rectilinear figures; the circle; proportion and similarity; areas of polygons; regular polygons and circles.[34] The present book deviates from Wentworth's fivefold way by treating circles after areas, which amounts to placing pieces of Euclid VI between Euclid II and Euclid III.

Until World War II American geometry texts continued to instil mental discipline, not computational skills or ability to analyze practical problems. Wentworth directed his texts 'not so much to the *attainment of information* as to the *discipline of the mental faculties*'; none of the 479 exercises in his textbook of 1895 concerns a practical

29. Tacquet, *Elementa* (1654), ff. a.2v–a.3r (quote), a.4.

30. Bouton, *School sci. math.*, *11* (1911), 332–6, 353–4; Bobynin, in Cantor, *Vorl.* (1907), 4, 323–4, 339, 343–4; Jones, *MT, 28* (1935), 419.

31. Bobynin, in Cantor, *Vorl.* (1907), 4, 381–2; Legendre, *Elements*, tr. Farrar (1825), vi (quote); infra, §5.2, at Fig. 5.2.37, and §6.1.

32. Jones, *MT, 37* (1944), 3–6; Davis, *Elements* (1872, 1874), iii; Bowser, *Geometry* (1891); iii–iv; Committee of Ten, *Report* (1894), 113–14 (recommending that geometry be done geometrically).

33. By Charles Davies; edn. of 1890 by J.H. van Amringe.

34. Wentworth, *Text-book* (1879, revised 1888); Wentworth and Smith, *Revised plane geometry* (1910), iii–v; Palmer and Taylor, *Plane geometry* (1915), 'Preface'. Similar arrangements in Byerly, *Chauvenet's treatise* (1887), 7, and Slaught and Lenne (1901), *Plane geometry* (1918), v.

Fig. 1.1.11 More native American designs. Finger-woven sashes from the Great Lakes, eighteenth and nineteenth centuries. From Wilson, *North American Indian designs* (1984), no. 41.

problem or hints that geometry can be of any use at all.[35] Despite recommendations by important national committees, the schools resisted fuller integration of algebra with geometry.[36] More calls by more reforming committees between the world wars had similar success; instruction in geometry stultified; and students, taking advantage of increasingly flexible curricula, avoided the subject despite the elitist

35. Wentworth, *Text-book* (1895), iv. Cf. Bowser, *Geometry* (1891), iii ('to discipline and invigorate the mind').

36. Wentworth, *Text-book* (1895), iv; NCTM, *History* (1970), 16, 19, 27, 39–40, 161–2, 173, 180–2; Bouton, *School sci. math., 11* (1911), 454, 512–14, 517. An exception is Palmer and Taylor, *Plane geometry* (1915).

argument of the National Committee on the Teaching of Mathematics that 'much of [geometry's] importance lies in the fact that it is part of the common background and experience of educated men'.[37] Between 1890 and 1920, the number of students graduating from American high schools increased eightfold; the schools managed to maintain and even to increase the proportion of students in geometry, from about one in five in 1890 to almost one in three in 1909 back to one in five in 1921. Between the wars, however, this fraction slipped to one in seven, and after 1945 it went into free fall. In 1955, American high school students took geometry at exactly half the rate they had in 1921 (11.4% as against 22.7%), which was about a third of what they had attained in 1909.[38]

Since after World War II the nation thought it needed large numbers of scientists and engineers, the disinterest of students in high-school mathematics courses became an acute national problem. The solution that won out, after much discussion, was the so-called 'new math' devised by the School Mathematics Study Group (SMSG), a set of university mathematicians and school teachers picked by the American Mathematical Society and supported financially by the National Science Foundation. The SMSG began to issue guidelines and texts in the early 1960s. They took a very high line: students

37. Beatley, *MT, 28* (1935), 335, 373, 375–6; NCTM, in *MT, 24* (1931), 229.
38. NCTM, *History* (1970), 48, 54, 234. Cf. Betz, *MT, 43* (1950), 377–8.

Fig. 1.1.12 A piece of sidewalk geometry. Ventilator grating, Stockton, California. Melnick, *Manhole covers* (1994), 174. Photograph © MIT Press.

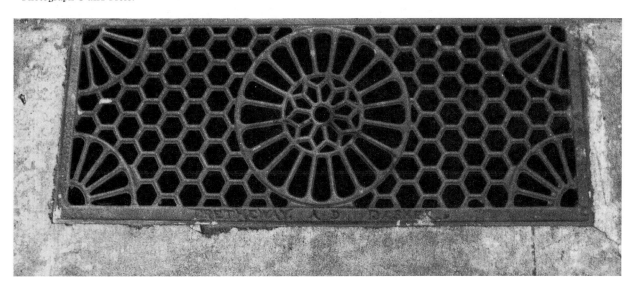

should be taught over-arching or foundational concepts, the true basis of all mathematics, such as set theory, the theory of functions, and generalized algebras.[39] The SMSG thus elected to treat geometry as science, which would open the way to an understanding of the range and beauty of mathematics. They ended close to Proclus in spirit, although they supported something more basic than Euclid and aimed at something less elevated than the Good.

Geometry as technique

Proclus knew perfectly well that Greek geometry had its origins in surveying. He did not try to cover up this compromising past; on the contrary, he made a long list of what he called the lower and humbler applications of his science. They included geodesy, mechanics, optics, astronomy, perspective, and fortification. Proclus conceded that these humble applications made the reputation of geometry in the wider society. He gave the example of the feat of the great Archimedes, who had built so large a ship for Hieron, the tyrant of Syracuse, that not all the Syracusans together could launch it. Archimedes thereupon contrived a machine that enabled Hieron alone to draw the ship to the shore. Knowledge of geometry conferred power over things. Hieron appreciated the performance and the performer. 'From this day forth [he said] we must believe everything that Archimedes says'.[40]

There therefore descended from Proclus a secondary justification of the study of geometry, which, like the first, attracted the attention of Euclid's editors during the Renaissance. Dee gives over most of the space in his extensive preface to the first English Euclid to applications, in order of their distance from the 'purity, simplicitie, and Immateriality of our Principall Science of *Magnitudes*'. There is surveying, mensuration, gauging, geography, cartography, and 'stratarithmetic', the art of military formations; also perspective, astronomy, astrology, statics, architecture ('the naturall subject of the peerless *Princesse Mathematica*'), navigation, and half a dozen more, down to 'Thaumaturgike', the art of making 'strange workes, of the sence to be perceived, and of men greatly to be wondered at', that is, stage tricks.[41]

Dee's litany of applications had no echo in the text that followed his preface. The first English Euclid gave out geometry as it had been systematized more than eighteen centuries earlier. Similarly the Italian edition, by the great mathematician Nicolò Tartaglia, published the

39. NCTM, *History* (1970), 68–75, 81–2, 278, 282–3.
40. Proclus, *Commentary* (1970), 17, 18, 31, 33, 51 (quote).
41. Dee, 'Preface', passim.

Fig. 1.1.13 Geometry on the floor. Tiles from a mosque at Edirne, Turkey, fifteenth century. From Wilson, *Islamic designs* (1988), no. 36.

year before the English one, and rendered so clear, he said, 'that every middling intelligence, without knowledge or support of any other science, will easily be able to understand it', extols all the applied arts dependent on geometry from astronomy to perspective, but does not ease the tasks, or raise the interest, of mediocre minds by bringing in practical problems.[42] The same disconnection occurs in the French edition by Jean de Tournes, dedicated to the noble youth of France because 'geometry is necessary and proper for the art of war, whether for fortification, or for placing, exercising, and using armies'. But there are no applications to military or any other practical problems in his traditional text, or in the texts to which the seventeenth century educator Henry Peachem refers the 'compleat gentleman' for instruction in the science whose principles inform the construction of ships, forts, palaces, and wheelbarrows.[43]

The need for gentlemen to know mathematics was a constant theme, or advertisement, of mathematical writers of the Age of Reason. 'Wisdom is certainly one of the main springs of human happiness', lays down an author in the middle of the eighteenth century. The twin fountains of wisdom are arithmetic and geometry; 'no state ever did or scarcely could flourish without them'. Therefore come buy my book.[44] 'The Nations that want [geometry] are altogether barbarous, as some Americans, who can hardly reckon above twenty'.[45]

Although French mathematicians of the later eighteenth century urged that students be introduced to geometry by constructions, measurements, and applications,[46] it was not until the end of the nineteenth century that authors who followed Proclus' second line allowed themselves to employ practical exercises or illustrations in textbooks of geometry. 'Geometry should be begun as early and as simply, in behalf of industrial life, as arithmetic is in behalf of business life'. Thus the British reformer of the turn of the century, John Perry, who led the successful effort to leaven Euclid with algebra, also urged the importance of practical problems to arouse interest in the otherwise 'soul-destroying, weary, worrying study' of school geometry.[47] It proved harder to insert practical problems than algebra into the textbooks, perhaps because the mathematicians and pedagogues who wrote them did not have the necessary knowledge or interest. A strong effort in the United States during the 1930s

    42. Tartaglia, *Euclide megarense* (1569), t.p., 3v, 5v.
    43. Tournes, *Les six premiers livres* (1611), [i]; Peachem, *The compleat gentleman fashioning him absolute in the most commendable qualities* (1622), cited by Watson, *Beginnings* (1909), 343–4.
    44. Stone, *Euclid* (1763), iii.
    45. Anon., *Essay* (1701), 30–2.
    46. Smith, *Teaching* (1903), 240.
    47. Warren, *Primary geometry* (1887), iii; Perry, *Neglect of science* (1900), 45 (quote), 51–2.

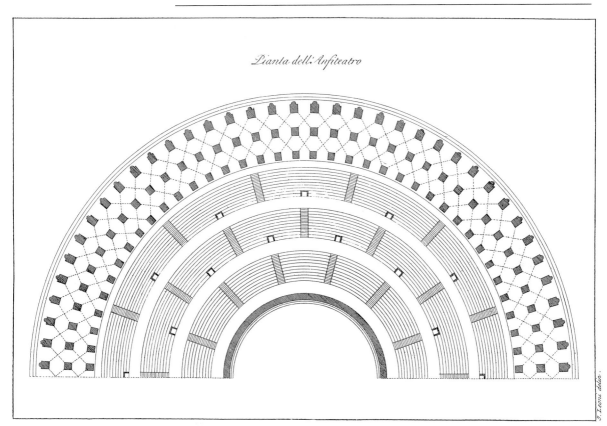

Pianta dell'Anfiteatro

Fig. 1.1.14 A typical architectural drawing from the Renaissance. Plan of an amphitheater. Alberti, *Ten books of architecture* (1755), 180.

made only a little headway with great difficulty despite the argument that the increasing number of students and the deepening economic Depression emphasized the need to make geometry attractive and useful to adolescents of all classes.[48]

Texts on applied geometry proliferated during the nineteenth century to serve working men eager to better themselves within the factories of the industrial revolution. Unlike the advanced industries of our time, which exploit electronics, those of the last century depended upon simple machines. The design of the gears, cams, cranks, screws, pulleys and the systems they made up depended in turn on the application of geometry. The texts devoted to would-be mechanics often included chauvinistic flights that read like contemporary exhortations to American youth to prepare for technological and commercial battle with Germany, Japan, and the five little dragons. An interesting early example, aimed at the workers of Metz, promised that assiduous study of applied geometry would make them excellent in their crafts, increase their pay and their happiness, perfect their morals ('it is extremely rare to see a drunk [geometer]'), and make them 'profoundly religious'. Not least, a horde of French

48. NCTM, *History* (1970), 48, 67, 234.

Fig. 1.1.15 Gothic geometrical tracery, from the cloister of Hauterive (1320s). Artmann, *Math. intelligencer, 13:2* (1991), cover illustration.

workers filled with geometry ('already they are rushing in crowds to courses in geometry and mechanics, and soon the lecture halls will not be large enough to contain them') would save the *patrie* from ruin in the industrial competition with England.[49]

But the geometry purveyed to the workers of Metz would not be that of ordinary schools. The author, a local academician, had composed his text expressly for his uneducated countrymen. Therefore he had omitted abstract ideas and preferred simplicity to rigor in his account of propositions and concepts. 'It would be a waste of time to demonstrate them to you so as to make them secure against all objections'.[50] Metz was not a place for Proclus. In any case, and with the best will and the greatest flexibility, it is difficult to present practical problems alongside proofs of propositions within a book of acceptable size. It is easier to assume the elements of geometry, or state

49. Bergery, *Géométrie appliquée* (1828), xii–xiv, xv (quote), xvi–xviii, xxi–xxii (quotes).
50. *Ibid.*, xxii.

them without proof, in texts on surveying or navigation or architecture, and refer curious or forgetful students to a standard geometry for what is missing. This book operates in reverse. It indicates applications, without describing the needed apparatus in detail, and refers to practical treatises for further applications and techniques.

1.2. Geometry in and as culture

Geometry and gender

'Nothing makes men more gentle than the cultivation of that heavenly philosophy [mathematics]. But, dear God, how rarely is this gentleness a quality of theologians! And how desirable it would be in this century if all theologians were mathematicians, that is, gentle and manageable men!'[51] This definition of the culture of the mathematician, apparently realized in Fig. 1.2.1, occurs in the preface to a text on trigonometry published in Germany in 1608. A similar message, though addressed to a different audience, occurs in the introduction to a French edition of Euclid published around the same time. There the author urges the young aristocrats at whom he aimed his work to act among themselves, and for their king, like

Fig. 1.2.1 A civilized theologian-geometer (probably Luca Pacioli) instructing a prince (the Duke of Urbino); another of Pacioli's students was Leonardo da Vinci. Painting by Jacopo de' Barbari (1495) in Museo e gallerie nazionali di Capodimonte, Naples. From Shea, *Storia delle scienze*, 2 (1992), 281.

51. Pitiscus, *Trigonometria* (1612), fol. a.3.

propositions and demonstrations in geometry, mutually confirming and supporting one another. Only mad or wild people act otherwise. 'You can and ought to do better, both for the obedience you owe to God and for the duty and obligation you owe to France'. Let the study of geometry tame you.[52] Or improve your morals. '[A geometer] is so conversant in truth, that he naturally hates falsehood, tho' perhaps [we deal here with an honest pedagogue] he sometimes is obliged, contrary to his inclination, to make use of it'.[53]

Geometers have repeated this self-serving theme down the centuries. Three hundred years after the French writer just quoted recommended study of his subject for ethical improvement, an American pedagogue found its most important consequence in the inculcation of honesty. 'The fact that a piece of mathematical work must be definitely right or wrong, and that if it is wrong the mistake can be discovered, may be made a very effective means of conveying a moral lesson'.[54] Tell the truth: your lies, like your mistakes, can be detected, exposed, and graded.

No doubt the study of mathematics can be civilizing. But it can also produce arrogance and exclusiveness. 'Mathematics is written for mathematicians', wrote Copernicus, on the doorpost of his revolutionary, and highly technical book of astronomy, which taught that the earth moves and the sun stands still. No doubt with this restriction he hoped to fend off criticism of his planetary geometry based upon what he considered to be ill-founded objections by the wider society. The arrogance and unsociableness of mathematicians may be better known than their gentleness. Benjamin Franklin wrote of one of his acquaintances: '[he] was not a pleasing companion, as like most great mathematicians I have met with, he expected universal precision in everything said, or was forever denying or distinguishing upon trifles to the disturbance of all conversation'.[55] Poets through the ages have condemned mathematical culture for (to quote but one of them), its 'arrogant sterility' and 'insolent tyranny'.[56] An old schoolmaster warned his students: 'Mark all mathematical heads, which be onely and wholly bent to those sciences, how solitary they be themselves, how unfit to live with others, and how unapt to serve the world'.[57]

---

52. Tournes, *Six premiers livres* (1611), [ii].

53. Stone, *Euclid* (1763), vi, after castigating 'tyrants and arbitrary governors, immersed in pride, and stain'd with madness' for being 'too stupid and brutal' to support the study of geometry (*ibid.*, iv).

54. G.B. Matthews, *School world, 1* (April 1899), 129, quoted by Smith, *Teaching* (1903), 239.

55. Copernicus, *De revolutionibus* (1543), pref.: Franklin, *Autobiographical writings* (1945), 259.

56. Lamartine, *Oeuvres* (1838), 398.

57. Roger Ascham, *Scholemaster* (1570), quoted in Watson, *Beginnings* (1909), 254.

The arrogance of Western mathematical culture has helped to exclude women from studying, and, after universal education, from practising, mathematics. The eighteenth century expressed itself directly on this subject. Mathematics was esteemed for giving 'a manly vigour to the mind', for preparing men for mastery of astronomy and other 'noble and manly sciences', and for permitting the completion of the century's greatest project in applied geometry, the determination of the meter, 'with the confidence of a male spirit'.[58] But this exclusiveness was only part of the story. In 1704, while Newtonians celebrated the manliness of geometry, there appeared the first number of an annual almanac, entitled the *Ladies diary*, that was to run for over a hundred years. The *Ladies diary* included puzzles, poems, and mathematical problems. These problems, many of them geometrical, were posed and solved by the readership. Many women took part in these questions and answers. For example, in 1754 an otherwise unknown Maria Atkinson supposed that a courier, riding on a straight road from A to B to C (Figure 1.2.2), took two days to do each segment, and each day rode a distance less than the previous day's by the same amount. She wanted to know the distance covered in each day if AB = 136 miles and BC = 104 miles. Among the correct answerers (the answer is 72, 64, 56, and 48 miles) was a gentleman, who took the liberty of asking the puzzler's age. Maria replied:[59]

Five times seven and seven times three

Add to my age, the sum will be

As many above six nines and four

As twice my years exceed a score.

It took no great mathematician to work out Maria's age. The gentleman was up to it: 18. He added to his solution a most patronizing remark: 'A very good age for matrimony, Miss'. She had spent enough time on mathematics; she should now get on with the serious business of being a woman. Even the early editors (all male) of the *Ladies diary* could not free themselves entirely from the prejudice that mathematics might render the doer unfeminine. In 1718, an editor wrote in praise of his women correspondents that they had 'as clear Judgements, as sprightly quick wit, as penetrating Genius, and as discerning and sagacious Faculties as [mens']', and betrayed himself by adding, '[they are] the *Amazons* of our Nation'.[60]

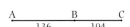

Fig. 1.2.2 Diagram for a puzzle from the *Ladies diary* (1754). Let x be the first day's journey, $x - a$ the second's, $x - 2a$ the third's, $x - 3a$ the fourth's; then $2x - a = 136$, $2x - 5a = 104$, whence $x = 72$, $a = 8$.

58. Quotes from, resp., Newton's physician (agreeing with Anon., *Essays* (1701), 7); one of Newton's disciples; and the official who accepted the prototype meter from the mathematicians who led the project. References in Heilbron, in Frängsmyr et al., *Quantifying spirit* (1990), 210.

59. Leybourn, *Ladies diary* (1817), 2, 88, 105–6. Let x be Maria's age. She tells us that $x + 35 + 21 - 58 = 2x - 20$.

60. Quoted in Schiebinger, *The mind* (1989), 42.

By the 1750s the *Ladies diary* had been taken over largely by gentlemen, who used it to promote mathematics in England. The *Ladies'* editor during the time Atkinson contributed to it wrote: 'For upwards of half a century, this small performance, sent abroad in the poor dress of an almanac (and that under a title not calculated to raise the highest expectations) has contributed more to the study and improvement of the mathematics than half the books professedly written on the subject.'[61] As will appear, many of the geometrical problems in the *Ladies diary* still make challenging exercises.

Although girls are now taught mathematics in all advanced countries, many still appear to feel, or are made to feel, that to excel at it is unfeminine. Insofar as this feeling rests on the notion that women lack ability for mathematics, or that its study will make them less feminine, there is no objective basis for it. As for the supposed want of ability, it can be refuted by pointing to women's contributions to mathematics both pure and applied.[62] Another persuasive argument is their performance on examinations. The Cambridge Mathematical Tripos affords a most striking case. In the 1880s, women had the right to take the examination and to receive grades, but not to appear on the winners' list. Instead, the positions they would have had had they been men were given separately. A great shock ran through the University when, in 1890, it read on the separate listing that a woman had placed above the Senior Wrangler. She was Philippa Fawcett, the daughter of a suffragette and a Cambridge professor. According to the *Dictionary of national biography*, Philippa's achievement 'materially advanced the cause of higher education for women'.[63]

As for the business about feminity, girls like geometry until they are taught that it is not for them. The recent discovery of this natural interest by the U.S. National Science Foundation recovered a result known long ago. The father of the famous philosopher Herbert Spencer wrote an unorthodox text around 1830 called *Inventional geometry*, which led students to work out problems by themselves by posing a series of gentle, graduated questions. 'Can you divide a line in two?' 'Can you make a rhombus?' In tune with this kinder, gentler tone, which contrasts favourably with the usual style ('Prove this', 'Demonstrate that'), the elder Spencer advised that 'it is better not to be anxious about keeping pace with others—indeed, all your efforts should be free from anxiety'. According to his son, the technique worked brilliantly with both girls and boys. 'I have seen it

61. Hutton, *Diarian miscellany* (1775), *1*, iii.

62. E.g., in Shiebinger, *The mind* (1989), 79–93, and Rossiter, *Women scientists* (1982), s.v. 'Mathematicians, women'.

63. *DNB*, *6* (1908), 1117 (Henry Fawcett), and *DNB, 1922–1930* (1937), 298 (Dame Millicent Fawcett).

Fig. 1.2.3 An unusual Gothic light at the Franziskanerkirche, Salzburg; the large empty circle may have structural importance. Courtesy of B. Artmann.

create in a class of boys so much enthusiasm that they looked forward to their geometry-lesson as a chief event in the week. And girls initiated in the system by my father have frequently begged him for problems to solve during their holidays'.[64]

64. W.G. Spencer, *Inventional geometry* (1877), 12 (quote), 34; H. Spencer, in *ibid.*, 4.

Fig. 1.2.4 Geometrical traffic interchanges discussed by the American Association of State Highway and Transportation Officials. AASHTO, *Policy on geometric design* (1984), 1008.

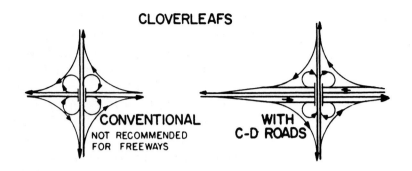

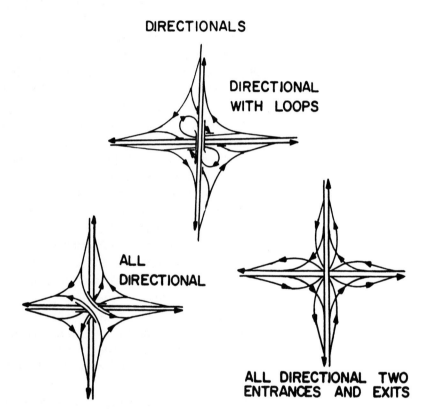

Non-Greek geometries

'Mathematical culture' may signify the characteristics of mathematics in a society as well as the culture of mathematicians. In Western geometry, these characteristics include a compulsion for proof, an exaltation of geometrical reasoning as the entry to the ideal world or as the basis of liberal education, and an exploitation of geometrical practice for the benefit of the arts and sciences. We know that other cultures did not feel the compulsion, and may have lacked the concept, of an axiomized deductive system; in consequence, perhaps, even those that had well-developed geometries, like the Egyptians,

Babylonians, Indians, and Chinese, treated the subject as a source of puzzles and as a practical art rather than as a mental discipline.

The geometrical problems that have descended from the Egyptians are severely practical. They have to do with the areas of rectilinear fields (connected with surveying and resurveying land plots) and the volumes of baskets, drums, and pyramids (connected with the collection of tribute, the storage of grain, and the disposal of pharaohs).[65] The most ancient Babylonian clay tablets concern economic matters, like inventories, tribute lists, and the size of fields. Tablets with rough sketches of fields, whose areas seldom agree with their dimensions, have a claim to being the oldest geometrical texts extant. Whether geometry in Mesopotamia descended from boundary problems, which seems likely from economic tablets and analogy to other ancient cultures, or whether it has a still more remote origin in artisanal practice, as suggested by the agreement of geometrical terminology with words used to describe designs on textiles and pottery, Babylonian geometry evolved out of concern with practical problems.[66] A good many of the later tablets estimate the volumes of containers of various shapes and the amounts of earth removed in digging or clearing canals. Facility with such problems did not lead to a deductive geometry of the Greek type, but to elaborate arithmetical exercises, at which the Babylonians excelled.[67]

Chinese manuals also relate closely to application, especially to problems of surveying. The Chinese depended upon their rivers for irrigation and commerce at least as much as the Egyptians and likewise developed their geometry with an eye to laying out fields after floods. At an early period, no later than the time of Christ, they applied their knowledge to military cartography, to manoeuvring armies, and to finding the distances to inaccessible objects, like besieged cities, pagoda roofs, and mountain peaks. The methods of handling these last problems differed significantly from those used in the West: whereas Western surveying relied on the theory of similar triangles, Chinese surveying made do with rectangles only.[68]

The Babylonian way. The cuneiform texts give recipes based on tacit general formulations or, sometimes, results with no directions whatsoever. Figure 1.2.5 is a transcription of a taciturn tablet dating from around 1700 BC. The marks that look like nails and birds, which were pressed into wet clay with a wedged-shaped reed, indicate

65. Cf. Joseph, *Crest* (1991), 81–9.

66. Neugebauer, *Vorl. (1934), 166–7*; Kilmer, in Gunter, *Artistic environments* (1990), 87–9. A triangle was called 'head of a nail'. Thureau-Dangin, *Textes* (1938), xvii.

67. E.g., Thureau-Dangin, *Textes* (1938), 124–30, and Neugebauer and Sachs, *Math. cuneiform texts* (1945), 59–99.

68. Swetz, *Sea Island manual* (1992), 2–11; infra, §4.3.

Fig. 1.2.5 A Babylonian field survey from ca. 1700 BC. From Neugebauer and Sachs, *Mathematical cuneiform texts* (1945), 42.

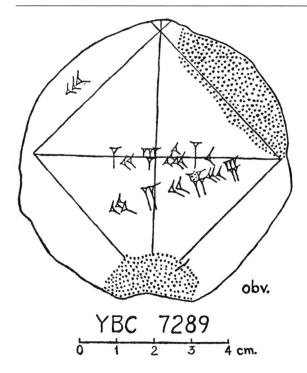

numbers. The three birds above the upper left side of the square indicate three tens, or thirty; the first nail along the horizontal diameter signifies 1; the rest of the line reads 24, 51, 10. If the Babylonians had used a decimal system like ours, the whole number might be read as 1.245110, that is, as $1 + 24/100 + 51/100^2 + 10/100^3$. But the Babylonians preferred to count in sixties, so the number above the diagonal is

$$1 + 2460 + 51/60^2 + 10/60^3 = 1.4140,$$

to four places of decimals. Now the ratio of the diameter of a square to its side is $\sqrt{2} = 1.4142$; Fig. 1.2.5 states this ratio along the horizontal diameter and adds beneath it the solution to the implied problem:

$$30\sqrt{2} = 42 + 25/60 + 35/60^2 = 42.4294.$$

A plausible conjecture about the Babylonian's route to their good approximation to $\sqrt{2}$ will be given later.[69]

The proposition that the square of the diagonal of a square is the sum of the squares of its sides is a special case of what the Greeks later knew as the theorem of Pythagoras. At some point the Babylonians found the general theorem, as appears from tablets

---

69. Neugebauer and Sachs, *Texts* (1945), 1–2, 42–3; infra, APS 4.1.7.

more recent than the one in Fig. 1.2.5. These tablets also have the merits of illustrating the style of Babylonian mathematical instruction and the type of pseudo-practical play problem that accompanied it. Here is a representative problem in its entirety:[70]

A reed stands vertically against a wall. If it falls three ells, it moves out nine ells. How long is the reed? How high is the wall [if, in the final state, the top of the reed meets the top of the wall]?

   Assuming that you don't know the answer, $3 \times 3 = 9$; $9 \times 9 = 81$. Add 9 to 81. Multiply 90 by 1/2: 45. The reciprocal of 3 is 1/3. Multiply 1/3 by 45: 15, the reed. What is the wall? $15 \times 15 = 225$; $9 \times 9 = 81$. Subtract 81 from 225 : 144. What must I multiply by to get 144? $12 \times 12 = 144$. The wall is 12.

The recipe is a special case of the algebra implicit in Fig. 1.2.6, where y stands for the reed, p for the distance the reed slips vertically, and q for the distance it moves horizontally. Then, by the Pythagorean theorem,

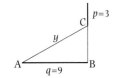

Fig. 1.2.6
A Babylonian Pythagorean problem.

$$y^2 = (y - p)^2 + q^2, \quad \text{or} \quad 2py = p^2 + q^2$$

$$y = \frac{p^2 + q^2}{2p} = \frac{3 \cdot 3 + 9 \cdot 9}{2 \cdot 3} = 15.$$

The scribe once started on his squares could not stop. Instead of finding the height of the wall immediately by subtraction (wall = reed − 3 = 12), he applied the Pythagorean theorem again to ΔABC:

$$\text{wall}^2 = \text{reed}^2 - 9^2 = 144, \text{ wall} = 12.$$

The triangle ABC allows an integral solution to the Pythagorean equation. The sides are in the proportion 3:4:5, the form in which the mathematicians of Babylon—and also Egypt, China, and India—liked to treat right triangles. Their preference for integral solutions may well obscure our appreciation of their knowledge of the general case.[71] When the problem of the slipped reed reappeared much later in the West, with a ladder as protagonist, it usually involved roots of numbers that are not perfect squares.

The Chinese way. Typically, the manuals delivering the technique of rectangles operate by very terse and often incomplete recipes; the student learned by watching the master, like an apprentice cook. Also like cooking, the ingredients in the recipes are always physical; the Chinese do not speak, as do the Greeks, of abstract line segments denoted by letters, but of 'distance from tree' or 'depth of water'. The terseness of the manuals made an opening for commentaries, which, by always leaving something out, made room for further glosses.

70. Thureau-Dangin, *Textes* (1938), 60–1. Cf. Coolidge, *History* (1940), 8.
   71. Neugebauer and Sachs, *Texts* (1945), 38–41, transcribe a tablet dating from between 1900 and 1600 BC that gives 15 sets of integral Pythagorean triplets.

Thus a commentator could state his purpose as writing well enough that 'he who reads can grasp 'more than half of it'. One of the most important early commentators, Liu Hui, who wrote around 300 AD, explained his allusive method by quoting Confucius: 'I showed him [the student] one corner and he understood the other three.'[72]

Despite their teaching by special cases, imitation, and innuendo, there can be no doubt that the Chinese knew why their recipes worked: the procedures imply mastery of the principles that the area of a rectangle is the product of its length and breadth, and that its diagonal splits it into equal parts. Furthermore, the recipes are so expressed that the student can see how to apply them to other, similar problems. Such recipes might be accounted the status of a generalization, but not, as some wish, of an implicit proof.[73] The concept of recipe fits the character and social context of Chinese geometry as delivered in the *Shu shu chiu chang*, which dates from the apogee of Chinese mathematics, around 1250 AD. This text survived because the editors of a later encyclopedia, in no fewer than 11 095 volumes, chose to include it as an authority. Its authoritative author, a man said to be as 'violent as a tiger or wolf and poisonous as a viper or scorpion', may have represented the standing of geometers as well as of geometry. This viper compiled his book 'for gentlemen of broad knowledge to peruse in their spare time, for although [mathematics] is a minor art it is worth pursuing'. The book gives many geometrical problems, some merely as pastimes, others as exercises in practical applications, none as an occasion for deductive reasoning.[74]

The pastime geometry of the Chinese seems to have descended from their discovery of the proposition known in the West as the theorem of Pythagoras. Stated in the Chinese manner, it reads that the square of the diagonal of a rectangle equals the sum of the squares on the other two sides. The approach is characteristic: in contrast to Euclid's, it begins numerically, with the special case of the (3, 4, 5) triangle $(25 = 16 + 9)$, and moves, centuries later, judging from the extant texts, to a general algebraic proof. The numerical statement appears from Fig. 1.2.7: ABCD is the original rectangle, with sides of 3 and 4, AC the diagonal; the rectangle is reproduced three times, as AEFG, FHIJ, and CKIL. We have

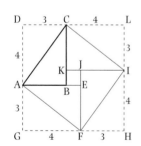

Fig. 1.2.7 A Chinese representation of the theorem of Pythagoras.

$$\square\text{AFIC} = \text{AC}^2 = \Delta\text{ABC} + \Delta\text{AEF} + \Delta\text{FJI} + \Delta\text{CKI} + \square\text{BEJK}$$
$$= 4\Delta\text{ABC} + \square\text{BEJK}.$$

72. Martzloff, *Histoire* (1988), 54, 67–68 (quotes), 70–1, 258–9.

73. Cf. Joseph, *Crest* (1991), 125–9, Needham, *Grand titration* (1979), 41–4, and Martzloff, *Histoire* (1988), 69.

74. Libbrecht, *Chinese mathematics* (1973), 2, 13, 31 (quote), 62 (quote); Needham, *Science and civilization, 3* (1959), 40–3, 48.

But $\Delta ABC = AB \cdot BC/2 = 4 \cdot 3/2 = 6$; $\square BEJK$, having a side of 1, has an area of 1; hence $AC^2 = 25$ and, as claimed, $AC = 5$. The text followed here, the *Chou pei suan ching*, the oldest extant Chinese mathematics manual, was drawn up probably between 100 BC and 100 AD, but transmits much older material. The information about the (3, 4, 5) triangle occurs in a dialogue whose characters alternately do mathematics and wonder at it. 'Great indeed is the art of numbering', says one. 'The combination of the right angle with numbers is what guides and rules the ten thousand things.'[75]

To obtain an algebraic generalization of the proposition exemplified in Fig. 1.2.7, and so to escape the need of remembering ten thousand things, commentators on the *Chou pei* around 200 AD replaced the numbers 3, 4, and 5 by kou ('leg') for the shorter side, ku ('thigh') for the longer side, and 'diagonal'.[76] Then they manipulated the generalization of Fig. 1.2.7 to read

$$\text{diagonal}^2 = 4\text{kou} \times \text{ku}/2 + (\text{ku} - \text{kou})^2$$
$$= \text{ku}^2 + \text{kou}^2.$$

(The last step is easily seen by writing $a = \text{kou}$, $b = \text{ku}$, $(b - a)^2 = b^2 - 2ab + a^2$, $4ab/2 = 2ab$). The generalization suggested many problems for the amusement of gentlemen of which that illustrated in Fig. 1.2.8 has an interesting cultural resonance. A bamboo stalk 10 units tall (AC) breaks in the wind at B and touches the ground at C, 3 units from A. How high above the ground did it break, that is, what is AB? The puzzler writes down the solution: the standing piece equals $(10 - 9/10)/2$ units, or, as we might write in general, setting $AB = x$, $AB + BC = b$, and $AC = a$, $x = (b - a^2/b)/2$. The justification for the recipe, not given in the Chinese text, is the Pythagorean theorem:

$$AB^2 = BC^2 - AC^2,$$

or

$$x^2 = (b - x)^2 - a^2,$$

whence

$$2bx = b^2 - a \text{ and } x = (b - a^2/b)/2.$$

This problem went west as Chinese and Indian civilizations came into contact. In the form the famous Indian mathematician Aryabhata gave it around 500 AD, using something more character-

75. Needham, *Science and civilization*, 3 (1959), 19–23.
76. Cf. *ibid.*, 96; Joseph, *Crest* (1991), 180–2, 186–8; Swetz and Cao, *Was Pythagoras Chinese?* (1977), 28; Lam and Shem, *AHES, 30* (1984), 87, 89, 110.

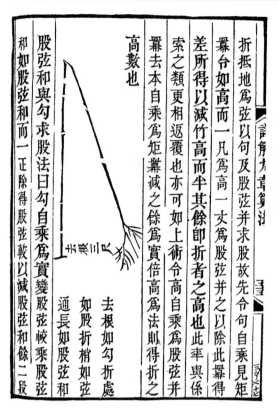

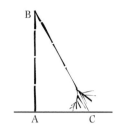

Fig. 1.2.8 The broken bamboo, an ancient Chinese brain teaser involving the Pythagorean theorem. Needham, *Science and civilization, 3* (1970), 28.

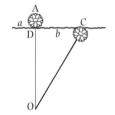

Fig. 1.2.9 The floating lotus, or the bamboo problem in India.

istic of his country than bamboo, a lotus plant rooted in water at O (Fig. 1.2.9) and carrying a full blossom AD bends in the wind so that the top of the blossom just touches the water at C. What is the depth of the water OD if AD is a quarter unit and DC is 2 units? Let OD = x, AD = $a = {}^1/_4$, DC = b = 2. Then, from Pythagoras, OC^2 = $(x + a)^2 = b^2 + x^2$, whence $x = (b^2/a - a)/2$. In fact, the Indians did not solve the lotus problem by the Pythagorean theorem but by what they called the method of arrows.[77]

The Indian way. The method and its name are illustrated in Fig. 1.2.10. Here ACBD is a circle, AB its diameter, CD a line (called a chord) making a right angle with AB and meeting the circle at C and D. Imagine that the arcs DBC and DAC are bow strings and that BE, AE, are their respective arrows. (In the Western tradition, only EB counts as an arrow). According to a proposition for which the Indians did not offer a proof, the product of the arrows equals the square of the half-chord ED, that is, AE × EB = ED^2. (Euclid of course

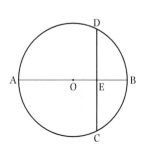

Fig. 1.2.10 The Indian method of bows and arrows.

77. Bag, *Mathematics* (1979), 157–8; Jha, *Aryabhata* (1988), 209–10.

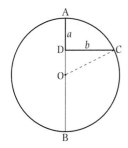

Fig. 1.2.11 The lotus problem solved by bows and arrows.

proved it, as we shall see). Now in the lotus problem, OA = OC, so that a circle centred on O will pass through A and C (Fig. 1.2.11). From the unproved proposition about the arrows and the half-chord, aDB = b^2. But DB = 2AO − a = 2(a + x) − a = a + 2x, so a(a + 2x) = b^2, or x = (b^2/a − a)/2 as before. The Indian text from which this problem comes delivers the answer, x = 7 and 7/8 units, without demonstration.

By the time they transformed the bamboo into the lotus, the Indians had had a practical geometry for about 1000 years. Their motivating problems arose from the implications of the Vedic religion, which required sacrifices on altars built to very exacting specifications in one of sixteen shapes determined by the objective of the sacrificer and proportioned to his height.[78] If, as appears likely, the implementation of these specifications preceded Pythagoras, the origins of geometry might better be sought in religious ritual than in land surveys. To those who believe that civilization progresses without backsliding and that religion represents a more primitive stage of thought than mathematics, the direction of advance must be from ritual to geometry, not the reverse. The insistence of priests on the precise fulfilment of their arbitrary rules lest the altar fail its purpose, which is reminiscent of their need to pronounce a mantra with absolute correctness to make it work, certainly indicates a geometrized ritual if not a ritualized geometry.[79] Each altar had to have 5 courses of bricks, 200 to each course; each of the 1000 bricks had to be cut into one of 16 standard triangles or rectangles; and the assembled whole had to be oriented precisely east–west and north–south. Figure 1.2.12 shows some of the allowed bricks; Fig. 1.2.13, their assembly into a complicated altar called the falcon; and

Fig. 1.2.12 Allowable shapes of bricks used in Verdic altars. After Bag, *Mathematics in India* (1979), 118–19.

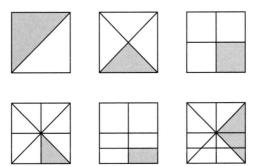

78. Kulkarni, *Layout* (1987), i, 5, 7.
79. Seidenberg, *AHES, 1* (1962), 520; Amma, *Geometry in India* (1979), 1–3, 16–17.

Fig. 1.2.13 The Vedic altar known as the falcon. Thibaut, 'Sulvasutras' (1875), **Fig.** 11.

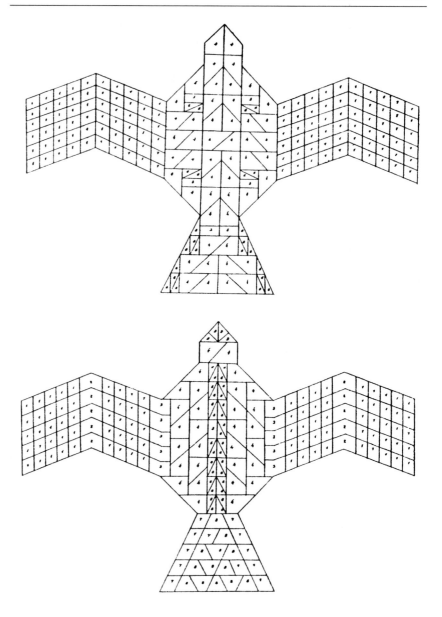

Fig. 1.2.14, the outline of a rudimentary altar that shows the method of orientation.

In Fig. 1.2.14, EW is laid out east–west by astronomical observation, and made 36 units long by measurement with a graduated rope. (The literature conveying the rules for constructing altars are called 'sulvasutras', or 'rules of the rope'.) The directions require that the altar be symmetric about EW; that AE = EB = 12 units and

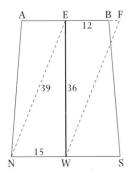

Fig. 1.2.14 Method of orientation of a Vedic altar using ropes and Pythagoras. After Thibault, 'Sulvasutras' (1975), **Fig. 2.**

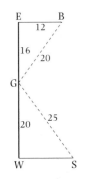

Fig. 1.2.15 Another method of orientation.

NW = WS = 15 units; and that AB and NS run rigorously north–south, so that EFSW is a perfect rectangle. (F is 15 units from E along EB extended.) First rule: take a cord 54 units in length, mark a point P on it 15 units from one end, and tie that end to a pole at W; tie the other end to a pole at E; take the mark P in your hand and walk around until the cord becomes taut; P now coincides with N, and you have found the north–west corner of the altar. The reason: since EP + PW = 54 and PW = 15, EP = 39; 39, 36, and 15 (or, casting out common factors, 13, 12, 5), make the sides of a right triangle, since $39^2 = 36^2 + 15^2$; NW runs north–south if EW runs east–west. The cord stretchers found S similarly; then they reversed the ends of the rope, stretched P out to locate F, subtracted 3 units from EF to find B, and repeated the process to get A.

Another recipe for achieving the same construction used a pole G placed 16 units along EW from E and two ropes, one 32 units long marked 12 units from an end, the other 40 units long marked 15 units from an end (Fig. 1.2.15). The first rope when tied at E and G and stretched located B (and A); the second, tied at G and W, S (and N). The right angle at E arose from the triplet (12, 16, 20), which, reduced to its prime factors, is the familiar (3, 4, 5); that at W from (15, 20, 25), another multiple of (3, 4, 5).[80]

These complicated rules for making right angles may have developed from simpler constructions also transmitted in the *sulvasutras*. For example, the perpendiculars to the east–west line EW may be found by tying the ends of a long rope at W and E, taking hold of the rope at its centre, and walking north until the rope is entirely taut (Fig. 1.2.16). That brings you to N, where you place a peg. A similar walk to the south locates S. NS is perpendicular to WE. A square may be constructed with only slightly greater trouble. Let EW = 2*a* be its side. Take a rope equal in length to EW, make a loop at either end and mark it at its midpoint and at its quarter points (Fig. 1.2.17). Now stretch the rope along EW and put pegs in the ground

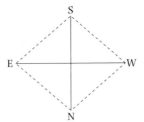

Fig. 1.2.16 An old way to make a perpendicular with a rope and two pegs.

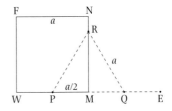

Fig. 1.2.17 A similar construction for a square.

80. Thibaut, 'Sulvasutras' (1875), 228–9, 235–6.

(a)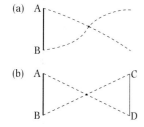

(b)

Fig. 1.2.18 Laying out a rectangle, in Mozambique manner.

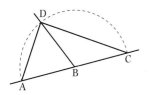

Fig. 1.2.19 Construction of a perpendicular in a circle, in the style of medieval masons.

at the ends (E, W), at the middle (M), and at the quarter points (P, Q). Loop the ends over the pegs P, Q; take hold of the rope at its middle, and walk north until the rope is fully taut; put a peg there (R). Next, loop both ends over the peg at M, again grasp the rope at its midpoint and walk north, stretching the rope over the peg at R to N—where you place a peg. Finally, loop the rope at N and at W, take the middle, stretch the rope, which brings you to F, a corner of the square.[81] As Confucius might have said, 'I've shown you how to find one corner, and you now know how to find the other three'.

Rectangular plots are still laid out with ropes in many parts of the world. For example, in Mozambique the villagers take a bamboo pole AB the length of one side of the plot and attach a rope to each end (Fig. 1.2.18). These ropes are equal in length and knotted together at their midpoints. When their free ends are pulled so that both ropes are taut, the ends locate the missing corners C, D of the rectangle. This technique depends not on the Pythagorean theorem, but on the knowledge that the diagonals of a rectangle bisect one another.[82] The masons who built the medieval Gothic churches made use of similar tacit knowledge. They would draw a line, AC say, and mark its midpoint B; and, from B, they would lay out in any direction a line BD = AB = BC (Fig. 1.2.19). By joining D to A and C, they would have the right angle ADC.[83] The masons used, no doubt without knowing, the information that the angles opposite equal sides in a triangle with two equal sides are equal, and that the sum of the angles in any triangle is two right angles; or, alternatively, that an angle inscribed in a semi-circle is a right triangle.

No ancient illustrations of Indian rope pullers seem to exist. We are better fixed for Egypt. Figure 1.2.20 shows a triplet of surveyors with their rope, as depicted on the walls of the tomb of a high priest responsible for fields whose harvest supplied an important temple.[84] Did their technique involve the (3, 4, 5) triangle, as the Greek

Fig. 1.2.20 Egyptian surveyors with their rope. From Clarke and Engelbach, *Ancient Egyptian Construction* (1990), 64.

81. Amma, *Geometry in India* (1979), 23, 31.

82. Gerdes, *Ed. stud. math.*, 19 (1988), 13, 14.

83. Shelby, *Gothic design* (1977), 114–15. That ΔADC is a right angle is proved infra, §5.1 at Fig. 5.1.21.

84. Clarke and Engelbach, *Egyptian construction* (1990), 64–8; Davies, *Tombs* (1923), p. 11 and plate X. There is another example in Campbell, *Theban princes* (1910), after p. 86.

philosopher Democritus intimated by calling Egyptian geometers 'rope-pullers'? Or did they limit themselves to the techniques described in the preceding two paragraphs? Certainly nothing in their extant writings indicates that they knew the Pythagorean theorem.[85] Here they fell short of the Indians, whose knowledge of the general case appears from their rules for increasing the size of previously constructed altars.

The Vedic priests placed a curious constraint on the rebuilding of altars. The second one on a given site had to exceed the first by a certain fraction, the bricks being increased in size so that no more were needed in the later than in the earlier constructions. Eventually they came to require an altar twice the area of a given one. The problem may be associated with the theological one of combining two gods into a single one with all the powers of its parts. The trick is easily done, with altars if not with gods,[86] if all the bricks are square. The Indians knew that the side of the bricks for the larger altar should be the diagonal of the bricks for the smaller one, that is, they knew the Pythagorean theorem for squares. They also stated, as usual without proof, the generalization of the theorem for rectangles, reasoning, perhaps, from the various special cases—(3, 4, 5), (5, 12, 13), (8, 15, 17), (7, 24, 35), and (12, 35, 37)—they had discovered. But since they also knew how to find a square brick equal in area to the sum of the areas of two different square bricks, and how to turn squares into rectangles and vice versa, they must have understood the Pythagorean theorem in general.[87]

The Islamic way. Another culture in which geometry played a major cultural role was Islam. On the theoretical side, Islamic mathematicians mastered Euclid and his successors both for intellectual discipline and practical application. In this endeavour they had the support of the caliphs. The first translation of the *Elements* into Arabic took place under Hārūn ar-Rashíd, the caliph for whom the romances of the *Thousand and one nights* also were composed.[88] (It takes less time to learn the *Elements* than to hear the stories). The Arabs applied their new learning in many ways. One of the more appealing was the design of tiles for mosques and palaces, as exemplified in Figs 1.1.6 and 1.1.13. Since Islam does not allow icons or images in its places of worship, it promoted abstract decorations; and it preferred the regularity in abstraction produced by geometrical figures. For intricacy and beauty of abstract design Islamic

85. Heath, *History* (1921), *1*, 121–2, 178. The inference from Democritus' term is contested; Gillings, *Mathematics* (1982), 238, 242.
    86. Seidenberg, *AHES, 1* (1962), 490–2.
    87. Thibaut, 'Sulvasutras' (1875), 230–5, 244–7; Kulkarni, *Layout* (1987), 10–11, 123.
    88. Busard, *Latin translation* (1984), ix.

Fig. 1.2.21 Detail from a Koran
produced in Iran in 1313. From
Wilson, *Islamic designs* (1988),
no. 81.

tiles, plates, and tapestries have few if any equals. Figure 1.2.21 and plates V–VI offer further examples.

1.3. No royal road

According to Proclus, Ptolemy I, King of Egypt, once asked Euclid for a shorter way to mastering geometry than working through the *Elements*. Euclid answered: 'There is no royal road to geometry'.[89] The learning and progress of twenty-three centuries have not much softened the way, despite the claims of textbook writers to have constructed (as one put it in the 1790s) 'a new royal way'. 'I will be bold to assert [wrote this hopeful pedagogue] that those who cannot, from this work, acquire a competent knowledge of all that is requisite, in common Geometry, will never be able to attain it at all.'[90] He was then but the latest, and certainly not the last, to brag that he had 'set out every Demonstration in such a regular, plain, and easie Manner ... [that] the Young Student might proceed with Ease and cheerfulness'.[91] Rare is the writer of textbooks who admits that he cannot remain true to his subject and, at the same time, 'deliver himself in such a manner as to be always intelligible ... to those whom he would have for readers'. Every honest pedagogue must worry, as did Newton's teacher, Isaac Barrow, that his 'Performance [might] not thoroughly please everybody'.[92]

The reasons that geometry puts off the average student were set forth neatly by an anonymous writer at the beginning of the Age of Reason. Here is his daunting list:[93]

The aversion of the greatest part of Mankind to serious attention and close arguing; Their not comprehending sufficiently the necessary or great usefulness of these in other parts of Learning; An Opinion that this study requires a particular Genius and turn of head, which few are so happy as to be Born with; And the want of ... able Masters.

Victorian pedagogues agreed that, as Wentworth expressed it, 'most persons do not possess and do not easily acquire, the power of abstraction requisite for apprehending geometrical conceptions, and for keeping in mind the successive steps of a continuous argument'. What to do about it? How to instil in refractory minds that 'knowledge of geometric facts and relationships [that] is as desirable in

---

89. Proclus, *Commentary* (1970), 57.
90. Malden, *New royal road* (1793), v.
91. Hill, *Elements* (1726), 'Preface'.
92. Resp. Williamson, *Elements* (1781), *1*, *2*, and Barrow, *Elements* (1660), sig. ∴ 3v.
93. Anon., *Essay* (1701), 2.

social life as the ability to recognize the most common quotations from Shakespeare'?[94]

Easing the path

Many devices have been invented to ease the presentation of geometry, some to good effect. The Jesuit Tacquet, who, as we know, thought freely about geometry, introduced signs and symbols to replace the verbiage that had cluttered earlier texts. The tradition that descended from Tacquet employed =, +, −, the form A:B = C:D for 'A is to B as C is to D', and aq ('a quadratus'), which we now write a^2, for 'the square of side a'. With such simplifications, it was hoped, Euclid would be made fit for the modern age.[95] That the use of symbols in the teaching of geometry was uncommon in the early eighteenth century is confirmed by an edition of 1722, which claimed a novelty in pandering to 'the Desires of those who are delighted more with symbolical than verbal Demonstrations'.[96]

Another important help, unfortunately not sufficiently employed in modern texts, is to distinguish typographically among elements in the diagrams that have different functions or standings. Heavy lines might be used for given or known elements in figures, lighter lines for unknowns, dashed lines for constructions, as in Wentworth's books, and in this one. Alternatively, the different functions might be distinguished by different colours.[97] Shadings to differentiate one triangle from another in a complicated figure came in (and, it appears, also went out) in the eighteenth century.[98] A very useful practice, which began early, is to reproduce a figure wherever it is discussed, so that the student need not flip about trying to match words with pictures. And, finally, the technique of placing every distinct assertion in a proof on a separate line together with its justification, which, though it has put off many students through the ages, has its value, entered geometrical texts in the middle of the eighteenth century.[99]

Perhaps the greatest helps to students are relaxation of unnecessary rigor, reduction of superfluous propositions, and introduction of practical examples.[100] No doubt, no one now knowingly writes a

---

94. Wentworth, *Plane and solid geometry* (1899), iii; W.R. Longley, in NCTM, *Yearbook, 5* (1930), as paraphrased in Beatley, *MT, 28* (1935), 373.

95. Tacquet, *Elementa* (1762), v–vi, viii, xii–xiii; this edition reprints prefaces by several previous editors.

96. Wilson, *Euclide's Elements* (1722), 'To the Reader'.

97. Wentworth, *Text-book* (1895), iii, and *Plane and solid geometry* (1899), iii; Deakin, *Euclid* (1903); Byrne, *First six books* (1847).

98. Malton, *New royal road* (1793).

99. Todhunter, *Elements* (1884), ix, xi; Lardner, *First six books* (1855), xviii–ix.

100. Cf. Henrici, *Geometry* (1879), ix; Blackhurst, *Humanized geometry* (1935), 3–7.

'model of soul-destroying systematization', as a reformer of 1899 described the school Euclids used in his time. Nonetheless, it is all too easy to slip into a pointless formalism, as do many current books that employ a complicated notation to distinguish line segments from lines and angle measures from angles, and make no effort to relate the abstract theorems to everyday experience. 'Strict proofs are not for boys', says an innovative German geometer of the 1920s. Boys consider belaboured foundations 'leeres Wortgeplapper', empty babble: 'do not injure their thinking prematurely by the constraint of a strict form, for which people are ready only later, if ever'.[101] I shall follow that advice and assume, with the innovative author of a textbook of 1904, that my readers know only algebra and how to sharpen a pencil.[102]

As for reduction of postulates and propositions, this book contains about 60 taken from Euclid and perhaps another 60 developed as problems. This compares favourably to the texts in use in the United States in the early 1930s, for which a detailed study exists. The average number of postulates in the six tests examined was 40 (range 29–52), of constructions and theorems, 195 (range 162–231). A set of nine texts published in the 1920s required on average 43 technical words about angles, 16 about triangles, 30 about polygons, 14 about circles, and 50 about things in general.[103] This book introduces around 80 altogether. We must practise continence. 'A complete account of all known elementary theorems regarding [only] the circle would be far beyond the strength of any writer, or reader'.[104]

Ad lectorem

In our century, four-colour printing, an infinite choice of fonts, attractive pictures of Egyptian pyramids, Gothic cathedrals, modern cubical skyscrapers and other geometrical forms exploited by architects, have enriched and enlarged texts. Furthermore, in our permissive democratic age, a multitude of approaches to purveying the content of geometry, from the deductive system of Euclid to algebraic paraphrases and weak informal proofs, compete for the attention of school boards and publishers. The author of a new geometry should therefore consider with Christopher Clavius, who brought out a fresh

---

101. Minchin, *Nature*, 59 (1899), 369; Kusserow, *Los von Euklid* (1985), 10, 8. Cf. Beatley, MT, 28 (1935), 403, quoting the National Committee of Mathematical Requirements' *Report* of 1923.
102. Pressland, *Introduction* (1904), 3, 6, 11, presents properties of figures by telling students how to draw them, an approach used earlier in German-speaking Switzerland.
103. Christofferson, *Geometry professionalised* (1933), 23, 43.
104. Coolidge, *Treatise* (1916), 4.

Fig. 1.3.1 Native American bowls, ninth to twelfth centuries. From Wilson, *North American Indian designs* (1984), no. 13.

one over 400 years ago, 'why, after so many excellent versions of Euclid have been published by famous and practised mathematicians, I should myself write a new one'.[105]

Clavius answered that Euclid as presented in his time could not be mastered *sine maximo labore, ac studio*, 'without very hard work and study'. He accordingly provided a help, in two volumes of commentary. I answer that the desire to render Euclid easy is not a good reason to write a text or commentary. Excellent texts on geometry have been plentiful for three centuries, notwithstanding their authors' tendency to find little merit in their predecessors.[106] It is

105. Clavius, *Euclidis elementa* (1574), *1*, 'Ad lectorem'.
106. Bobynin, in Cantor, *Vorl.*, 4 (1908), 331.

reported by a Spanish-speaking bibliographer, perhaps reliably, that the efforts of editors and bowdlerizers have made Euclid's *Elements* the most frequently printed work after the Bible and *Don Quixote*. It is certain that an Italian bibliographer, writing at the end of the last century, itemized over 200 significant pedagogical reworkings, which, in their manifold editions, amount to many thousands of different hopeful presentations of the ancient text.[107] Although I hope that my approach will make geometry more pleasant and useful to learn and, to that extent, make it easier, I cannot suppose that, after two millennia of failed attempts by experienced teachers to find a royal road, I have stumbled upon one.

My reason for writing is to continue to reap, and in a measure to repay, the pleasure that studying geometry has given me. This pleasure is in part aesthetic—delight in finding a clear tidy proof, in seeing a powerful application of simple principles, in perceiving unexpected spatial relationships in buildings, in patterns on ceramics and textiles, in highway interchanges. Part of the pleasure is in advancing beyond the *Elements* to higher geometries, spherical astronomy, and geodesy. A further satisfaction is gaining confidence in systematic reasoning. Architects of the French Revolution thought that the best guarantee of liberty was the ability to reason independently about the quantitative transactions of everyday life. '[We want] to insure that in the future all citizens can be self-sufficient in all calculations related to their interests; without which they can be neither really equal in rights ... nor really free.'[108]

Another source of pleasure is the integration of the pictorial and the verbal. Few other studies require close attention to diagrams and illustrations and also to precision and economy of language. The requirement about illustrations is plain enough; that about language may appear implausible to those who like the abbé Desfontaines, a rhetorician of the eighteenth century, hold that 'geometry uses up all the force of an ordinary mind and renders it incapable of any other thing'.[109] But as the leading American geometry textbook at the turn of the century put it, writing out proofs makes 'a valuable exercise as a language lesson'. At the same time, around 1900, the board set up by American mathematical societies to reform the teaching of geometry urged oral recitation, the presentation of theo-

---

107. Duarte, *Bibliografiá* (1967), 7; Riccardi, *Saggio* (1887).
108. Condorcet, quoted in Heilbron, in Frängsmyr, *Quantifying spirit* (1990), 213.
109. P.F.G. Desfontaines, *Le nouveau Gulliver* (1745, 1787), quoted by Cajori, *Mathematics* (1928), 67. Cajori gives much to the same effect from other literary men; ibid., 53–4, 69–72, 77–80.

Fig. 1.3.2 Geometry applied to the earth and the heavens. Even with such simple instruments, astronomers and surveyors of the Renaissance could determine to useful accuracy the angular separation of the moon and a star, the distances to castles, and the heights of towers. Johannes Werner and Peter Apianus, *Introduction geographica* (1533), in Bennett, *Divided circle* (1987), 56.

rems before the class, as the obvious remedy to 'the vicious habit of slovenly expression'.[110] The long-time secretary of the Paris Academy of Sciences, Bernard Le Bovier de Fontenelle, a literary man as well as (or, rather, more than) a mathematician, taught that 'a work on morality, politics, criticism, perhaps even eloquence will be more elegant, other things being equal, if it is shaped by the hand of geometry'. Jean Le Rond d'Alembert, a leading mathematician of the eighteenth century and, with Denis Diderot, one of the two original editors of the great French *Encyclopédie* of all respectable knowledge, pointed to Fontenelle and a half dozen others to prove that the phrase, 'dull as a mathematician', already a chestnut in his time, admitted exceptions. A person properly reared in geometry and language 'will know how to solve a problem, and to read a poet; to calculate the movements of the planets, or to enjoy a play'.[111]

Finally the pleasure, or my pleasure, has been cultural. Pursuing geometry opens the mind to relationships among learning, its applications, and the societies that support them. Some of these relationships, which run from high philosophy through surveying, have

110. Resp., Wentworth, *Plane and solid geometry* (1910), v, and Committee of Ten, *Report* (1894), 114. Cf. Gillet, *Euclidean geometry* (1896), 15, and Bouton, *School sci. math.*, 11 (1911), 516.
111. Fontenelle, 'Preface sur l'utilité des mathématiques' (1699), and d'Alembert, 'Géometrie', in the *Encyclopédie*, ed. d'Alembert and Denis Diderot (1753), quoted in Cajori, *Mathematics* (1928), 61, 62.

Fig. 1.3.3 A classic cloverleaf interchange. AASHTO, *Policy on geometric design* (1984), 952.

been indicated. The text that follows contains many more. They are chosen for their relevance and interest to the student of geometry, not from considerations of political correctness.[112] These geometries receive attention either because, as deriving from traditions uninfluenced by the systematization of the Greeks, they suggest less formal or more intuitive presentation of useful concepts, or because they enrich and vary the range of problems and exercises. Once admitted on these bases, however, the non-Western material may be accompanied by historical information designed to awaken further interest in, and respect for, the societies that created it.

To convey a sense of all these pleasures, as well as a grasp of geometry sufficient to excel on college aptitude tests, I have adopted an eclectic approach. I emphasize the earthy origins of geometry and its applications to practical problems. That does not mean that the text neglects theorems or their proofs. Quite the contrary. The theorems and proofs do not appear, however, as arbitrary theoretical impositions, but as generalizations of relationships illustrated in special practical problems. Some of these problems have to do with geodesy and astronomy. Students who work their way through this book will have a sturdy knowledge of the traditional content of plane geometry and also some elements of the 'globes', as the old astronomers called their models of the earth and the celestial sphere.

Typically, geometry texts either do not enter into trigonometry or do so only toward the end. That is to miss an opportunity. The principle and purpose of trigonometry ('measure of a triangle') follow immediately from the application of the concept of similar triangles to the measurement of heights and distances. The basic ideas of trigonometry are introduced early in this book and used freely

112. Cf. Nelson et al., *Multicultural mathematics* (1993), 7–9, 18–19, 31–41.

throughout. Ratios of the lengths of lines are the soul of Euclidean geometry. The so-called trigonometric functions are just such ratios, and can always be exhibited in geometrical proofs by dropping perpendiculars and invoking the Pythagorean theorem.[113] But the power of trigonometry lies in the treatment of the ratios as algebraic elements. That is another reason—besides the simplicity it admits in some derivations—for the unrestricted use of algebra in derivations and problems. I have tried, however, to bear in mind the advice of a committee of the (British) Mathematical Association, that 'the use of trigonometry in elementary geometry should not pass the level at which an *immediate* reinterpretation by means of similar right-angled triangles is possible; human nature takes refuge only too readily in a formula'.[114]

An unusual feature of this book is the range of its exercises and their close integration into the expository text. The problems come from all times and cultures; the *Ladies diary* proved an especially rich source. Solutions are always provided, usually in detail and often with an indication of the rationale for the approach taken. Few readers will wish to attempt them all, although, at 150, they fall short of the 479 offered in Wentworth's *Text-book*.[115] The justification for this cornucopia is that geometry possesses no general method; to master it requires seeing and solving many diverse problems. Hence the great French mathematician Jacques Hadamard took as the single guiding rule in his course on elementary geometry to multiply exercises excessively. His countryman, who shielded himself under the initials 'F.G.M.', perhaps to escape the wrath of overworked students, practised Hadamard's rule more strictly than anyone before or since: his *Exercises de géométrie*, in its sixth edition published in 1920, presents over 2500 problems and 1000 theorems in a closely printed volume of 1300 pages.[116] The only effective restraints on the multiplication of problems are the calculations of the publisher and disabilities of the sort suffered by my predecessor of the 17th century, William Oughtred, whose wife was so penurious that she would not allow him a candle after dinner, 'whereby,' he said, 'many a good notion was lost and many a problem unsolved.'[117]

113. Cf. Harrison, *Nature*, 67 (1903), 578.
114. Text of 1923, quoted in Beatley, *MT*, 28 (1935), 405.
115. Wentworth, *Text-book* (1895), vii, preface to first (revised) edition of 1888.
116. Hadamard, *Leçons* (1931), preface to the first edition; G.-M., *Exercises* (1920), iii. Cf. Committee of Ten, *Report* (1894), 113, 115: lack of a general method is 'the weakness of elementary geometry as a science'.
117. Ball, *History* (1889), 30–1.

2 From points to proof

Can you 'point to the center of the earth'? No? You have all the necessary apparatus at hand. Stand up straight; hold your arms to your sides along your legs; extend your forefingers. They are pointing to the center of the earth. If you prefer to use an apparatus other than yourself, you can tie a weight to one end of a string and hold the other end in your hand so that the weight hangs freely. When it stops swinging, the string will point directly at the center of the earth. The string then is said to be 'vertical': that is, to lie along the direction from where you are toward or away from the center of the earth.

2.1. Necessary ingredients

Point, line, plane

The string plus weight is called a 'plumb line', from *plumbeum*, the Latin word for lead. Bring a plumb line to the corner of your room where two walls join. If you could bring the string all the way to the joint, the two would run together: the joint between the walls also is vertical. Pretend that the string, remaining vertical, is moved along the wall. In the motion, the string maps out a vertical surface the length and breadth of the wall and one string in thickness. Imagine now that this wall of strings decreases in thickness, to become a wall of threads, then of single hairs, then of split hairs, and so on, until it has no thickness whatsoever. This is the same process by which, in *Alice in Wonderland*, the grinning Cheshire cat slowly fades away, leaving nothing but its smile. You will recall that the author of *Through the looking glass* and *Alice in Wonderland*, Lewis Carroll, was a determined teacher of geometry.

We come to two essential geometrical definitions. The breadthless string that in its motion constructs a wall without thickness is called a *line*. The vanishingly thin string wall is called a *plane*. As the joint between two walls suggests, the intersection of two planes is a line. Two lines intersect in a *point*. A plane has length and breadth but no thickness; a line, length, but no thickness or breadth; a point, neither length, nor breadth, nor thickness. You cannot have one

Fig. 2.1.1 Opening page of the first printed edition of Euclid (Venice, 1482). The diagrams illustrate the unpicturable: a breadthless line, a dimensionless point, a depthless plane. From Beretta, *Starry messenger* (1995), no. 2, p. 63, taken from Copernicus' copy, which a Swedish army stole from Poland in 1626 as a spoil of war.

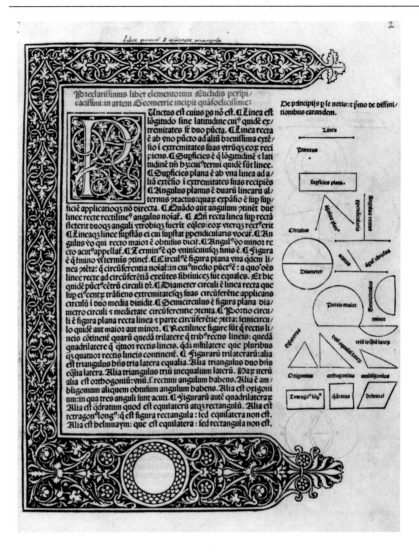

point bigger than another point: all have the same size, namely, none at all. Sometimes the old geometries began with illustrations of the fundamental geometrical concepts; but they could give no plausible picture of a point (Fig. 2.1.1). None of these things—points, lines, or planes—exist in the world of immediate experience. And yet they exist. By the powerful act of your imagination, you have been able to conceive them clearly and distinctly.

We can go further. The line we conceived by abstraction from the string can be *extended* indefinitely in either direction; the extended vertical runs through the earth's center and out the other side and up to the stars at both ends (Fig. 2.1.2). Similarly, the plane with

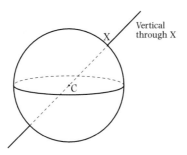

Fig. 2.1.2 The vertical at a point on the earth's surface.

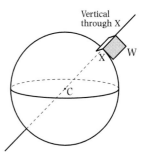

Fig. 2.1.3 A plane containing the vertical splits the earth in half. W is a wall containing the vertical through X; the plane of the shaded wall is the plane of the paper, which divides the earth into the hemisphere facing you and one behind the paper.

length and breadth equal to the wall's can be imagined to split the earth, and the heavens, into two hemispheres (Fig. 2.1.3). A point, however, is always a point, none bigger than any other.

That does not mean that all points have the same significance. A singular point is the center of a circle, which is a region of a plane entirely enclosed by a curved line. What makes the curved line the boundary or circumference of a circle is that every point on the boundary is the same distance from that singular point, the circle's center. In practice you draw a circle by opening a compass to the constant distance (the 'radius' of the circle) and following your intuition. The sharper your pencil, the closer your circle will resemble its definition.

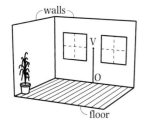

Fig. 2.1.4 The horizontal at the earth's surface.

Angles in general

Like the joint between two walls, the joint between either of them and the floor may be abstracted as a line. Figure 2.1.4 represents this floor; it also shows a vertical drawn through a point V in the middle of one wall and the intersection O of this vertical with the floor. Draw another line through O in the plane of the wall—or, better, on the blackboard or on paper. This new line can be drawn as you please provided that it does not run on top of (does not 'coincide with', as geometers say) either of the other lines in the diagram. Label the end of the new line 'P'. You have a picture that looks like Fig. 2.1.5. VO is the vertical. The inclination of OP to OV is called the *angle* VOP, usually written as ∠VOP; the point at which the lines constituting the angle meet, O in this case, goes between the other letters. Figure 2.1.5 contains five other angles: ∠POH (indicated by

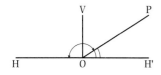

Fig. 2.1.5 Angles in the vertical plane.

Table 2.1.1   Words about angles

Word	Definition	Example
Right	Half a straight angle	$\angle$VOH', $\angle$VOH
Straight	The inclination that one segment of a single straight line has to another	$\angle$HOH'
Acute	Less than a right angle	$\angle$POH'
Obtuse	More than a right angle	$\angle$POH
Complementary	Two angles that sum to a right angle	$\angle$H'OP, $\angle$POV
Supplementary	Two angles that sum to a straight angle	$\angle$H'OP, $\angle$POH
Vertical	Pointing toward or away from the earth's centre	
Horizontal	Perpendicular to the vertical	

the arc with double arrows), $\angle$POH' (indicated by the double arc), $\angle$VOH', $\angle$VOH, and (though it may appear an odd angle) $\angle$HOH'.

The angles VOH' and VOH (which, you recall, are abstractions of the intersection between the vertical defined by a plumb bob and the joint between the wall and the floor) are very special angles. First, we shall take it for granted that they are equal, that VO divides equally, or 'bisects', the 'angle' defined by the straight line HOH'. Angles like $\angle$VOH are called 'right angles'; we agree further that all right angles, wherever found, are equal. We take the equality for granted because it seems obvious; no one has yet been able to prove it, except by assuming things less obvious. Euclid makes the equality a 'postulate', the fourth of five he put at the beginning of his text, just after the definitions of point, line, surface, and so on. Angles like $\angle$HOH' are called 'straight angles'; since $\angle$HOH' = 2 right angles, all straight angles are equal.

The Greeks liked to give a name to everything. An angle like $\angle$POH', less than a right angle, is called 'acute'; one like $\angle$POH, more than a right angle but less than a straight angle, is 'obtuse'; pairs of angles like $\angle$H'OP and $\angle$POV, which sum to a right angle, are 'complementary'; pairs of angles like $\angle$H'OP and $\angle$POH, which sum to a straight angle, are 'supplementary'. Table 2.1.1 repeats these definitions. Finally, lines meeting at right angles are said to be 'perpendicular', and any line, like HOH', perpendicular to a vertical, is said to be 'horizontal'.

Parallels

Suppose two lines OV, O'V' perpendicular to a third OO' (Fig. 2.1.6). The line OO' is the perpendicular distance between the lines. Euclid took as a fundamental assumption of his abstract geometry that no matter how far the lines OV, OV' might be extended on either side of

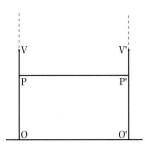

Fig. 2.1.6 Parallels in the vertical plane.

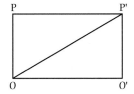

Fig. 2.1.7 The rectangle.

OO′, they never meet. The perpendicular distance between them remains the same no matter where it is measured. Lines that lie in the same plane but never meet no matter how far extended are called 'parallel'. The assumption with which Euclid grounds the theory of parallels, his so-called 'fifth postulate', considers a more general case than two lines perpendicular to a third line. It will be presented in §2.3.

Suppose another perpendicular distance between OV and OV′, say PP′. Since OO′ and PP′ are both perpendicular to OV, they are parallel. Geometers call a figure like OO′P′P, which has four sides and four right angles, a 'rectangle'. The opposite sides of the rectangle, being the perpendicular distances between pairs of parallel lines, are equal in length: OP = O′P′, OO′ = PP′. Note that the sum of the angles in a rectangle is four right angles. This observation leads to a far-reaching conclusion.

Draw a line, called a 'diagonal', from one corner of the rectangle to the opposite corner (the one farthest away). You have the situation indicated in Fig. 2.1.7, which contains two spaces, OO′P′ and OP′P, each entirely enclosed by three straight lines; such spaces are called 'triangles' and indicated by '△'. Evidently △OO′P′ and △OP′P have much in common: an equal angle (∠P′O′O = ∠OPP′, since all right angles are equal), and three pairs of equal sides (OO′ = PP′ and PO = P′O′, since POO′P′ is a rectangle, and OP′, the common diagonal). We might conclude that the triangles themselves are equal, by which we mean that for every part of one (parts being lines and angles) there is a corresponding equal part of the other. Do the remaining parts satisfy the condition, that is, do ∠P′OO′ = ∠OP′P and ∠P′OP = ∠OP′O′?

Several ways of answering the question suggest themselves. One is to declare the equality obvious: it appears to hold and the imagination can scarcely picture it to be otherwise. But we often deceive ourselves in our imaginings and perhaps a surer or solider way exists. Take a solider way. Cut out the triangle OO′P′ and lay it on OPP′ so that the right angles coincide and O′O lies on PP′. The triangles fit perfectly—or would, we suppose, if the drawing and cutting had been more exact. Still, the little misfits might indicate that the equality does not hold under all circumstances; and, in any case, demonstration by cutting and pasting can scarcely be allowed in Euclidean geometry, in which plane figures have no solidity. A third way is to superpose the two triangles in your mind; imagine it done; admire the perfect fit; and persuade yourself that all the corresponding parts of the triangles are equal. You might say that these demonstrations are almost as identical as the triangles. Indeed, the choice is largely one of taste. Euclid uses the third; no fastidious Euclidean can use the second; and many connoisseurs, thinking that Euclid's argument of superposition does not fit Euclidean geometry any better than cutting and pasting, prefer the first. Euclid himself appears not to have liked his own argument, since he uses demonstration by superposition only rarely, and in our century a committee of the British Mathematical Association advised against using it at all.[1]

Congruence

Euclid gained more from his reluctant application of superposition than we have. In his fundamental proposition I.4,[2] he considered any two triangles, not just ones containing a right angle. The triangles ABC and A′B′C′ in Fig. 2.1.8 are drawn with sides AC = A′C′ and CB = C′B′, and ∠ACB = ∠A′C′B′; note the convention on the figure for indicating the equality of the parts. By placing C′ on C so that A′ coincides with A, you make B′ coincide with B, etc.; the triangles are equal because they were drawn with two sides and the *included* angles equal, or, as geometers say, because of 'side-angle-side' (*s.a.s.*) equality. Note that the angle in question *must* be included between the equal sides. As appears from Fig. 2.1.9, if this condition is not met the triangles generally will not be equal. If you put ∠A′ on its equal ∠C so that C′ coincides with B, B′ will not fall on A except in the special case AC = A′B′.

The 'equality' of triangles appears as something more extensive than the equality of parts. Two triangles can have some parts equal but not all; two right triangles, for example, have one equal part, namely the right angle, but they need have no other parts equal. However, if the triangles themselves are 'equal', all the corresponding parts must be so too. Geometers have introduced a special name and symbol to indicate the wide meaning and implication of equality in the case of triangles and rectilinear figures that can be composed

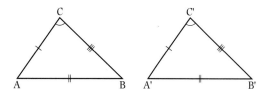

Fig. 2.1.8 Congruence of triangles by *s.a.s.*

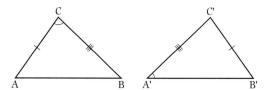

Fig. 2.1.9 The equal angles referred to in *s.a.s* must lie between the equal sides.

1. Health, in *Euclid, 1,* 225, 249; *The teaching of mathematics in the schools* (1923), in Beatley, *MT, 28* (1935), 404.

2. The roman numeral refers to the book, the arabic to the proposition, in Euclid's *Elements of geometry.*

of triangles. The name is 'congruence'. They write $\triangle ABC \cong \triangle DEF$ and read 'triangle ABC is congruent to triangle DEF'.

The notion of congruence suggests a technique that is of the first importance in solving geometrical problems. Often one can test the possibility of a demonstration by asking whether enough information is given to allow the unambiguous construction of a figure. Only if figures are uniquely determined can they be proved congruent. Giving two sides and the included angle of a triangle is such a specification. You see intuitively that the specification amounts to fashioning a unique and rigid structure; the ends of the given lines, being fixed, need only to be joined to complete the triangle.[3] But if the given angle is not included between the given sides, the triangle is not only not determined, it may not be possible. This arresting claim will be clear from Fig. 2.1.10.

Let CAD be the given angle and AC one of the given sides. Open your compass to the separation x equal to the length of the second given side. Place the compass point at C and draw a circle of radius x. All possible locations of the third angle of a triangle with given sides AC and x, and given (non-included) angle CAD, must lie on this circle. If x is less than the perpendicular CE on AD, no triangle is possible, since the circle of radius x centered at C will not intersect AD. If $x = $ CE, one triangle may be drawn, the right triangle AEC. If x lies between AE and AC, or exceeds AC, two solutions are possible, for example, $\triangle ABC$ and $\triangle AB'C$, and $\triangle ACD$ and ACD', respectively. If $x = $ AC, there is again but one possibility, the triangle with sides equal to AC and base equal to twice AE.

The same sort of reasoning indicates that a triangle can be made perfectly rigid, and so be given completely, by specification of all three of its sides. Let the sides be AB, BC, AC. Lay out AB (Fig. 2.1.11). With A as center, draw a circle of radius $a = $ AC; with B as center, draw a circle of radius $b = $ BC. If the circles intersect, they do

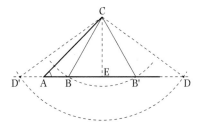

Fig. 2.1.10 Further demonstration that the angles in *s.a.s.* must be included between the sides.

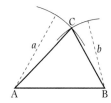

Fig. 2.1.11 Congruence by *s.s.s.*

3. Cf. Frank Land, *Language* (1963), 185.

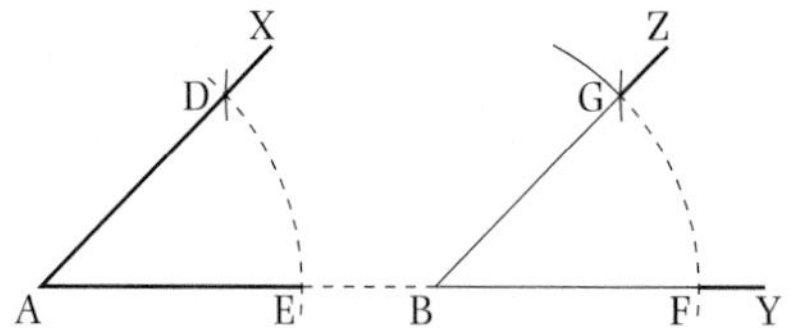

Fig. 2.1.12 To copy an angle.

so at the unique point C (or its mirror image below AB). From this argument, you would expect that two triangles with all three sides respectively equal would be congruent. You would be right. Euclid's proof of congruence by side-side-side (s.s.s.), his I.8, is deferred to §3.1.

Congruence by s.s.s. suggests a way to copy an angle and so to draw parallels. Let ∠XAY (Fig. 2.1.12) be given. An equal angle can be drawn at B as follows. Describe circles of equal radius around A and B intersecting AX at D and AY at E and F. Now place the compass point at E and open it to the radius ED; move the point to F and draw around F a circle of radius ED. Let the intersection of this circle with that previously drawn around B be G. Draw ZGB through G. You have ∠ZBY = ∠XAY since ΔDAE ≅ ΔGBF by s.s.s. (DA = GB, AE = BF, DE = FG by construction). We have further AX∥BZ. A proof will be offered shortly.

Congruence by s.s.s. also suggests a way to draw a right angle without copying and so to drop a perpendicular from a point onto a line. Let C in Fig. 2.1.13 be the point, AB the line. Open your compass to a convenient distance for drawing a circle around C that will intersect AB twice, as in D and E. Then with D and E as centers, and the same opening, draw arcs intersecing at F. From the construction, DC = CE = EF = FD. Draw the line CF; it intersects the line AB at right angles at G as required. To show that ∠CGE and ∠CGD are rt.∠s, we need only to show that they are equal, since they sum to the st.∠DGE. The equality can be demonstrated in two steps. First, ΔFCD ≅ ΔFCE by s.s.s., whence ∠1 = ∠2. Second, ΔCGE ≅ ΔCGD by s.a.s., whence ∠CGE = ΔCGD = rt.∠.

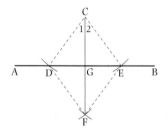

Fig. 2.1.13 To drop a perpendicular from a point onto a line.

The important construction just given allows you not only to drop perpendiculars from points to lines, but also to erect a perpendicular at any point on a line. For let that point be G in Fig. 2.1.13. Then locate points D and E equidistant from G with the compass and with any convenient constant setting, draw arcs about D and E that intersect at C (Fig. 2.1.14). CG then is perpendicular to AB. Furthermore, since by construction DG = GE, CG divides the segment DE into two equal parts at G. Geometers say that CG is the 'perpendicular bisector' of DE.

Table 2.1.2 lists geometers' words and symbols encountered so far.

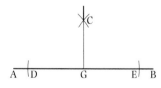

Fig. 2.1.14 To raise a perpendicular at a point on a line.

Table 2.1.2 Some useful symbols

Symbol	Significance
∠	Angle
=	Equals or is equal to
Δ	Triangle
≅	Is congruent to
‖	Is parallel to
⊥	Is perpendicular to

Alternate and vertical angles

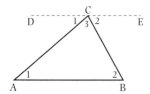

Fig. 2.1.15 Alternate interior angles.

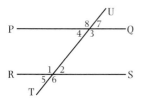

Fig. 2.1.16 The sum of the angles in a triangle is a straight angle (Euclid I.32).

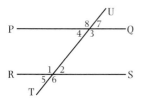

Fig. 2.1.17 Alternate and vertical angles.

Let us return to our rectangle (Fig. 2.1.7), redrawn as Fig. 2.1.15 with its angles labelled more conveniently. Since ΔOO′P′ ≅ ΔOPP′, ∠1 = ∠3 and ∠2 = ∠4. But we know that ∠2 and ∠3, and ∠1 and ∠4, are complementary. It follows that ∠1 and ∠2, and ∠3 and ∠4, are complementary too. Therefore the sum of the angles in each of the right triangles OO′P and OPP′ is two right angles, or one straight angle. In fact, the proposition (Euclid I.32) that the angles of a triangle sum to one straight angle is true of all triangles, as appears from Fig. 2.1.16, where DCE is parallel to AB. Notice that ∠DCA = ∠CAB; they stand in the same relation as ∠1 and ∠3 in Fig. 2.1.15. Similarly, ∠BCE = ∠ABC (as you can persuade yourself, if you need persuasion, by considering the angles formed by the diagonal PO′). Hence, by adding ∠ACB to ∠ACD (= ∠CAB) and ∠BCE (= ∠ABC), we have the straight angle DCE. This proposition, that a triangle's angles sum to a straight angle, is one of the most important and useful in geometry.

The angles formed by lines (or 'transversals', as they are called) cutting parallel lines have special names. In Fig. 2.1.17, the parallels PQ, RS, are cut by the transversal TU. 'Alternate' signifies that the angles under consideration fall on opposite sides of the transversal. The angles 1, 2, 3, and 4 are called 'interior', because they fall within the parallels. Since ∠1 = ∠3 and ∠2 = ∠4, geometers say that 'when two parallel lines are cut by a transversal, the alternate interior angles are equal' (Euclid I.29). The angles 5, 6, 7, 8 are 'exterior'. Certain equalities hold for them too. These equalities are easily demonstrated with the concept of 'vertical angles'.

Vertical angles are formed by and at the intersection, or vertex, of two straight lines. Figure 2.1.18 shows two lines AB and CD forming the pairs of 'vertical' angles 1, 3 and 2, 4. We want to show that ∠1 = ∠3 and ∠2 = ∠4. We have ∠1 + ∠2 = ∠2 + ∠3 = a straight angle, whence ∠1 = ∠3, and similarly for ∠2 and ∠4. This is the

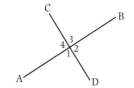

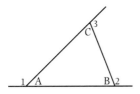

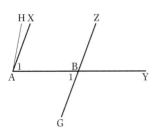

Fig. 2.1.18 Vertical angles are equal (Euclid I.15).

Fig. 2.1.19 Exterior angles.

Fig. 2.1.20 *Reductio ad absurdum*, according to Euclid I.27 and 1.29.

content of Euclid I.15: alternate vertical angles are equal. They are extremely useful.

As for the exterior angles, 5, 6, 7, 8, we have $\angle 6 = \angle 8$ and $\angle 5 = \angle 7$. These equalities follow because $\angle 5 = \angle 2$ and $\angle 7 = \angle 4$ (they are pairs of vertical angles); but $\angle 2 = \angle 4$ (alternate interior angles); whence $\angle 5 = \angle 7$. To deduce $\angle 5 = \angle 7$ from the equalities $\angle 5 = \angle 2$, $\angle 7 = \angle 4$, and $\angle 2 = \angle 4$, we must invoke what Euclid calls the 'common notion' that things equal to the same things are equal to one another. The same argument applied to angles 6, 1, 3, and 8 delivers $\angle 6 = \angle 8$. We say that alternate exterior angles are equal.

One more kind of angle needs a brief notice. In Fig. 2.1.19, angles 1, 2, 3, each of which is made between a side of the triangle and an adjacent side extended, are called 'exterior' angles. Each is equal to the sum of the opposite 'interior' angles (Euclid I.32). For example, $\angle 1 + \angle A = \text{st.}\angle = \angle A + \angle B + \angle C$ (because the sum of the angles of a triangle equals one straight angle), whence $\angle 1 = \angle B + \angle C$. Similarly, $\angle 2 = \angle A + \angle C$, $\angle 3 = \angle B + \angle A$. That brings us to a second table of angles. (Table 2.1.3) and a few words about plane figures (Table 2.1.4).

The promised proof that AX ∥ BZ in Fig. 2.1.12 may now be given. The essential parts are redrawn as Fig. 2.1.20, where $\angle XAB = \angle ZBY$ by construction. Extend ZB to G. Then, by vertical angles, $\angle ABG = \angle ZBY$; but $\angle ZBY = \angle XAB$ by construction. We therefore have a transversal cutting two lines so that the alternate internal angles are equal. We know that if the lines are parallel, the angles are equal. Is the reverse (which geometers call the 'converse') true? It must be. For, if XA is not parallel to BZ, draw a line through A, AH say, that is parallel. Now since, by hypothesis, AH ∥ BZ, $\angle HAB = $

Table 2.1.3 More words about angles

Word (subject)	Definition	Example
Interior (∥)	Any angle formed between parallels by a transversal; or	∠s1,2,3,4, Fig. 2.1.17
Interior (Δ)	Any angle of the three angles of a Δ	∠sA,B,C, Fig. 2.1.19
Exterior (∥)	Any angle formed outside parallels by a transversal; or	∠s5,6,7,8, Fig. 2.1.17
Exterior (Δ)	Any angle formed between a side of a triangle and an adjacent side extended	∠ s1,2,3, Fig. 2.1.19
Vertical	Any angle formed by the intersection of two straight lines	∠ s1,2,3,4, Fig. 2.1.18
Alternate	Falling on opposite sides of a line or a vertex	

Table 2.1.4 Words about plane figures

Word	Definition
Rectangle	Four-sided figure with four interior right angles
Square (□)	Rectangle two adjacent sides of which are equal
Diagonal	Line drawn between opposite corners of a rectangle
Transversal	Line intersecting parallel lines
Radius	Distance from the centre of a circle to any point on it
Diameter	Line, equal to twice the radius, from one point to another on a circle that passes through its centre
Equilateral	Describes a plane figure all of whose sides are equal

∠ABG = ∠1. But ∠HAB = ∠HAX + ∠1; hence, if AH were parallel to ZB, ∠HAB would have to be both equal to and greater than ∠1, which is impossible. Similarly if AH ∥ BZ were drawn inside ∠XAB, ∠HAB would have to be both equal to and less than ∠1. Therefore AH must coincide with AX, and the converse is true. Indeed, Euclid first proves that if the alternate interior angles are equal, the lines are parallel (I.27) and thence demonstrates as the converse that parallel lines make equal alternate interior angles (I.29). But, as will appear, not all converses are true.

Euclid's argument in obtaining I.27 parallels that used here, that is, he assumes a proposition counter to what he wants to prove and shows that the assumption leads to a contradiction. This method of reasoning is called 'reduction to absurdity'. We shall have to invoke it often.

Ad pleniorem scientiam ('for fuller understanding,' as the scholastic philosophers used to say, when multiplying examples) of the concepts so far introduced:

APS 2.1.1. Figure 2.1.21 contains several pairs of supplementary, complementary, and vertical angles. Can you name them all?

APS 2.1.2. Write down all pairs of vertical, and alternate interior and exterior angles that occur in Fig. 2.1.22. AB, CD, and EF are supposed to be parallel.

APS 2.1.3. In Fig. 2.1.23, which angles are external? Which internal? Which complementary? Which supplementary? Which acute? Which obtuse? Which right? Are there any parallel lines? Which ones? The sign ⌐ indicates a right angle.

APS 2.1.4. Which of the triangles in Fig. 2.1.24 are congruent, if any?

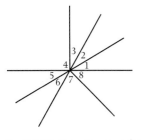

Fig. 2.1.21 Exercise on angles.

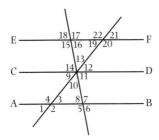

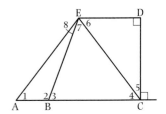

Fig. 2.1.22 Another exercise on angles.

Fig. 2.1.23 Yet another exercise on angles.

Fig. 2.1.24 Exercise on congruence.

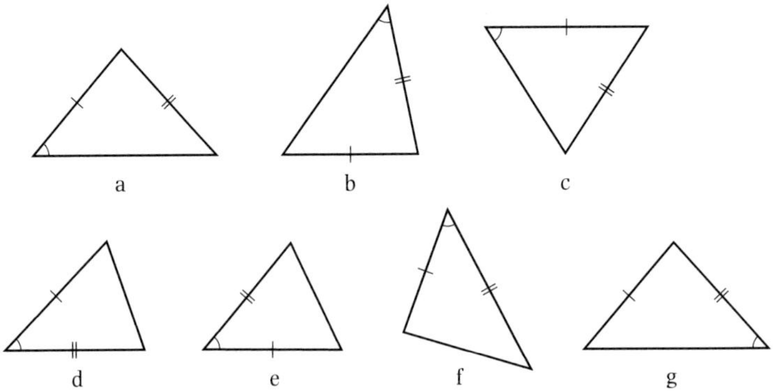

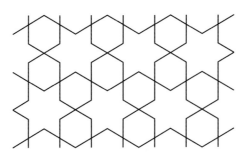

a   b   c

d   e   f   g

Fig. 2.1.25 Islamic interlace pattern to serve as exercise on angles and parallels.

APS 2.1.5. Describe the various systems of parallel lines, and of vertical and complementary angles, in Fig. 2.1.25, which represents a standard interlace pattern in Islamic art. You might want to introduce letters into the diagram to refer to in your description of representative parts of the pattern.

APS 2.1.6. Here is how to draw a line from a given point making a given angle with a given line. Let F (Fig. 2.1.26) be the point, DE the line, ∠ABC the angle. Copy ∠ABC at D by the method described in the text (in Fig. 2.1.12). That produces the situation in Fig. 2.1.27. Now join D and F and draw ∠DFC = ∠A′DF. Then,

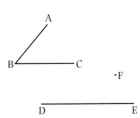

Fig. 2.1.26 To draw a line from a given point making a given angle with a given line.

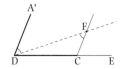

Fig. 2.1.27 The task of
Fig. 2.1.26 solved.

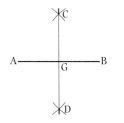

Fig. 2.1.28 To draw the
perpendicular bisector of a line.

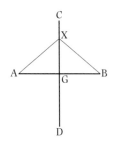

Fig. 2.1.29 The locus of all
points equidistant from two
given points.

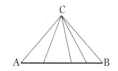

Fig. 2.1.30 Possible definitions
of the 'distance' between a point
and a line.

by alternate interior angles (Euclid I.27), FC∥A′D and ∠FCE = ∠A′DC = ∠ABC, which completes the task at hand.

APS 2.1.7. In Fig. 2.1.14, CG is the perpendicular bisector of the line segment DE. Can you draw the perpendicular bisector of any line AB? You can employ a modification of the method used earlier to construct CG. Set your compass at any convenient opening greater than half of AB. Draw intersecting arcs centered on A and B above and below AB, as in Fig. 2.1.28. Connect the intersections C, D. Show that CD is perpendicular to AB and bisects it at G.

Loci

A point may be given in the plane, like F in Fig. 2.1.26 or unambiguously defined by the intersection of lines or circles, as in Fig. 2.1.28. Geometers also like to define a set or collection of points according to some rule, for example, the set of points equidistant from a given point. Such a set is called a locus; in the example just given, the set of points equidistant from a given point, the locus is a circle with the given point as center and the equal distance as radius. The locus therefore differs essentially from a given point: any number of points make up the locus but a given point is singular and unique. However, the locus has this in common with the point, that all points on it uniquely satisfy a certain condition, namely the rule that defines the set.

We are already acquainted with an important locus, that of all points equidistant from two given points. Let the two points be A, B in Fig. 2.1.28. By construction, both C and D are equidistant from A and B. The obvious generalization, that every point on CD (and CD extended) is equally distant from A and B, holds. For consider any point X on CG (Fig. 2.1.29): since AG = GB (G is the midpoint of AB), CG is common, and ∠XGA = ∠XGB = rt.∠, ΔXGA ≅ ΔXGB by *s.a.s.* The corresponding parts XA, XB therefore are equal. We can say that the locus of points equidistant from two given points is the perpendicular bisector of the line joining the given points.

Another important locus is that of all points equidistant from two intersecting straight lines. You may well object that the locus so stated is indeterminate, since the distance of a point from a line may be any value you please (Fig. 2.1.30). By 'the distance from a point to a line' geometers mean the perpendicular, and so make the definition unique. Let us assume that point P in Fig. 2.1.31 satisfies the definition of the locus, that is, that PX = PY, AB and CD being the given straight lines intersecting at O. Join O and P. Do you see that ΔPOX ≅ ΔPOY? You have PX = PY and ∠PYO = ∠PXO = rt.∠ by definition of the locus under consideration, and OP is common.

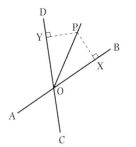

Fig. 2.1.31 Congruence by *h.s.*

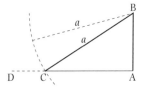

Fig. 2.1.32 Further to congruence by *h.s.*

'But', you object, and rightly, 'the equal angle is not included between the equal sides'. Consequently congruence does not follow via *s.a.s.* However, in the special case of right triangles, congruence does hold if the sides opposite the right angle, and corresponding legs adjacent to the right angle, are equal.

You can make this assertion plausible by imagining yourself drawing a right triangle given the side opposite the right angle (which is called the 'hypotenuse') and either of the other two sides or legs (Fig. 2.1.32). Let AB be the leg; make a right angle at A and extend its side indefinitely toward D, open your compass to the length *a* of the given hypotenuse; place the compass point at B and describe an arc cutting AD at C. ΔABC is the unique solution to the problem. Compare the discussion of Fig. 2.1.10, in which a unique solution to the construction of a triangle given two sides and an angle not included between them is a right triangle.[4]

If we admit the congruence of right triangles by hypotenuse-side (*h.s.*), we have, in Fig. 2.1.31, ΔPYO ≅ ΔPXO. Hence the corresponding parts ∠YOP and ∠XOP are equal; each of them must therefore be half of the angle between the given lines; the line OP therefore bisects the angle DOB. Now the point P was any point that satisfied the definition of the locus. Hence all points on the line OP produced constitute the desired locus. We can say that the locus of points equidistant from two intersecting straight lines is the bisector of the angle between them.

APS 2.1.8. You will have no trouble showing that the locus of points equidistant from two parallel lines is a line parallel to both and halfway between them.

APS 2.1.9. The concept of locus is sometimes useful in solving problems that otherwise might be very tedious. A well-known theorem, not in Euclid, relates the angle bisectors of two external angles and one internal angle of a triangle as follows. The external angles at A and B in ΔABC are bisected as indicated in Fig. 2.1.33. Let the bisectors XB, YA extended meet at a point D (we skip the important question whether they meet at all). Join D to the remaining vertex C. The line CD bisects ∠ACB. This extraordinary property follows directly from the definition of the angle bisector as the locus of points equidistant from two intersecting straight lines. For drop the perpendiculars from D onto AB and onto CA and CB extended; call the feet of these perpendiculars E, F, G, respectively. Since by construction the external angles at A and B are bisected by YA and XB, so are the angles FAE and EBG

4. This amounts to a proof by superposition. Cf. Hadamard, *Leçons* (1931), 36–7.

Fig. 2.1.33 The locus of all points equidistant from two given lines.

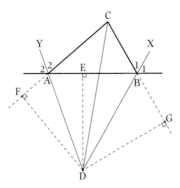

by AD and BD. (By the theorem of the equality of vertical angles, Euclid I.15, ∠CBX = ∠DBG = ∠1, etc.) Since D is common to the two angle bisectors BD and AD, the locus theorem requires that DE = DF = DG. Now, DF and DG are the distances of D to the sides of ∠ACE; since they are equal, then, according to the locus definition, D lies on the bisector of ∠ACB, that is CD splits the internal angle of C into two equal parts, as asserted. An alternative way of arriving at the same result is to see that ΔDFC ≅ ΔDGC (by *h.s.*) and hence that the corresponding angles ACD and BCD are equal.

We have accumulated a few more words (Table 2.1.5):

Table 2.1.5 Terms mainly about relationships

Locus	A set of points each of which satisfies the same condition
Bisector	A line that cuts a line or an angle in half
s.a.s.	Congruence of triangles via the equality of two pairs of corresponding sides and the included angles
s.s.s.	Congruence of triangles via the equality of three pairs of corresponding sides
h.s.	Congruence of right triangles (and right triangles only) via the equality of the hypotenuses and one pair of sides
Hypotenuse	The side opposite the right angle in a right-angled triangle
Leg	Either side of a right triangle not opposite the right angle

2.2. The size of the earth

The small corner of geometry so far explored does not yet allow us to measure the land in the manner of the old Egyptian tax collectors or design altars in the style of the Vedic priests. But it enables us to undertake a measurement very much grander, a measurement of the entire earth. Here is how to do it. Find a wall running north–south and imagine that the vertical plane through it is

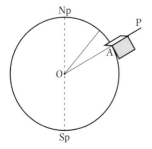

Fig. 2.2.1 The meridianal plane.

extended so as to include the north pole of the earth and all the stars that you can see directly overhead. The intersection of this plane with the earth is a circle, represented by NpASp in Fig. 2.2.1. You stand at A, beside a long stick AP that marks the vertical (it would coincide with a plumb line suspended from P). Now suppose that the sun also lies in the plane AONp—which happens once every day at every place on earth. This special time, when the sun comes into the vertical plane running north–south at any place, marks and makes noon there: noon is defined by sundials, though not by watches, as the moment the sun stands due south or due north of the observer.

By Greek measurement

At some places on the earth's surface—any place between the tropics of Cancer and Capricorn, to be precise—the sun can appear directly overhead. Suppose that A lies between the tropics. When the sun stands directly overhead at A at noon, its rays fall on the earth parallel to the stick AP, which therefore casts no shadow.

Now suppose that on the very day that the sun is overhead at A, it casts a shadow equal to some fraction, say, $1/f$ of a straight angle, at a more northerly point B in the plane AONp (Fig. 2.2.2). Since the plane AONp is the north–south plane at B, it is noon at B when it is noon at A. Because the sun is so far from us, its rays arrive at the earth almost in parallel; they are so drawn in the figure.

Let us imagine that you have hiked from A to B and measured the distance roughly by counting your steps. You have a friend and collaborator at B, who had the foresight to measure the angle Q at noon on the day that the sun stood overhead at A; your friend is more advanced in geometry than you, and knows how to obtain $\angle Q$ by measuring the shadow cast by the long pole BQ; a little later, you will know how to do it too. Now you reason as follows: $\angle Q = \angle BOA$ (they are alternate interior angles owing to our treating the sun's rays at the earth as parallel); $\angle BOA$ is the separation of A and B as seen from the earth's center. Now $\angle Q$, according to friend B's estimate, is $1/f$ of a straight angle, and the distance corresponding to it, as you discovered after a very long walk, is 10 000 paces; had you walked f times as far, you would have covered a distance corresponding to a straight angle as seen from the earth's center, that is, a distance equal to that from the North pole to the South pole. Therefore an entire circuit of the earth, its circumference, is $2 \times 10\ 000/f$ paces.

The method just outlined is very ancient. A friend of Archimedes named Eratosthanes and nicknamed 'Beta' because he was second-best at everything (beta is the second letter of the Greek alphabet)

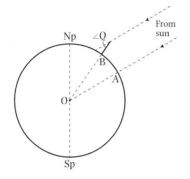

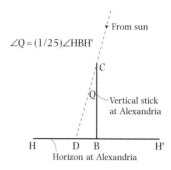

Fig. 2.2.2 Sun's rays in the meridianal plane, marking and making local noon.

Fig. 2.2.3 Shadow of a vertical stick, for finding the size of the earth.

first applied the method in Egypt. He lived in Alexandria, where he headed the famous library containing thousands of scrolls filled with the learning of the Greeks and the Egyptians. He knew a place due south of the Library at which the sun came into the zenith (stood directly overhead) at noon on midsummer day; this place, the modern Aswan, corresponds to point A on Fig. 2.2.2. To complete his measurement, he had to estimate the distance AB from Aswan to Alexandria and to observe the angle Q made by the shadow of a vertical stick erected in the library's grounds at noon on midsummer day. He estimated the distance at 5000 stades, a number so round as to create the suspicion that it was more a guess than a measurement; and he observed the angle to be about a twenty-fifth of a straight angle (Fig. 2.2.3). As the ancient chronicler Cleomedes, from whom we have the report of Eratosthenes' achievement, wrote, we need only to invoke the rule 'proved by the geometers, [that] straight lines falling on parallel straight lines make the alternate angles equal', to transform the data into a value for the earth's circumference. Eratosthenes made the earth $2 \times 25 \times 5000 = 250\,000$ stades around.[5]

How good was this result? Apparently, far better than the rough ingredients in the calculations. According to the Roman writer Pliny, who died in the eruption that destroyed Pompeii in 79 AD, Erastothenes' stade equalled 300 royal Egyptian cubits; and since the length of the cubit is known from existing Egyptian monuments and other sources to be 0.525 meters, or 1.72 feet, Eratosthenes' earth was about 40 000 km or 25 000 miles in circumference.[6] This result falls astonishingly close to the modern value. Pliny and his contemporaries ascribed to Eratosthenes the number 252 000

5. Cleomedes, in Health, *Greek astro.* (1932), 109–12: Health, *Greek math.*, (1921) 2, 104.
6. Ibid., 2, 107.

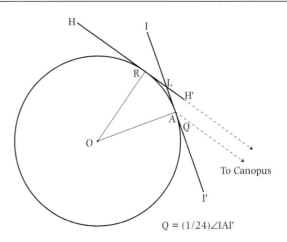

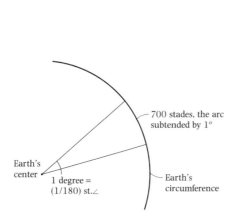

Fig. 2.2.4 Eratosthenes' measurement of the earth's circumference.

Fig. 2.2.5 Posidonius' measurement of the same.

stades, probably because it is more convenient than 250 000 for people who, like the ancient geometers and astronomers, used a numerical system based on 60 rather than 10, and divided a straight angle into 180 parts or degrees (written 180°). With a circumference of 252 000 stades, each part, or degree, of a straight angle would cut off, or 'subtend', an arc of 252 000/2(180) = 700 stades on the earth's surface (Fig. 2.2.4). Now 252 000 stades corresponds to 24 662 miles, about 150 miles shorter than the modern value.

The Greeks had a second method of obtaining the earth's size that used a fixed star rather than the sun. The method, for which the Greek philosopher Posidonius, the teacher of Cicero, has garnered the credit, is illustrated in Fig. 2.2.5. Posidonius noticed that when the bright star Canopus can be seen just above the horizon to the south at Rhodes, it stands at 1/24th of a straight angle above the horizon at Alexandria. Assuming, what is more or less true, that Rhodes and Alexandria lie on the same noon circle or meridian (that is, that Rhodes, Alexandria, and the center of the earth lie in the same plane), take HH′ as the horizon at Rhodes (R) and II′ as the horizon at Alexandria (A); note that the horizon lines are perpendicular to the earth's radii, that is, to the verticals, at each place, as the definition of 'horizontal' requires. Now ∠I′ACanopus = ∠ROA, the angle subtended at the earth's center by the arc connecting Rhodes and Alexandria, as you can see from the following argument.

In Fig. 2.2.6, which re-presents the relevant part of Fig. 2.2.5, ∠I′LH′=∠I′ACanopus since HH′‖ACanopus. We have further that ∠RLA equals the straight angle HLH′ less the elevation of Canopus ∠I′LH′, that is, ∠RLA = st.∠ − ∠I′LH′. Another value for ∠RLA may be obtained from the right triangles OAL and ORL, namely, ∠RLA = st.∠ − ∠ROA. To arrive at this second value, invoke Euclid I.32 (the

Fig. 2.2.6 Further to
Posidonius' method.

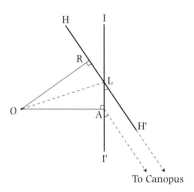

sum of the angles in a triangle—any triangle—equals a straight
angle). The four-sided figure ROAL consists of two triangles, ROL and
LOA; hence the sum of its angles, ∠ROL + ∠RLA, and the rt.∠'s
ORL and OAL, are 2st.∠'s. Since 2 rt.∠'s = 1st.∠, ∠RLA = st.∠ –
∠ROA. If we equate our two values of st.∠ – ∠RLA, we obtain
st.∠ – ∠RLA = ∠ROA = ∠I'LH' = ∠I'ACanopus. Posidonius' estimate that ∠I'ACanopus = 1/24th st.∠, or 7.5°, made ∠ROA =
1/24th st.∠ and the arc AR 1/48th of the earth's circumference.

It remained only to measure the arc AR from Alexandria to
Rhodes. Since they are separated by water, the distance between
them could not be paced out. With inspired inconsistency, Posidonius
took the distance from Eratosthenes' calculations, which, as we
know, made the earth's circumference over 24 000 miles. Since,
according to Eratosthenes, arc AR = 3750 stades, Posidonius' earth
had a circumference of (48)(3750) = 180 000 stades, or a little
under three-quarters of the size of Eratosthenes' earth.[7] Posidonius
erred in his observation, or supposition, of the elevation of Canopus,
which he should have made 5.25° rather than 7.5°.

By Columbus' speculations

The value of 180 000 stades, transmitted to posterity through the
widely read work of the Greek geographer Strabo, has had more profound consequences than any other geometrical error ever commit-
ted. Columbus made decisive use of it in arguing the practicability of
his proposed voyage to the East Indies by sailing to the West.
Columbus argued his case for support before learned geometers and
geographers who knew the world is round. They were charged to
decide whether the proposed voyage over the immense ocean in
small wooden ships had a chance of success. They did not worry
that the ships might fall off the edge of a flat earth but that they

7. Heath, *Greek math.* (1921), 2, 219–20, and *Greek astr.* (1932), 121–3.

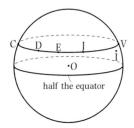

half the equator

Fig. 2.2.7 Columbus's conception of the distances to the Indies. C,D, and E are the easternmost points of China according to Ptolemy, Marinus, and Marco Polo, respectively, and V is the westernmost point of Europe.

might sink from rotting timbers or that the crews might die from thirst or hunger. The decision hung on estimates of the likely length of the voyage, which, in turn, hung on estimates of the size of the Earth.

Columbus argued before the learned advisors of the king and queen of Spain as follows. According to the great Greek astronomer and geographer Ptolemy (who lived long after Eratosthenes, Posidonius, and Strabo), a traveler who began his journey at the westernmost point in Europe and continued east until returning to his starting point would do the first half of his journey over land and the second half over water. In other words, according to Ptolemy the Eurasian land mass occupies about 180° of a parallel of latitude in the North temperate zone. In Fig. 2.2.7, O is the earth's center, V the westernmost point in Europe (Cape St. Vincent in Portugal), C the easternmost point of China, according to Ptolemy (that is, arc VC = 180°). Ptolemy's estimate implied half land, half sea, around the parallel VCV. That left too much sea between Portugal and China going West to suit Columbus. He therefore seized on the baseless speculation of another Greek geographer, Marinus of Tyre, whom Ptolemy had mentioned only to criticize. Following Marinus, Columbus supposed the distance from Portugal to the easternmost point of China proceeding East to be 225°.

Since the distance from Cape St. Vincent to Shanghai going East along the parallel VC is about 130°, Columbus, by assuming it to be 225°, shrank the distance from Portugal to China proceeding West from the true value of 360°–130° = 230° to 360°–225°=135°. (The angular distance around a circle, as seen from its center, is two straight angles, or 2 × 180° = 360°.) Seizing on Marinus was only the first of several steps Columbus took to reduce the earth to a manageable size. By his time there existed a popular travel book about China written by the Venetian merchant Marco Polo. Owing to the invention of printing in the middle of the fifteen century, Columbus could afford to acquire his own copy of Polo's book of travels, on which he made some notes. This book, together with others Columbus marked up, have survived. The jottings show how he went about decreasing the estimate of the West–East distance he would have to travel to reach the Indies. According to Polo, using the licence to exaggerate given to travelers and fishermen, the east-ernmost point in China lay 28° further from Portugal (going East) than even Marinus had made it; and the Indies came somewhere around the great island of 'Cipangu' (Japan), which Polo put 30° further East than the end of China.

Polo's placement of Cipangu is marked J on Fig. 2.2.7; arc CJ = 45° + 28° + 30° = 103°. That left 180° – 103° = 77° from Cape St. Vincent to Japan (or the Indies) going West, that is, arc VJ = 77°.

Columbus planned to begin his voyage not from Cape St. Vincent, but from the Canary Islands (I on Fig. 2.2.7), some 9° to the West and a little South of V. Hence he would have only 68° to go. That still seemed a little far to him. Moreover, he disliked complicated numbers. So he arbitrarily knocked off another 8° and announced to the astonished committee of experts summoned by the queen that he would find the Indies 60° to the West of the Canaries, about a third of the angular distance that the highly respectable Ptolemy had supposed.

To judge the practicability of such a voyage, the distance in degrees had to be converted into distance in miles. It was here that Posidonius' error entered. Since Columbus wanted to minimize distances, he much preferred Posidonius' value for the earth's circumference to Eratosthenes'. Hence in his opinion 60° at the equator would amount to

$$\frac{60}{360} \times \frac{180\,000}{252\,000} \times 24\,660 \approx 3000 \text{ miles.}$$

(The first factor reduces Eratosthenes' circumference of 24 660 miles to a 60° arc; the second factor reduces Eratosthenes' earth to Posidonius'.) But Columbus did not intend to sail along the equator, but along a parallel of latitude, the circumference of which is smaller than that of the equator. Later (§3.2) we shall consider how to calculate the circumference of any latitude circle. The latitude of the Canary islands is about 23° N, which implies a circle with a circumference about 0.92 that of the equator (Fig. 2.2.8). Hence Columbus expected to find Japan and the Indies a little more than 2700 miles West of his starting point.

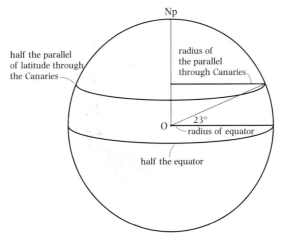

Fig. 2.2.8 Radius of the circle of latitude on which Columbus planned to sail.

Fig. 2.2.9 Columbus' conception
of the position of the Indies
superposed on a modern map.
From Fernández-Arnesto,
Columbus (1991), [xxi].

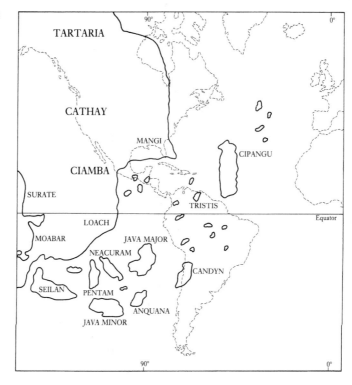

Columbus no doubt concluded his lecture to his examiners with
the opinion recorded in one of the marginal comments in his copy of
a popular book on cosmography. 'The end of Spain and the beginning of the Indies are not very distant...the sea [between] is naviga-
ble in a few days with a favorable wind.' He calculated he could do it
in 30 days. The experts laughed at him. Relying on Ptolemy, dis-
counting Polo, and preferring Eratosthenes (whose result had been
confirmed by more recent measurements), they would have calculated the gap at close to its true value of 12 000 miles. No sailing
ships of Columbus' time could have crossed so immense an expanse
of ocean. The committee so advised the king and queen. But Isabella,
harkening to those who, for various reasons, pushed Columbus' view
of the world, decided to take the modest risk of underwriting part of
the project. Private speculators subscribed the rest. Their admiral set
forth from the Canaries and, 33 days out, he struck land almost
exactly where he had expected, some 57° West of his starting point
(Fig. 2.2.9). He had not reached Japan, as he had expected, but a
New World.[8]

After presenting Eratosthenes' method of measuring the world by
using a stick and the principle of the equality of alternate interior

---

8. Fernández-Arnesto, *Columbus* (1991). chapt. 2.

angles, the eighteenth century author of a *New royal road to geometry* exclaimed: 'If such extraordinary Things can be done by the knowledge of this simple Property of Right lines only, which is, in a manner, self-evident, how much greater Matters may be expected, from a thorough knowledge of the whole [of geometry]? [This question] cannot but occur to the mind of every inquisitive and speculative Enquirer after Truth; it must determine him to lose no Time acquiring it.'[9]

Inquisitive and speculative enquirers, lose no time. Read on.

2.3. The point of proof

The tournament of Raoul and Raimbeau

The last flicker of the glory of ancient Greece and Rome went out for a time in western Europe about 750 years after Euclid's death. His legacy had been greatly perfected by then; but almost all of it disappeared when the Huns and the Goths and others who did not admire geometry ransacked the Roman Empire. The pathetic state of the subject in northern Europe around 1025 AD, over five centuries after the deposition of the last Roman emperor, appears from an exchange of letters between two of the most learned men of the time, Raimbeau (pronounced something like 'Rambo') of Cologne and Raoul of Liège. This correspondence amounted to a geometrical tournament, in which Raoul, who was much the younger man, attempted to earn his spurs by defeating Raimbeau. At Raoul's invitation, Raimbeau shot off the most difficult problem he had run across—and to which, as it turned out, he did not know the answer. The problem was one with which we are acquainted: to prove that the sum of the interior angles of a triangle equals a straight angle (Euclid I.32). Raimbeau may have taken it from a book that gave a few simple Euclidean propositions without demonstrations or, indirectly, from the texts of the Roman agrimensores. These manuals could still be found in Italy. The famous Gerbert, who became a Pope, came across one there about 50 years before Raoul and Raimbeau went to battle, and, thus inspired, composed a manuscript on geometry that contained the very proof sought by the ignorant gladiators beyond the Alps.[10]

They began by throwing dust in one another's eyes. It occurred to Raimbeau that if triangles have interior angles they must also have exterior ones. What might they be? They asked all their friends. No

9. Malton, *Royal road* (1793), 61.
10. Günther, *Geschichte* (1887), 114–16.

one knew. No one they consulted could put his hands on the first book of Euclid's *Elements*. Raimbeau ended by taking 'interior' to mean 'acute' and 'exterior' to mean 'obtuse'. With this understanding he found it impossible to prove anything at all. He would have had trouble even if he had his definitions right. Neither he nor Raoul had the slightest conception of a geometrical proof. Raoul did manage to show that the sum of the interior angles is a straight angle in the special case of a triangle with two equal sides, formed by drawing the diagonal of a square. But the general case eluded him. All he could suggest was either to declare the proposition true by intuition or to draw a triangle on parchment, cut out its angles, and place them together to form a straight angle (Fig. 2.3.1).[11]

Raoul's technique cannot satisfy a geometer. For one reason, it is not general: to prove, or make plausible, the proposition in his way you would have to cut up every triangle you wanted to analyze. The principal reason for constructing formal proofs in geometry is to demonstrate, once and for all, a property common to *all* geometric figures defined in the enunciation of the proposition. Secondly, Raoul's technique violates the rules of Euclidean geometry. Euclid did not include scissors in the tool kit of geometers. He allowed only three tools: a pencil for making points; a straight edge for lines; and a compass for circles, which define planes. You may use a ruler for a straight edge provided that you do not use it to measure; in pure Euclidean geometry, you do not express the lengths of lines as numbers or prove propositions by measuring things. Also, you may employ an ordinary compass, although, strictly speaking, you should get one that collapses every time you raise it from the plane on which it has drawn a circle or circular arc. That is because Euclid does not allow you to know anything about the behavior of a compass outside the plane (the straight edge can be moved about *in* the plane as required). The limitation to figures constructible by a straight edge and compass may be taken as a definition of the scope of Euclidean plane geometry.[12]

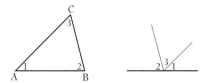

Fig. 2.3.1 The point of proof, according to Euclid I.32.

11. Tannery *Mémoires*, 5 (1922), 90–93, 237–93.
12. Cf. Committee of Ten, *Report* (1894), 113.

The spirit of Euclid

The machinery of proofs. What is at stake in proofs may be gathered from the 'postulates', or suppositions that Euclid put at the head of the *Elements* without any proof, because, as he rightly recognized, they rest upon nothing more elementary: postulates are ingredients, not objects, of proof (Table 2.3.1).

Table 2.3.1 Euclid's postulates

1. That a straight line can be drawn from any point to any point.
2. That a finite straight line can be produced continuously (carried further) in a straight line.
3. That a circle can be described with any centre and radius.
4. That all right angles are equal to one another.
5. That, if a straight line falling on two straight lines make the interior angles on the same side less than two right angles, the straight lines, if continued indefinitely, meet on that side on which lie the angles less than the two right angles.

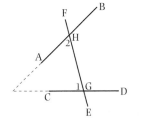

Fig. 2.3.2 The axiom of parallels and the proof of Euclid I.28.

Postulates 1–3 ensure that we can always draw what lines and circles we please; postulate 4 allows us to build up the geometry of perpendiculars; and the jabberwocky postulate 5 does the same for parallels. Figure 2.3.2 will make its content evident. The first straight line is the transversal EF; the two lines, AB, CD; the angles on the same side, ∠1 and ∠2. The postulate says that if ∠1 + ∠2 is less then a st.∠, AB and CD extended will meet on the side of the transversal containing ∠1 and ∠2.

If ∠1 + ∠2 = st.∠, AB and CD extended will either not meet on either side of EF, or, by symmetry, will meet on both sides; but, in this second case, two straight lines would enclose an area, which violates the spirit if not the letter of postulate 1. We have it from Proclus that Euclid meant that between two points one *and only one* straight line can be drawn.[13] Postulate 5 therefore implies Euclid I.28: if the interior angles on the same side of a transversal cutting two straight lines sum to a st.∠, the lines are parallel. Euclid proves I.28 by observing that, if ∠2 = st.∠ – ∠1 = ∠DGH, then AB and CD are parallel via alternate interior angles (I.27).

Perhaps you can see how the fifth postulate grounds the discussion of parallels in §2.1. There we assumed that two lines perpendicular to a third line would never meet. Since both internal angles in this case are right angles, they sum to a straight angle, and so, by

13. Heath, *Elements* (1926), *1*, 195.

inference from the postulate, are parallel. Our assumption allowed us to declare the distance between the lines OV, O'V' (Fig. 2.1.6), each perpendicular to OO', to be the same wherever taken, and so to define the rectangle, from which the rules of alternate angles followed. Many attempts have been made to reduce the fifth postulate to a theorem, that is, to deduce it from the rest of the Euclidean apparatus.[14] All have failed. The fifth postulate must be left as an independent assumption; as mathematicians discovered only much later, it is an essential part of the definition of plane geometry.

Other geometries, with other postulates in place of Euclid's fifth, are possible. For example, a consistent system can be built on the assumption that two lines can enclose a space, as do two meridians running from pole to pole on the surface of the earth. This discovery, made during the nineteeth century, surprised and even shocked most mathematicians of the time. They soon conceded the logical possibility of the new system. Non-Euclidean geometries are not mere logical games, however. They are at the heart of the mathematics of Einstein's theories of relativity.

Euclid's third postulate has not had so distinguished a history as his fifth. Nonetheless it is no less characteristic of his system. It seems to specify very little—only that a circle can be drawn. One needs only to open the compass to a constant distance and turn the instrument. But—and here is the character-building part—the postulate does not say how to draw a circle around a given point with a radius equal to a given line that does not have the given point as one of its extremities. Nothing in the postulate allows the discriminating geometer to carry a distance with a compass—that is, to assume that a compass exactly preserves its setting when lifted from the plane—and use it as a radius of a circle around a point elsewhere on the plane (Fig. 2.3.3). But the postulates do say enough to tell you how to construct such a circle without carrying the distance. That is the content of the second proposition in the *Elements* (Euclid I.2).

In Fig. 2.3.4, A is the given point, BC the given line. The problem is to apply BC at A as a radius. The method is as follows. Join A and

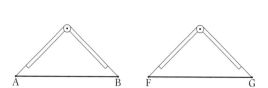

Fig. 2.3.3 Use of a compass to carry a distance.

Fig. 2.3.4 The problem of carrying a distance (Euclid I.2).

Fig. 2.3.5 To construct an equilateral triangle (Euclid I.1).

B (postulate 1). Erect on AB as base the 'equilateral' triangle (a triangle all of whose sides are equal) ABD. Euclid I.1—the very first geometrical theorem—tells you how to make the equilateral triangle. Draw a circle with A as center and AB as radius and a second circle with B as center and BA as radius (Fig. 2.3.5). The process does not exceed the authorization of postulate 3; the compass could collapse when removed from A and be reset to the radius AB = BA when replaced at B. The two circles meet at D. (The geometrical purist might object that nothing in the postulates obliges us to admit that the two circles meet anywhere; that is correct, but we shall not allow this scruple to distract us.) Evidently ΔABD is equilateral, since each of its sides was constructed equal to AB.

To return to our argument, extend DA and DB as indicated in Fig. 2.3.6 (postulate 2), place the compass point at B and open the other leg to BC; cut off BE = BC on DB extended. Place the point at D, open the compass to DE, and cut off DF = DE on DA extended. Now DF = DA + AF = DE = DB + DE = DB + BC, or DA + AF = DB + BC. But by construction, DA = DB. Therefore AF = BC: the line BC has been applied to the point A, as required. This tedious demonstration brings a great gain, which is the avoidance of even greater tedium in the future. From now on we can carry distances with a compass (and will do so) because, if challenged, we can always return to the construction of Euclid I.2. That is very comforting.

Proofs may be elaborate or spare, comprehensive or sketchy, and, of course, wrong. An elaborate form, which has consumed myriads of unnecessary student-hours, places every step, together with its justification, on a separate line. This form, which had the endorsement of Edmund Halley, who used geometry to plot the courses of comets, was invented early in the eighteenth century by a textbook writer named Henry Hill. '[It] is wholly different', he claimed, 'from any that has hitherto been made choice of by any of the Interpreters of Euclid', and it had the further merit, according to its inventor,

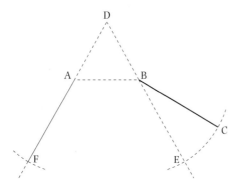

Fig. 2.3.6 To carry a distance (Euclid I.2).

that it 'does not so much require Intent and severe Thinking, as a bare and easie Inspection'.[15] Further to relax thinking, Hill printed the text of each proposition in italics and the assertion to be proved in boldface type. His proof of Euclid I.2 may be paraphrased as follows.

1. Join A, the given point, to B, one end of the given line   Post. 1
2. With AB as base, raise the equilateral $\triangle$ABD   Prop. I.1
3. Extend DA and DB indefinitely   Post. 2
4. With center B, draw the circular are EC   Post. 3
5. With center D, and opening DE, draw a circle cutting DA extended in F   Post. 3
6. DF = DE, BE = BC, DA = DB   Construction
7. AF = DF − DA = DE − DB = BE = BC   C.N. 3
8. $\therefore$ AF = BC, Q.E.D.   C.N. 1

The symbol '$\therefore$' in step 8 signifies 'therefore' and the abbreviation 'Q.E.D.' stands for 'quod erat demonstrandum', 'that which was to be proved'. The justification for steps 7 and 8 are two of the five 'common notions' (abbreviated 'C.N.') without which neither Euclid nor anyone else could get through the daily grind. You are already familiar with the first of these notions. The full five are given in Table 2.3.2.

Table 2.3.2   Euclid's common notions

1. Things which are equal to the same thing are also equal to one another.
2. If equals be added to equals, the wholes are equal.
3. If equals be subtracted from equals, the remainders are equal.
4. Things which coincide with one another are equal to one another.
5. The whole is greater than the part.

Most of the arguments given in this book are intended as adequate demonstrations, not as formal proofs. Whenever, however, because of the importance of the proposition or the neatness of the construction, it has seemed desirable to proceed formally, the proofs exhibit explicit steps in Hill's manner and the diagrams show the different elements of the argument in different ways. But in most cases the obvious and the picayune will be left aside; neither time nor patience in the twentieth century allows the teacher of geometry to follow the

15. Hill, *Elements* (1726). 'Preface'

counsel of perfection of the old Jesuit professors of the subject, 'nihil enim Tyronibus offerri debet, quod severe demonstratum non fit', 'nothing should be presented to beginners that is not formally demonstrated'.[16]

Nonetheless we should take time to examine a few Euclidean demonstrations of constructions that we already know. An instructive set is I.9–I.12 and I.23, in which, in succession, Euclid shows how to bisect an angle, bisect a line, erect a perpendicular, drop a perpendicular, and copy an angle. Our earlier demonstrations of these constructions appealed to congruence by s.s.s. and treated each independently of the other. The order therefore could be chosen at will; ours was copying the angle, dropping a perpendicular, erecting a perpendicular, bisecting a line, and (by loci) bisecting an angle (at Fig. 2.1.12, 2.1.13, 2.1.14, 2.1.28, 2.1.29). Euclid arranges matters so that each construction depends upon preceding ones.

To bisect an angle, Euclid (I.9) takes any point D on one leg of the given angle BAC and cuts off a length AE = AD on the other (Fig. 2.3.7). He joins D and E and erects on DE an equilateral triangle DEF (I.1). The line AF now bisects ∠BAC because ΔADF ≅ ΔAEF by s.s.s. (I.8). He moves on to bisect a line (I.10). Again he raises an equilateral triangle, this time on the given line AB (Fig. 2.3.8), and bisects ∠ACB by the line CD according to I.9. Since ΔADC ≅ ΔBDC by s.a.s (AC = AB, ∠ACD = ∠BCD, CD is common), AD = DB, as required.

We come to perpendiculars. To raise one on a line at a given point (I.11), say C on AB (figure 2.3.9), take any point D on AC; make CE = CD; construct an equilateral triangle DEF on DE (I.1); join C and F. Since ΔDCF ≅ ΔECF by s.s.s (I.8), ∠DCF = ∠ECF = rt. ∠. To drop a perpendicular (I.12) from a point C onto a line AB outside it, draw a circle with C as center and radius large enough to cut AB in two places, say G and E (Fig.2.3.10). Bisect GE at H and joint G,H, and E with C; CH is the desired perpendicular, since ΔGHC ≅ ΔEHC by s.s.s.

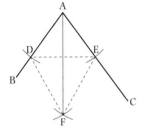

Fig. 2.3.7 To bisect an angle (Euclid I.9).

Fig. 2.3.8 To bisect a line (Euclid I.10).

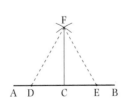

Fig. 2.3.9 To raise a perpendicular (Euclid I.11).

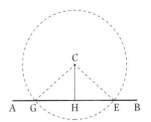

Fig. 2.3.10 To drop a perpendicular (Euclid I.12).

16. Tacquet, *Elementa* (1762), iii.

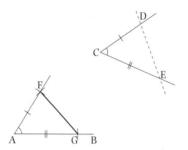

Fig. 2.3.11 To copy an angle (Euclid I.23).

Euclid I.23 copies an angle by drawing an arbitrary line across the given angle to create ΔDCE (Fig. 2.3.11). He then lays out $AG = CE$ on the given line AB and describes circles around A and G with radii CD and ED, respectively. Their intersection gives the point F. Since $\Delta FAG \cong \Delta DCE$ (*s.s.s.*), $\angle FAG = \angle DCE$ as required.

The Euclidean elements so far considered are listed in Table 2.3.3.

A handsome by-product. The construction of the equilateral triangle, which figures so prominently in Euclid's proofs, leads to a figure of great beauty and importance. Instead of drawing the triangle as in Fig. 2.3.12, draw the full arcs as in Fig. 2.3.13. The curvilinear shape ACB is that of the basic Gothic arch. You can see it everywhere in churches built during the later middle ages. Figure 2.3.14, showing the back end of Bayeux cathedral, begun around 1230, indicates its use in openings (the towers), windows of all sizes, and blind arcades (the lowest level). Often the interior of the arch is filled with other geometrical shapes, like squares, circles, and smaller arches, as in the nave of Saint-Denis (Fig. 2.3.15). We shall study some of these shapes later after meeting the Euclidean principles needed to draw them. Here the point to seize is that at the threshold of the study of geometry, in connection with the very first proposition of Euclid, you have encountered one of the most important and versatile elements of architecture.

Fig. 2.3.12 An equilateral triangle.

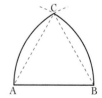

Fig. 2.3.13 An equilateral Gothic arch.

Table 2.3.3 Euclidean elements in Chapter 2

Definitions

Def. 1	Point
Def. 2–4	Line
Def. 5–7	Plane
Def. 8–12	Angles
Def. 15–18	Circles
Def. 19–21	Triangles
Def. 22	Rectangles (and other four-sided figures)
Def. 23	Parallel

Postulates

Post. 1–2	Drawing straight lines
Post. 3	Drawing a circle
Post. 4	All right angles are equal
Post. 5	Lines if not parallel meet

Common notions

C.N. 1	Things equal to the same thing equal one another
C.N. 2–3	Addition and subtraction of equals
C.N. 4	Things coincident with one another equal one another
C.N. 5	The whole is greater than the part

Propositions

Prop. I.1	Construction of equilateral triangle
Prop. I.2	Application of a straight line to a point
Prop. I.4	Congruence by *s.a.s.*
Prop. I.8	Congruence by *s.s.s.*
Prop. I.9	Bisecting an angle
Prop. I.10	Bisecting a line
Prop. I.11	Erecting a perpendicular
Prop. I.12	Dropping a perpendicular
Prop. I.15	Equality of vertical angles
Prop. I.23	Copying an angle
Prop. I.27	Equality of alternate angles implies parallelism
Prop. I.28	Interior angles on the same side summing to a st. $\angle$ imply parallelism
Prop. I.29	Equality of alternate interior angles
Prop. I.32	Exterior angle equals sum of opposite interior angles; the three interior angles sum to a straight angle

Fig. 2.3.14 The Gothic arch as deployed in the cathedral of Bayeux. From Bony, *French Gothic architecture* (1983), no. 297.

Fig. 2.3.15 The arch filled with other geometrical shapes in the cathedral of Saint Denis. From Bony, *French Gothic architecture* (1983), 372.

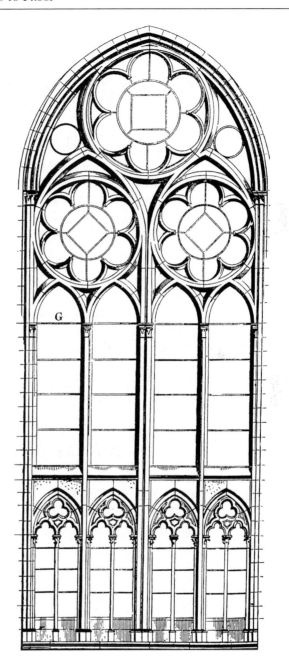

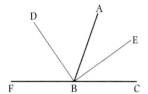

Fig. 2.4.1 Exercise 2.4.1.

2.4. Exercises

2.4.1. In Fig. 2.4.1, FBC and AB are any straight lines meeting at B. If now the angles ABF and ABC are bisected by the lines BD and BE, respectively, the bisectors are perpendicular. Prove that ∠DBE = rt.∠.

[2∠ABE + 2∠ABD = st.∠; ∠ABE + ∠ABD = ∠DBE = rt.∠.]

2.4.2. Given (in Fig. 2.4.2) ΔABC equilateral and CD⊥AB, AE⊥BC. What fraction of a straight angle is ∠CPE?
[ΔADC ≅ ΔBDC (*h.s.*), so ∠FCP = ∠ECP = ∠ACB/2 = st.∠/6. (Since an equilateral triangle obviously is also equiangular, each of the angles in ΔABC is st.∠/3.) Therefore ∠CPE = st.∠ – ∠CEP – ∠DCB = st.∠ – rt.∠ – st.∠/6 = st.∠/3.]

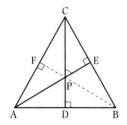

Fig. 2.4.2 Exercise 2.4.2.

2.4.3. Matters being as they are in the preceding exercise, show that PE = PD.
[Drop the perpendicular PF from P onto AC; PF = PE since, as we just showed, CD is the bisector of ∠ACB and the locus theorem (APS 2.1.9) applies. You can show similarly that PA bisects ∠FAD and that, therefore, PF = PD. We now have PD = PF = PE, and the desired result PE = PD. Now draw BP; BPF is a straight line, since ∠FPC = ∠CPE = ∠EPB = (st.∠/3), whence their sum, ∠BPF, = st.∠. The three perpendiculars from the vertices to the sides of an equilateral triangle meet in a point. Later (§3.3) we shall see that the property holds for all triangles.]

2.4.4. In Fig. 2.4.3, CE bisects ∠ACF and BE bisects ∠ABC. Show that ∠BEC = ∠BAC/2.
[The angle 2β being external to ΔBCA, 2β = 2α + ∠BAC; similarly, in ΔBCE, β = α + ∠BEC; hence ∠BCE = ∠BAC/2.]

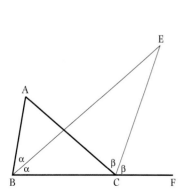

Fig. 2.4.3 Exercise 2.4.4.

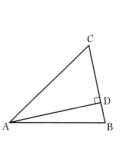

Fig. 2.4.4 Exercise 2.4.5.

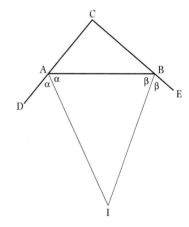

Fig. 2.4.5 Exercise 2.4.6.

2.4.5. ΔABC is any acute triangle (Fig. 2.4.4). If AD⊥BC, can you show that ∠BAD − ∠CAD = ∠C − ∠B? (When no ambiguity results, angles can be labelled by their vertices, for example, ∠ACB = ∠C, ∠ABC = ∠B.)

[∠BAD = rt.∠ − ∠B; ∠CAD = rt.∠ − ∠C; ∠BAD − ∠CAD = ∠C − ∠B.]

2.4.6. If the external angles made by the base and two sides of a triangle are bisected, show that the angle made by the bisectors is half the sum of the interior angles made by the base and the sides.

[Referring to the Fig. 2.4.5, we must show that ∠BIA = (∠ABC + ∠BAC)/2. Let BI divide ∠ABE into the equal angles β, and AI divide ∠DAB into the equal angles α. Then ∠BIA = st.∠ − α − β. But ∠ABC = st.∠ − 2β, whence ∠BIA = st.∠ − (st.∠ − ∠ABC)/2 − (st.∠ − ∠ACB)/2 = (∠ABC + ∠BAC)/2.]

3 Tricks with triangles

We come to the Bridge of Asses. That is the nickname—the *pons asinorum*—that medieval students of geometry, who knew even less than Raimbaud and Raoul, gave the fifth proposition of the first book of Euclid's *Elements*. In explanation of the nickname, some say that the shape of the figure resembles a trestle bridge; but, even if it did, that would not explain the asses. Others say that since Boethius, an erudite Roman aristocrat, who wrote Latin commentaries on Greek authors shortly after the final collapse of the Western Roman Empire and transmitted most of the scraps of Euclid known in the early middle ages, gave proofs only through I.4, I.5 separated the—let us change species—the sheep from the goats. No doubt, however, the great Jesuit geometer Christopher Clavius had it right in the annotated *Elements* he published over 400 years ago when he identified the asses with students who could or would not cross the bridge. No doubt also, the crossing presented difficulties. 'This theorem', wrote Clavius, 'usually appears hard and dark to beginners because of the number of lines and angles, to which they are not yet accustomed.'[1] Once you have crossed this bridge, you will have access to much new grazing land, and, most immediately, via Euclid I.7, to a most important practical application.

3.1. The bridge of asses

Euclid I.5

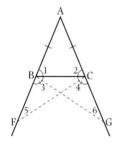

Fig. 3.1.1 The 'Bridge of Asses' (Euclid I.5).

The *pons asinorum* concerns triangles that have two equal sides. Unfortunately, the Greeks chose a non-English name for triangles with this property: isosceles, meaning 'the same legs', and the word has stuck. The theorem of the *pons asinorum* holds that the angles opposite the equal sides of an isosceles triangle are equal, and that the exterior angles under the equal angles also are equal. The proof rests on congruence by *s.a.s.*(I.4). In Fig. 3.1.1, AB and AC are the given equal sides of the isosceles triangle ABC; the proposition requires proof that $\angle 1 = \angle 2$ and $\angle 3 = \angle 4$. (Since $\angle 1 + \angle 3 = $ st.$\angle$, $\angle 1 = \angle 2$ if $\angle 3 = \angle 4$). To make the footing across the asses' bridge

1. Clavius. *Elementa* (1574), *l*, 26; cf. Heath. *Elements* (1926) *l*, 415–16.

as secure as possible, each step will be exhibited in the manner of Henry Hill. To make writing easier, now and in the future, we shall use the symbol '>' to signify 'greater than'. To avoid clutter, and proving the obvious, Euclid I.3, required at step 2, will not be demonstrated. It merely shows how to use his definition of a circle (Euclid, Def. 15) to cut off from a greater line a portion equal to a given lesser line. Now the steps:

1.	Extend AB and AC indefinitely	Post. 2
2.	From A as center, cut off AF = AG > AB = AC	I.3
3.	AB = AC	Given
4.	AF = AG	Construction
5.	∠FAC = ∠GAB	Same angle
6.	∴ ΔFAC ≅ ΔGAB	I.4 (*s.a.s.*)
7.	∴ ∠5 = ∠6, FC = BG	Step 6
8.	BF = CG (differences of equal lines)	C.N. 3
9.	ΔBFC ≅ ΔCGB	I.4 (*s.a.s.*)
10.	∴ ∠3 = ∠4	Step 9
11.	∴ ∠1 = ∠2 (differences of equal angles)	C.N. 3

Q.E.D., 'quod erat demonstrandum', which, you will recall, is the slogan with which geometers celebrate their success in demonstrating what Euclid proved long ago.

You might think that this demonstration goes to much unnecessary trouble. You would be in good company. Pascal exemplified the third fault that he found in geometers, 'far-fetched demonstrations,' by the bric-a-brac of the trestle below the bridge of asses. 'This defect is very common among the geometers ... [Could anything] be more ridiculous than to imagine that this equality [I.5] depended on these foreign triangles?'[2] As Pascal observed, there are other ways. A textbook writer of the early seventeenth century suggested flipping the isosceles triangle ABC to make ΔACB (Fig. 3.1.2); then, since AB = AC, AC = AB, and BC = CB, ΔABC ≅ ΔACB by *s.s.s.*[3] A good idea, were it not that Euclid does not give us the right to flip a triangle out of the plane. A more common alternative, employed in many Euclidean texts deriving from Legendre, avoids the flipping and appeals to *s.s.s.* (in the form of *h.s.*). In Legendre's simple proof,

Fig. 3.1.2 Alternative demonstration of Euclid I.5.

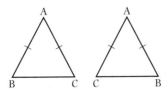

2. Arnauld, *Logic* (1851), 341.
3. Tournes, *Euclid* (1611), 31.

Fig. 3.1.3 Legendre's demonstration of Euclid I.5.

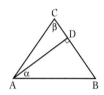

Fig. 3.1.4 Exercise on isosceles triangles.

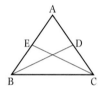

Fig. 3.1.5 Exercise on medians.

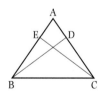

Fig. 3.1.6 Exercise on altitudes.

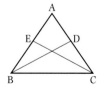

Fig. 3.1.7 A fourth, on angle bisectors.

the demonstrator drops a perpendicular from A onto BC at D (Fig. 3.1.3). Then, since AB = AC (given) and AD is common, $\triangle ABD \cong \triangle ACD$ (h.s.) and $\angle ABD = \angle ACD$, Q.E.D. The existence of these and other alternative proofs of I.5 inspired another explanation of its pet-name: 'because the asses who taught it took the greatest trouble to destroy the bridge which connects it with the real world.'[4]

Whatever you think of these proofs, they put you in a position to solve many interesting problems about isosceles triangles. Here is a trivial one. Triangle ABC is isosceles (Fig. 3.1.4). From the vertex of one of the base angles (the 'base' is not one of the equal sides) drop a perpendicular to the opposite side. Let this perpendicular make an angle α with the base. Can you show that α equals half the angle opposite the base? If you label that angle β you can express your chore as proving that $\alpha = \beta/2$. You have from $\triangle ADB$, $\alpha = \text{rt.}\angle - \angle ABC$, and from $\triangle ABC$, $\beta = \text{st.}\angle - 2\angle ABC$. Therefore $\beta = \text{st.}\angle - 2(\text{rt.}\angle - \alpha)$, or $\beta = 2\alpha$.

APS 3.1.1. Let D and E be the midpoints of the sides of the isosceles triangle ABC (Fig. 3.1.5). Prove that BD = EC. The obvious route is to prove $\triangle BCD \cong \triangle CBE$. Nothing could be easier. Since, by construction, BE = AB/2 and CD = AC/2, and AB = AC ($\triangle ABC$ is isosceles), BE = CD. Also (I.5), $\angle EBC = \angle DCB$. Finally, BC is common to both $\triangle BCD$ and $\triangle CBE$. Consequently $\triangle BCD \cong \triangle CBE$ by s.a.s. and their corresponding parts, BD and EC, are equal.

APS 3.1.2. A similar property holds for perpendiculars dropped from the vertices of the base angles to the equal sides. In Fig. 3.1.6, BD = EC since $\triangle BCD \cong \triangle CBE$, this time by h.s.

The properties just demonstrated have an analogy in the bisectors of the base angles of an isosceles triangle. In Fig. 3.1.7, BD, EC are the bisectors, which, as the analogy suggests, are equal. To prove the equality expeditiously, we require a third method of proof of congruency, by two angles and the included side (a.s.a.), Euclid I.26. Once again, the best way to persuade yourself of the validity of the proof is to imagine the construction of a triangle using the given parts and to see that only one solution is possible. So, in Fig. 3.1.8, XY is the given line, QXY and RYX the given angles; XQ and YR extended meet in at most one point (they would not meet only if QX∥RY), Z, making one and only one triangle, the unique $\triangle XYZ$.

To return to the subject illustrated in Fig. 3.1.7, BD = EC because $\triangle BCD \cong \triangle CBE$. The congruency, by a.s.a., follows from $\angle EBC = \angle DCB$ (I.5), BC common, and $\angle BCE = \angle DBC$ (halves of equal angles). A more elaborate application of the same set of ideas

---

4. Jones, *MT*, 37 (1944), 10; Hogben, *Mathematics for the million* (1937), 147 (quote).

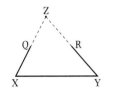

Fig. 3.1.8 Congruence by *a.s.a.*

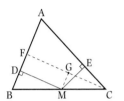

Fig. 3.1.9 Exercise on isosceles triangles.

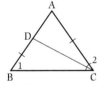

Fig. 3.1.10 Converse of the asses' bridge (Euclid I.6).

involves the perpendiculars dropped from *any* point M on the base BC of an isosceles triangle onto the equal sides (Fig. 3.1.9). We are to show that the *sum* of the perpendiculars, DM + ME, is the same no matter where M lies on BC. To prove this improbable claim, drop the perpendiculars CF from C onto AB and MG from M onto CF. The route is to show ME = GC. Since DMGF is a rectangle by construction, FG = DM; hence, if ME = GC, ME + DM = CF, the distance of C from AB, which is independent of the position of M. We have ΔMGC ≅ ΔMEC by *a.s.a.* For one angle, ∠MGC = ∠MEC = rt.∠; for another, ∠GMC = ∠MCE. (Since GM‖AB, ∠GMC = ∠B = ∠C = ∠MCE.) If two pairs of corresponding angles of a triangle are equal, so must the third: ∠GCM = ∠EMC. Since both triangles have the same hypotenuse, MC, and the angles on either end of the hypotenuse are correspondingly equal, ΔMGC ≅ ΔMEC and, as required, ME = GC.

Reduction to absurdity

A demonstration arrives at a proposition from certain givens. It occurs to inquiring minds to ask whether the process can be reversed. Assuming the truth of the proposition, can a logician deduce the givens? As we know, the new proposition derived by interchanging the givens and the old proposition is called a 'converse'. The converse of I.5 requires that the sides opposite equal angles of a triangle are equal; happily it is true and so proved in I.6, by the method we met in connection with I.27, the 'reductio ad absurdum', the demonstration of a proposition by showing that its negation leads to a formal, frontal, unresolvable contradiction, a 'reduction to the absurd'.

The converse of I.5 begins with ∠1 = ∠2 (Fig. 3.1.1 again) and requires a proof that AB = AC. Assume that AB ≠ ('does not equal') AC; let it therefore be greater, that is, let AB > AC, if that is possible. Of course, it is not possible, as the following argument shows. Mark off from B along AB a segment DB = AC (note that to do so you require Euclid I.2). It will be best to redraw the triangles as in Fig. 3.1.10. Since ∠1 = ∠2 (given), DB = AC by construction, and BC is common, ΔDBC ≅ ΔACB (*s.a.s.*). But, if D lies between A and B, as it does by hypothesis, we have just proved that the smaller ΔDBC is congruent to the larger ΔACB that contains it; which is an absurdity, since it violates C.N. 5. Therefore our supposition AB > AC must be false. Similarly AB < ('less than') AC can be ruled out: the demonstration interchanges the roles of AB and AC in the preceding argument. Since AB cannot be larger or smaller than AC, the lines must be equal. AB = AC. Q.E.D.

Euclid proves congruence by *s.s.s.*, for which we accepted provisionally an argument based on construction and plausibility (at

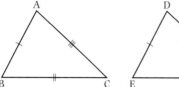 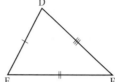

Fig. 3.1.11 Congruence by *s.s.s* (Euclid I.8).

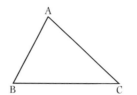 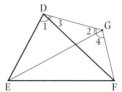

Fig. 3.1.12 A *reductio ad absurdum*, by *s.s.s* (Euclid I.7).

Fig. 2.1.11), by *reductio ad absurdum*. In his I.8, presented in our Fig. 3.1.11, we are given that AB = DE, BC = EF, AC = DF, and required to show that any angle in one of the triangles equals the corresponding angle in the other; for if we can show that, the triangles must be congruent by I.4 (*s.a.s.*). Imagine that B be placed on E (although he disliked doing so, Euclid could find no alternative to demonstration by superposition in the most obvious cases of congruence of triangles) and BC on EF, so that (BC = EF, a given) C falls on F. Suppose that A does not fall on D, but rather on a point G, where EG = AB, AC = GF. But the construction cannot be done, as Euclid had already shown, in I.7, as follows.

Join D and G (Fig. 3.1.12). Now by hypothesis DE = EG; therefore (1.5), ∠1 = ∠2. But also, again as a given, DF = FG, whence ∠3 = ∠4. But these two equalities cannot both hold at the same time; we have, by subtraction, ∠1 − ∠3 = ∠2 − ∠4. But ∠1 > ∠3 and ∠2 < ∠4. A quantity larger than zero cannot equal a quantity smaller than zero. Since the assumption that G does not coincide with D leads to a manifest absurdity, D does coincide with G, ∠ABC = ∠DEF (Fig. 3.1.10), and ΔABC ≅ ΔDEF by the trustworthy *s.a.s.* You may rightly ask, why have recourse to absurdity after appealing to intuition, which shows plainly enough that if one pair of equal sides are superposed, so are the other two. Many expositors content themselves with superposition.[5] But Euclid, always wanting to minimize its use, contrived a way to rest his proof of I.6 on I.4, and so to restrict most of the offense to the earlier demonstration.

APS 3.1.3. Euclid I.20, the theorem that the Epicureans thought evident even to an ass, lends itself to proof by *reductio ad absurdum*. (Euclid proved it directly after showing, in I.19, by reduction, that, in any triangle, the greater angle is subtended opposite the greater side.) In Fig. 3.1.13, we suppose BC to be the longest side of ΔABC and must show that it is less than the sum of the other two sides, that is, BC < AB + AC. Assume first that BC = AB + AC. Then cut

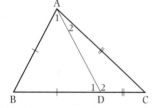

Fig. 3.1.13 Another *reductio*, by Euclid I.20, the proposition obvious to a donkey (Euclid I.19).

5. It was the French way: Combette. *Cours abrégé* (1898), 14–15; Hadamard, *Leçons* (1931), 25–6.

Table 3.1.1 Some words and symbols used in proofs

Word/Symbol	Definition
Converse	A proposition made by interchanging the givens and conclusion of another proposition
Reductio ad absurdum	A proof made by demonstrating that an assumption leads to an impossibility or a formal contradiction with a known truth
$\neq$	'Does not equal'
$>$	'Is greater than'
$<$	'Is less than'
a.s.a.	'Angle-side-angle': refers to triangles with two corresponding angles and the included side equal
s.a.s.	'Side-angle-side': refers to triangles with two corresponding sides and the included angle equal
s.s.s.	'Side-side-side': refers to triangles all of whose corresponding sides are equal
$\cong$	'Is congruent to'
$\therefore$	'Therefore'

Fig. 3.1.14 Still another *reductio* needed for Euclid I.20 (Euclid I.19).

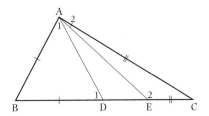

off BD = AB on BC, leaving DC = AC. That makes ΔABD and ΔACD isosceles and, via 1.5, the angles equal as indicated in the figure. But they cannot be equal; if they were, we would have ∠BAC = ∠BDC = st.∠, which is absurd; therefore BC ≠ AB + AC. Well, then, perhaps BC > AB + AC. That may be proved impossible with the help of Fig. 3.1.14. Cut off from one end of BC, BD = AB and, from the other, EC = AC. Again we have isosceles triangles, as shown in the figure. Now ∠BAC = ∠1 + ∠2 + ∠DAE. But, from ΔDAE, (st.∠ − ∠1) + (st.∠ − ∠2) + ∠DAE = st.∠, or ∠1 + ∠2 = st.∠ + ∠DAE. We have finally that ∠BAC = st.∠ + 2∠DAE > st.∠, which is absurd.[6]

6. After Heath, *Elements* (1926), *I*, 288, after Proclus.

3.2. Practice

Graduating an arc

Euclid devotes I.9, the proposition immediately following the proof of congruence by *s.s.s.*, to the bisection of an angle, which we took up earlier in connection with the locus of a point equidistant from two intersecting straight lines (APS 1.1.8). His way of doing it (see Fig. 2.3.7) may be rephrased for exploitation in the multiple subdivision of an arc, a matter of first importance for making and mounting telescopes and other instruments for angular measurement. Let ∠A (Fig. 3.2.1) be the angle to be halved. Place the point of a compass at A and mark off equal lines AB, AC. From B and C as centers, draw arcs of any convenient, equal radius (Euclid I.2) intersecting at D. Join A and D. The line AD bisects ∠A, that is, ∠CAD = ∠BAD. For draw CD, BD (Fig. 3.2.2): AC = AB, BD = CD, by construction, and AD is common, whence ΔCAD ≅ ΔBAD via *s.s.s* (Euclid I.8); the corresponding parts, ∠1 and ∠2, therefore are equal. To 'graduate' (mark equal divisions on) the 'limb' (the curved edge) of instruments like those depicted in Figs 3.2.6–3.2.8, we need only apply this construction over and over again.

Now draw a circle with radius OB = 8 inches. This circle will be the graduated limb. Erect the perpendicular EO (Euclid I.11) and divide the right angle EOB by making equal arcs from B and E (Fig. 3.2.3). Repeat, making equal arcs from B and F; that produces one-quarter of a right angle. Repeat, arriving at rt.∠/8, etc., until you get tired or obtain rt.∠/32. You can duplicate this small angle all over the limb by the method indicated in Fig. 3.2.4. Put the point of the compass at K and open it to KB, the arc subtending ∠KOB = rt.∠/32; then draw an arc intersecting the limb at H, so that ∠HOK = ∠KOB = rt.∠/32, and so on, until, after you have done it 31 times, you come to E. If in fact you do arrive close to E after 31 repetitions, you have a good hand with a compass.

This graduation will make your circle a useful quadrant, such as that pictured in Fig. 3.2.5(a), when you paste it on cardboard and attach a pivoted ruler, or 'alidade', at its center. A useful form of alidade appears in Fig. 3.2.5(b). With it you look at a distant object

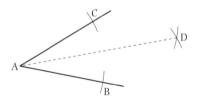

Fig. 3.2.1 To bisect an angle, practically.

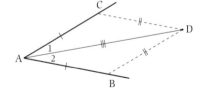

Fig. 3.2.2 Proof of the construction in **Fig.** 3.2.1.

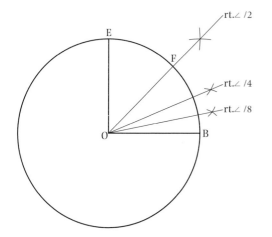

Fig. 3.2.3 Repeated halving of a right angle.

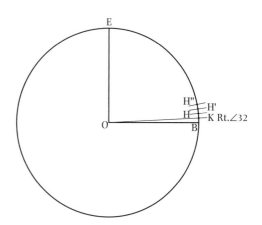

Fig. 3.2.4 Graduation of an arc to 1/32 of a right angle.

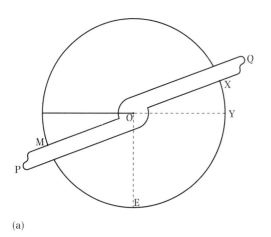

(a)

Fig. 3.2.5 Plate and alidade of an instrument of angular measure: (a) a home-made affair, in which the limb is graduated as in Figure 3.2.4 (b) a professional model, worked by an angel, as depicted in Moxon, *Tutor* (1686), 49.

(b)

and find its 'elevation' ∠XOY above the horizontal OY by reading the mark M where the line XO extended cuts the limb. (You will need a plumb bob to keep OE vertical during the observation.) The word 'alidade' is a Spanish version of an Arabic term for a pivoted ruler; while people like Raimbaud and Raoul in Europe were breaking their heads over the simplest geometrical theorems, the Arabs were perfecting an elaborate instrument for angular measurement and

Fig. 3.2.6 A surveyor with a theodolite (an instrument for taking angles in the plane) is at work between the globes; below him, between the muses of arithmetic and geometry, another practitioner uses a plane table. Aaron Rathborne, *The surveyor* (1616), in Bennet, *Divided circle* (1987), 48.

computation known as an astrolabe. When the Spanish became familiar with the astrolabe through the learned Arabs in their midst, they adopted it completely, down to the names of its parts. Indeed, they had little choice, since, owing to their relative backwardness in mathematics, they had no appropriate words of their own.

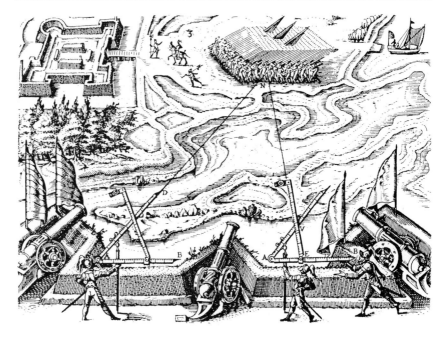

Fig. 3.2.7 The artillery, always a promoter of science, used two angular measurements (and a human baseline between the observing stations) to obtain the distance to a target. This device, by the famous instrument maker Joost Bürgi, is pictured in Benjamin Bramer, *Bericht zu Meister Jobsten ... Triangularinstrument* (1648), in Bennet, *Divided circle* (1987), 46.

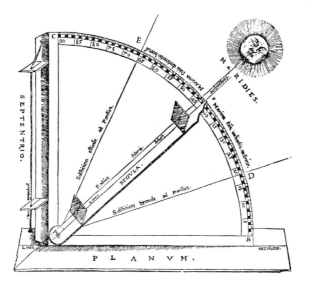

Fig. 3.2.8 An astronomical quadrant for taking the altitude of the sun. Oronce Fine, *De mundi sphaera* (1542), in Bennet, *Divided circle* (1987), 19.

When the alidade's edge at X does not fall precisely on a division, you can estimate easily where it comes between the two nearest marks. To ease the task you might wish to put in more divisions, not by marking smaller angles, for they would not be easy to draw, but by introducing submultiples of a third, rather than a half, of a right angle. In general it is not possible to divide an angle in thirds with straight edge and compass; but the trick can be done with a right

Fig. 3.2.9 Graduation of an arc to 1/12 of a right angle.

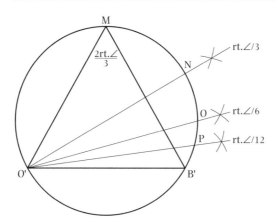

angle, as indicated in Fig. 3.2.9. Draw the equilateral triangle O'B'M (Euclid I.1). Since, by the theorem of the bridge of asses (I.5), all the interior angles are equal, and since, by I.32, their sum is two right angles, each is two-thirds of a right angle. Hence the continuous bisecting of ∠MO'B' will produce the angles rt.∠/3, rt.∠/6, rt.∠/12, and, a good place to stop, rt.∠/24. If you undertake this improvement, make O'B' about 6 inches and do the work on tracing paper. You can transfer the smallest angle, rt.∠/24, to the limb of your quadrant by placing O' on O (of Fig. 3.2.7) and O'B along OB. Then put your straight edge on O'P (∠PO'B' = rt.∠/24) and mark where it cuts the quadrant's limb. You can now divide the whole quadrant into intervals of rt.∠/24 in the same way that you divided it into intervals of rt.∠/32.

The designation of an angle as a fraction of a right angle is very reasonable and serviceable. However, most people use other conventions. One, which comes down from the Egyptians and Babylonians, was already ancient when Euclid was a boy. It assigns 90 parts or degrees to a right angle, and 60 parts called minutes to each degree. It is not easy to divide a right angle into 90, let alone 90 × 60 = 5400 parts. The closest you can come to 1° by bisections of the right angle is 90/64° or, very closely, 1°26' ('one degree, twenty-six minutes'); the closest by bisections of two-thirds of a right angle, 90/96°, or just about 56'' ('fifty-six seconds', a second being one-sixtieth of a minute). A right angle divided in this second way would have 96 parts rather than 90—certainly a good approximation to division by degrees for the sort of measurements we shall make, but inadequate for better instruments of the kind shown in Figs 3.2.4 and 3.2.5.

The Egyptians and Babylonians settled on their method of measurement by dividing the four right angles that complete the circle (Fig. 3.2.10) into 360 parts. The choice seems to have been motivated by their mathematical systems (the Babylonians used one based

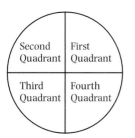

Fig. 3.2.10 Babylonian division of angles by degrees (1/360 of a circle).

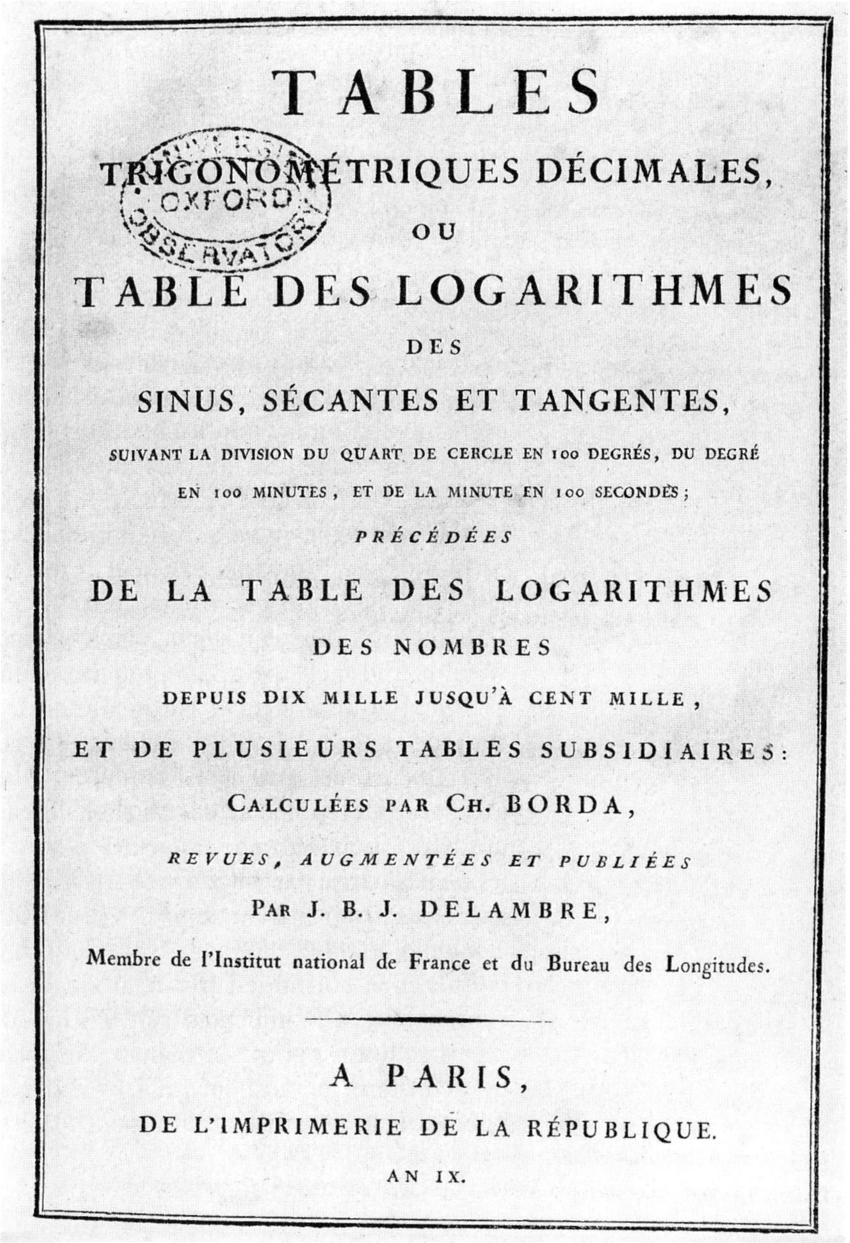

Fig. 3.2.11 French Revolutionary division of angles by grades (1/100 of a circle).

TABLES

TRIGONOMÉTRIQUES DÉCIMALES,

OU

TABLE DES LOGARITHMES

DES

SINUS, SÉCANTES ET TANGENTES,

SUIVANT LA DIVISION DU QUART DE CERCLE EN 100 DEGRÉS, DU DEGRÉ

EN 100 MINUTES, ET DE LA MINUTE EN 100 SECONDES;

PRÉCÉDÉES

DE LA TABLE DES LOGARITHMES

DES NOMBRES

DEPUIS DIX MILLE JUSQU'À CENT MILLE,

ET DE PLUSIEURS TABLES SUBSIDIAIRES:

CALCULÉES PAR CH. BORDA,

REVUES, AUGMENTÉES ET PUBLIÉES

PAR J. B. J. DELAMBRE,

Membre de l'Institut national de France et du Bureau des Longitudes.

———————

A PARIS,

DE L'IMPRIMERIE DE LA RÉPUBLIQUE.

AN IX.

on the number 60) and by the apparent motion of the sun around the earth during the year. It takes the sun—or, if you prefer, the earth—365 days, 5 hours, and 48 minutes, more or less, to complete this journey. That is not nearly so easy a number to work with as 360. On the Babylonian system, which assigns 360° to a full circuit, the sun seems to move about a degree a day in its annual motion. Although the minutes and seconds of angular measure may have originated in connection with the sun's motion, they no longer have anything to do with time. 'Minute' comes from a Latin word meaning something very small; 'second' is short for 'second minute,' something as much smaller than a minute as a minute is of its parent quantity.

In 1792 the French inflated the Babylonian system by assigning not 90 degrees, but 100 'grades', to a right angle (Fig. 3.2.11.). They did so as one of many reforms in weights, measures, and coinage that were to advance the cause of the French Revolution. A chief feature of the revolutionary measures was the decimalization of everything. For example, revolutionary time divided the day into 10 revolutionary hours each subdivided into 100 minutes; naturally, every revolutionary minute had 100 revolutionary seconds. Watches showed the new time and its relation to 'servile time', that is, the old scheme of 24 hours by which the English, Germans, and other 'enslaved' peoples who still had kings arranged their affairs. French revolutionary time was in effect from 1792 until 1806, when the Emperor Napoleon put an end to it.[7]

A more enduring reform was the decimalized metric system of weights and measures. The French revolutionaries defined the meter as one ten-millionth of the distance from the North pole to the equator (Fig. 3.2.12). By dividing the right angle into 100 grades and the grade into 100 gradettes, they brought their angular measure into conformity with their linear measure. An angle of one gradette, shown greatly enlarged as ∠AOB in Fig. 3.2.12, intercepts a portion of the earth's surface AB = (10 million meters)/(number of gradettes in a right angle) = 10 000 000/10 000 = 1000 meters = 1 kilometer. So a kilometer is the length of an arc of the earth's circumference contained in an angle of one gradette at the earth's center. If ∠AOB indicated not a gradette in French revolutionary measure but a minute in Babylonian measure, the arc AB would be, by definition, a nautical mile (= 1.16 regular or statute mile). A quarter of the earth's circumference therefore equals (90)(60) nauti-

Fig. 3.2.12 The meter, or 1/40 000 000 of the earth's circumference.

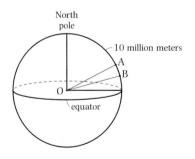

7. Heilbron, *Weighing imponderables* (1993), 249–57.

Table 3.2.1 Words about dividing arcs and angles

Word	Definition
Graduate	Divide an arc or angle into equal parts
Limb	The graduated edge of an instrument
Alidade	A pivoted sighting bar
Degree (°)	One-ninetieth of a right angle
Minute (′)	One-sixtieth of a degree
Second (″)	One-sixtieth of a minute
Grade	One-hundredth of a right angle
'Gradette'	One-hundredth of a grade
Kilometer	The length of an arc of a meridian subtending an angle of one gradette at the earth's centre
Nautical mile	The length of an arc of a meridian subtending an angle of one minute at the earth's centre

cal miles or (100)(100) kilometers, from which you see that a kilometer is 0.54 nautical mile or 0.625 statute mile.

Dividing a right angle into 100 parts offers no greater difficulty than dividing it into 90 parts. The continuous bisection of two-thirds of a right angle again gives 96 parts, which is closer to 100 than to 90. The scheme of grades and gradettes (the subdivisions of the grade in fact were not called gradettes, but, to make confusion unavoidable, minutes) is still in use, primarily by the French.

Keep your graduated quadrant in a safe place. You will need it soon.

Finding heights

A common problem for the practical geometer or forester was the determination of the heights of towers and trees without the bother of climbing them. Figure 3.2.13 presents the problem as drawn by a Renaissance architect-engineer, who wanted to deduce a tower's height from measurements made on the ground. To proceed, he needed the concept of similar triangles—triangles like GHI, JKL (Fig. 3.2.14) that have all their corresponding angles equal. Whereas in congruent triangles, the sides opposite the corresponding angles are equal, in similar triangles the sides opposite the corresponding angles are proportional. That means that, in Fig. 3.2.14, GH = pJK, HI = pKL, IG = pLJ; here p is some number which can be interpreted as an indicator of the relative size of the two triangles. If $p = 1$, the triangles are congruent; congruency is a special case of similarity.[8]

8. Cf. G.D. Birkhoff and Ralph Beatley, *A new approach to elementary geometry* (1930), in Beatley, *MT*, 28 (1935), 375.

Fig. 3.2.13 Height of a tower,
by similar triangles, in Peter
Apianus, *Quadrans astronomicus*
(1532). From Bennet, *Divided
circle* (1987), 42.

Fig. 3.2.14 Schematic of
similitude.

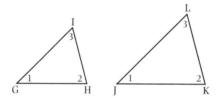

Even from this small beginning, something useful can be obtained.
The 'perimeter' of a plane figure is the sum of the lengths of its sides;
for example, the perimeter of a square is four times the length of
any of its sides. Add together the three preceding relations: GH + HI
+ IG = p (JK + KL + LJ), or perimeter ΔGHI = p × perimeter ΔJKL. As
you might guess, a triangle each of whose sides is p times the corresponding side of a similar triangle has an area p^2 times that of the
second triangle. We shall prove and use this relationship later.

The Greeks preferred to do without the number p. That is easily
managed by solving all the preceding equations for p:

$$p = \frac{GH}{JK} = \frac{HI}{KL} = \frac{IG}{LJ}.$$

The fractions GH/JK, etc., are called 'ratios'; the geometer says that
the ratio GH/JK equals the ratio HI/KL, or that 'GH is to JK as HI is
to KL', or that 'GH bears the same proportion to JK as HI does to KL'.
Soon we shall prove by Euclid's method that the corresponding sides
of similar triangles are proportional. (The method rests on theorems

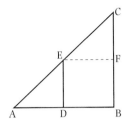

Fig. 3.2.15 A special case of similarity in triangles.

about parallelograms.) But first the proportionality can be made plausible, and ideas clarified, by considering a special case.

Figure 3.2.15 indicates the sorts of similar triangles involved in determining heights: BC might be a tower or tree, E the eye of the observer, ED a plumb line. Since D and B are right angles and ∠A is common, ΔABC and ΔADE, having all their corresponding angles equal, are similar. In the special case that AD = DB, the sides of ΔABC are each twice the corresponding sides of ΔADE. You can show this proportionality directly by drawing EF parallel to AB. To prove the proposition that BC = 2ED, etc., you need to show that ΔEFC ≅ ΔADE. That is the business of I.26, which is Euclid's proof of congruence by *a.s.a.*

Euclid proceeds by *reductio ad absurdum*. In triangles GHI and JKL (Fig. 3.2.16), ∠IGH = ∠LJK, ∠GHI = ∠JKL, and GH = JK. Suppose that GI does not equal JL, but exceeds it. Then cut off on GI a length GM = JL and join M to H. Then ΔMGH ≅ ΔLJK by *s.a.s.* and the corresponding parts, ∠GHM and ∠JKL, are equal. But ∠GHM < ∠JKL since ∠GHM = ∠GHI – ∠MHI and we are given ∠GHI = ∠JKL. Since a quantity cannot be both equal to and less than another, GI cannot be larger than JL. Nor can it be less. So, as we knew all along, GI = JL and ΔGHI ≅ ΔJKL by *s.a.s.*

If two triangles have two of their corresponding angles equal, they have all three equal, since the sum of the angles of any triangle is the same (I.32). The proof just given therefore works no matter what pair of corresponding sides you know to be equal: for if, in Fig. 3.2.16, it were given that GI = JL rather than that GH = JK, you need merely observe that ∠GIH = ∠JLK and that, as before, the equal sides come between two pairs of equal angles. You might think you have caught Euclid in an illogicality in using a later theorem (I.32) to fill out his proof of congruence by *a.s.a.* in an earlier theorem (I.26). It is not so easy to catch him. He completed the proof by redoing the *reductio* for the case that the equal lines do not come between the equal angles.

Applying *a.s.a.* to Fig. 3.2.15, you have ΔEFC ≅ ΔADE. Therefore, CF = ED = FB, and AE = EC. By construction, AD = DB; therefore, AB = 2AD, BC = 2ED, AC = 2AE. We have that the sides opposite the equal angles in similar triangles ADE and ABC are proportional; in this special case, $p = 2$. Assuming what has yet to be proved, that

Fig. 3.2.16 Congruence by *a.s.a* proved by *reductio* (Euclid I.26).

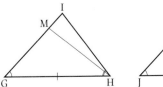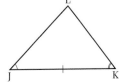

Fig. 3.2.17 Height of a tree by
similar triangles.

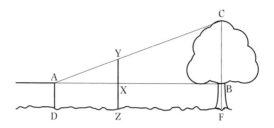

similar triangles have their corresponding sides proportional, we can
find an answer to the sort of problem illustrated in Fig. 3.2.13.

Stand at D (Fig. 3.2.17) with your eye at A. You will need the help
of two friends. Ask one of them to stand 6 feet away at Z holding a
plumb line. The bob should hover just above the ground and the
hand holding the line should be at Y, collinear with the top of the
tree at C and your eye at A. Your collaborator may have to climb on
a chair. Your second friend should measure the length of the plumb
line YZ, the height of your eye above the ground AD, and the distance DZ between your feet and those of your other helper. Use a
tape measure. Now $\triangle AXY \sim \triangle ABC$ ('$\sim$' signifies 'is similar to');
hence $YX/AX = CB/AB$. We know the value of $YX = YZ - AD$ by
measurement; also the value of $AX = DZ$. If we knew AB, we would
know the tree's height above your eye since $CB = AB (YX/AX)$. You
can find $AB = DF$ with your tape measure. Then the tree's height
above the ground, $CB + BF = CB + AD$, is known.

You can find heights without measuring the entire distance
between yourself and the base of the object of interest. The technique not only saves physical effort—at the cost, however, of a little
mental effort—it also avoids the difficulty of finding the exact center
of the bottom of trees or towers. Figure 3.2.18 indicates the technique. You do the measurement of Fig. 3.2.17 twice, first at A,
obtaining the relation $AH/GH = AB/CB$, and next at D, obtaining
$DJ/IJ = DB/CB = (AB - AD)/CB$. We have the following bit of
algebra:

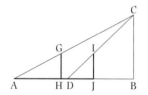

Fig. 3.2.18 The same, without
knowing the distance to the tree.

$$\frac{DJ}{IJ} = \frac{AB}{CB} - \frac{AD}{CB} = \frac{AH}{GH} - \frac{AD}{CB}$$

$$\frac{AD}{CB} = \frac{AH}{GH} - \frac{DJ}{IJ}. \tag{3.2.1}$$

You know the numerical values of everything in this last equation
except CB. If you call the numerical value of the right side of the
equation q, you have your answer: $CB = AD/q$.

APS 3.2.1. Another approach to the height of a tree makes a
double use of the concept of similarity. Figure 3.2.19(a) shows the

Fig. 3.2.19 Height and breadth of a tree by a ruler held at a distance. From Land, *Language* (1963), 189.

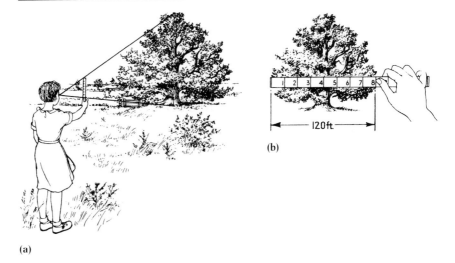

(b)

(a)

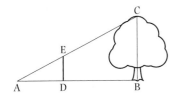

Fig. 3.2.20 Schematic for **Fig. 3.2.19**.

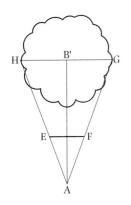

Fig. 3.2.21 Another schematic for **Fig. 3.2.19**.

usual arrangement, in which the triangle including the eye and the ruler is similar, more or less, to the triangle including the eye and the tree. (The ruler must be held parallel to the tree to achieve similarity.) Say the length of the ruler included in the triangle is 10 inches. How can that be translated into the tree's height without knowing the distance from the observer to the tree? The trick may be pulled by remaining in place and holding the ruler horizontally. As we see from Fig. 3.2.19(b), the tree appears to the observer to be 8 inches wide. She walks to the tree and measures its width to be 120 feet. She now knows that 120 feet of the tree corresponds to 8 inches at the place of observation and, consequently, that the tree's height is $(10/8)(120) = 150$ feet.

It will be well to rehearse this measurement making explicit the symmetrical triangles and the ratios employed. Figure 3.2.20 represents the vertical measurement: A the eye, BC the tree, DE the length of ruler that just covers the height. From the similar triangles ADE, ABC, we have BC = height = AB (DE/AD). Figure 3.2.21 represents the horizontal measurement: the similar triangles AHG, AEF, produce GH = AH (EF/AE). In the preceding paragraph, we took AH, the distance from the observer to the extremity of the branch, to be equal to her distance AB' from the center of the trunk; also, we took the distance of her eye from the top of the ruler, AE, in the horizontal measurement to be equal to its distance AD to the bottom of the ruler in the vertical measurement. With AB taken for AH and AD taken for AE, the equations BC = AB (DE/AD) and GH = AH (EF/AE) yield

$$BC = \text{height} = DE(GH/EF).$$

But GH/EF = 120 feet/8 inches; with DE = 10 inches, BC = 150 feet.

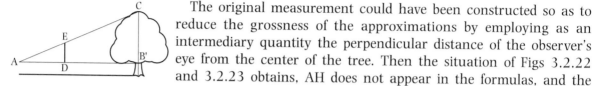

Fig. 3.2.22 A third schematic
for Fig. 3.2.19.

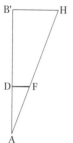

Fig. 3.2.23 A fourth schematic
for Fig. 3.2.19.

The original measurement could have been constructed so as to reduce the grossness of the approximations by employing as an intermediary quantity the perpendicular distance of the observer's eye from the center of the tree. Then the situation of Figs 3.2.22 and 3.2.23 obtains, AH does not appear in the formulas, and the distance from the eye to the ruler is the same in both measurements. Of course, the observer would have to add her own height to the result of her calculations of B'C to get the height of the tree. In general, it is best whenever possible to arrange measurements so as to invoke perpendicular distances.

APS 3.2.2. Another way to use right angles is illustrated in Fig. 3.2.24. You want to measure the height of the sheer cliff without climbing it. Put a stick vertically into the sand and draw a circle around its base with a radius equal to the stick's height. Then wait for a time on a sunny day when the stick's shadow just reaches the circle. At that instant, measure the cliff's shadow, which will equal its height. The implicit symmetry is indicated in Fig. 3.2.25, where AB is the cliff's shadow, AD the stick's. The symmetry holds in general, not just when the stick's shadow equals its height.

Fig. 3.2.24 Height of a cliff from
its shadow. From Hogben,
Mathematics for the million
(1937), 145.

Fig. 3.2.25 Schematic for
Fig. 3.2.24.

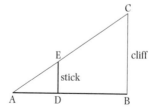

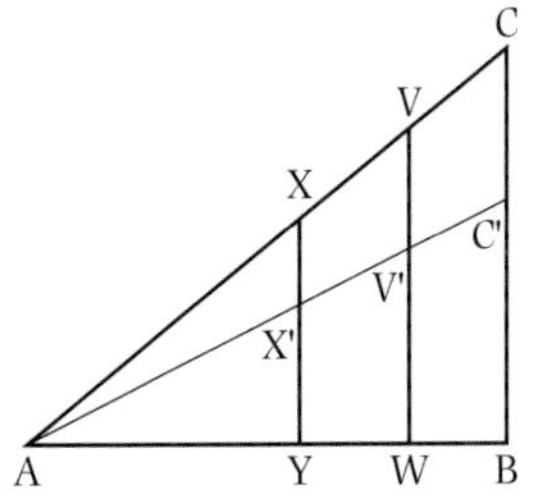

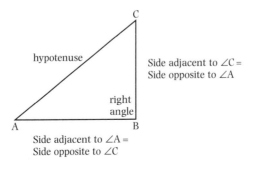

Fig. 3.2.26 From similarity to trigonometry.

Using trigonometry

Let us return to the situation of figure 3.2.18, redrawn in Fig. 3.2.26. We have the set of similar right triangles AYX, AWV, ABC. Hence XY/AY = VW/AW = CB/AB. The value of these ratios is determined by the size of ∠CAB; with another angle, such as ∠C'AB, the value would diminish, since the numerators CB, VW, XY would be replaced by their lessers, C'B, V'W, X'Y, while the denominators AB, AW, AY would remain the same. We might then say that the ratio CB/AB, or VW/AW, or XY/AY is a measure of ∠CAB. Evidently none is a unique measure; any of the equal ratios AB/AC, AW/AV, AY/AX, or of the equal ratios CB/AC, VW/AV, XY/AX might with equal reason be chosen. Geometers have opted not to take any of these ratios as the lone or even as the primary measure, but, rather, to accept them all, and, in accordance with the ordinary custom, to give each a special name. The following definitions of the names refer *only to a right triangle*.

Definitions of the functions. As we know, the longest side of a right triangle, which always lies opposite the right angle, is called the 'hypotenuse'. Each of the acute angles, ∠A ≡ ('is defined as') ∠CAB and ∠C ≡ ∠ACB, has the hypotenuse as one side and its 'side adjacent' as the other; the third side of the triangle is called 'the side opposite'. Figure 3.2.27 expresses these relationships. From the standpoint of ∠A, AC = hypotenuse, AB = side adjacent, CB = side opposite. From the standpoint of ∠C, CB = side adjacent, AB = side opposite. Now the first measure of ∠A discussed in connection with Fig. 3.2.19 was CB/AB and its equals, that is, the side opposite/side adjacent. This ratio is called 'tangent'; geometers write tan A = CB/AB. The ratio BC/AC and its equals, the side opposite/hypotenuse, is the 'sine'; we write sin A = BC/AC. Finally, AB/AC and its equals, the side adjacent/hypotenuse, is the 'cosine'; we write cos A = AB/AC. It follows immediately from the definitions that sin A = cos C, cos A = sin C. *Recall: all this holds only for a right*

Fig. 3.2.27 The trigonometric functions.

hypotenuse

Side adjacent to ∠C = Side opposite to ∠A

right angle

Side adjacent to ∠A = Side opposite to ∠C

triangle. Further to fix ideas, if △ABC were isosceles, that is, if ∠A = ∠C = rt.∠/2, then tan A = tan C = 1. *Note: The abstract functions have no definite values,* although we can say that (can you prove it?) the sine and cosine must lie between 0 and 1. Thus, in general, tan = side opp/side adj.; in particular, tan ∠45° = 1.

The utility of the 'trigometrical functions', as these ratios are called, appears immediately from equation (3.2.1). We have from the relevant figure, 3.2.18, GH/AH = tan A, IJ/DJ = tan D. Now the values of the trigonometric functions are fixed by the size of the angles that appear in them; if you were to meet the same angles A and D again in another problem, you would immediately know their tangents if you had had the foresight to write down the values of the measured ratios GH/AH and IJ/DJ where you could recover them. Since all practical geometers need to know the values of the trigonometric functions for the angles they encounter, they have made and published tables of these functions for every minute of every degree from 0° to 45°. Figure 3.2.28 presents a sample. Recall that sin A = cos C (Fig. 3.2.27). Therefore the tabulations need run only to 45°, since the sine of an angle between 45° and 90°, say $(45 + x)°$, equals the cosine of $(45 - x)°$.

A glance at the entry for 23° will allow the fulfillment of an obligation incurred in connection with Fig. 2.2.8, which depicts Columbus' calculation of the length of his projected voyage. The obligation was to reveal how he obtained the result that the circumference of the parallel of latitude through the Canary Islands is about 0.92 of the circumference of the earth at the equator. Since the circumferences of circles are as their radii (so as not to tell all at once, the proof of this plausible fact is put off to §5.4), we have (see Fig. 2.2.8) radius of parallel through Canaries/radius of the earth = cos 23° = 0.9205.

You have computed CB (the height of a tree or tower) in accordance with the scheme of Fig. 3.2.18 by measuring the lengths AH, GH, AD, IJ, DJ. Now try the same calculation by measuring ∠A and ∠D with your homemade quadrant and looking up their tangents in any book of trigonometric tables. According to equation (3.2.1),

$$\frac{AD}{CB} = \frac{1}{\tan A} - \frac{1}{\tan D}, \quad \text{or} \quad \frac{CB}{AD} = \frac{\tan A \tan D}{\tan D - \tan A}. \quad (3.2.2)$$

It will be interesting to see how close the two answers agree. You might consider whether the blame for the difference between them falls more on errors in the linear measurements AH, GH, IJ, and DJ or on errors in determining the angles at A and D.

We now have still more words about angles (Table 3.2.2). The last three functions have been added merely to complete the list. Since

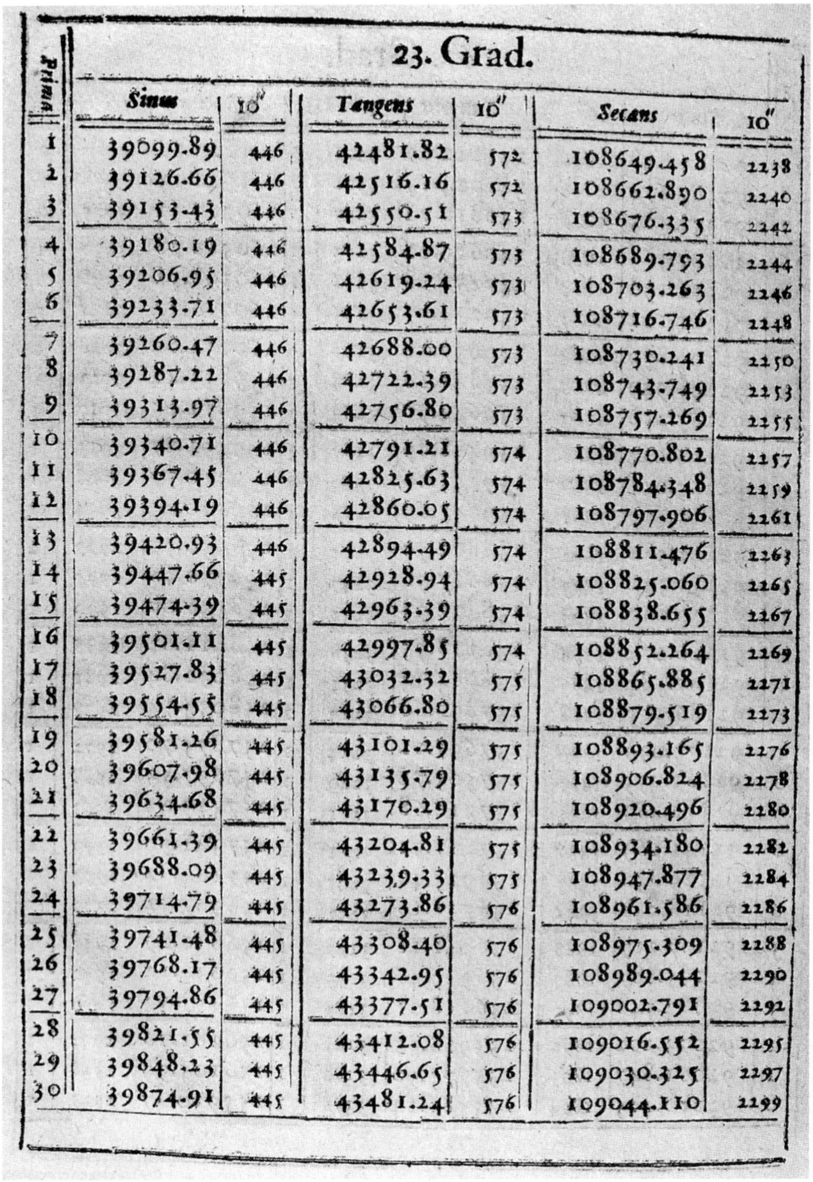

Fig. 3.2.28 A piece of a table of trigonometric functions. From Pitiscus, *Trigonometria* (1612), one of the earliest books on the subject; the tables have not changed much in format or values in the last 300 years.

23. Grad.

Prima	Sinus	10″	Tangens	10″	Secans	10″
1	39099.89	446	42481.82	572	108649.458	2238
2	39126.66	446	42516.16	572	108662.890	2240
3	39153.43	446	42550.51	573	108676.335	2242
4	39180.19	446	42584.87	573	108689.793	2244
5	39206.95	446	42619.24	573	108703.263	2246
6	39233.71	446	42653.61	573	108716.746	2248
7	39260.47	446	42688.00	573	108730.241	2250
8	39287.22	446	42722.39	573	108743.749	2253
9	39313.97	446	42756.80	573	108757.269	2255
10	39340.71	446	42791.21	574	108770.802	2257
11	39367.45	446	42825.63	574	108784.348	2259
12	39394.19	446	42860.05	574	108797.906	2261
13	39420.93	446	42894.49	574	108811.476	2263
14	39447.66	445	42928.94	574	108825.060	2265
15	39474.39	445	42963.39	574	108838.655	2267
16	39501.11	445	42997.85	574	108852.264	2269
17	39527.83	445	43032.32	575	108865.885	2271
18	39554.55	445	43066.80	575	108879.519	2273
19	39581.26	445	43101.29	575	108893.165	2276
20	39607.98	445	43135.79	575	108906.824	2278
21	39634.68	445	43170.29	575	108920.496	2280
22	39661.39	445	43204.81	575	108934.180	2282
23	39688.09	445	43239.33	575	108947.877	2284
24	39714.79	445	43273.86	576	108961.586	2286
25	39741.48	445	43308.40	576	108975.309	2288
26	39768.17	445	43342.95	576	108989.044	2290
27	39794.86	445	43377.51	576	109002.791	2292
28	39821.55	445	43412.08	576	109016.552	2295
29	39848.23	445	43446.65	576	109030.325	2297
30	39874.91	445	43481.24	576	109044.110	2299

Table 3.2.2 Still more angle words

Word	Abbreviation	Definitions
Sine	sin	Side opp./hyp.
Cosine	cos	Side adj./hyp.
Tangent	tan	Side opp./side adj.
Cotangent	ctn	Side adj./side opp. = 1/tan
Secant	sec	Hyp./side adj. = 1/cos
Cosecant	csc	Hyp./side opp. = 1/sin

they are the reciprocals of the first three, they bring nothing new except occasional convenience: for example, they permit writing equation (3.2.2) as CB/AD = 1/(ctn A − ctn D), which some geometers may regard as more beautiful. We shall make very little use of the cotangent, secant, and cosecant.

APS 3.2.3. The term 'sine' comes from the Latin *sinus*, meaning 'bosom' or 'bay'. It came into trigonometry by a mistranslation by the Arabs of the word Aryabhata coined to signify 'side opposite/hypotenuse'. Aryabhata defined trigonometric functions with reference to a circle, as in Fig. 3.2.29. By definition, sin ∠AOB = AB/OA; Aryabhata observed that AB lay in the direction of the arc AC and that the extended arc ACD resembled the string of a drawn bow. He therefore called AB/OA 'Jya', or 'bow string', for which the Arabs substituted a word, 'Jaib', which resembled the original in spelling and in the concept of a curve, since it means 'bosom'; whence *sinus*, whence sine.[9] The Arabs introduced the remaining five trigonometric ratios.[10]

APS 3.2.4. All the trigonometric functions can be represented in the same diagram, each by a single line, by use of a 'unit circle'. (The unit circle has a radius set equal to 1; that way the ratios with the radius as denominator may be expressed misleadingly as straight lines.) Representation by the unit circle (Fig. 3.2.30) is about 500 years old.[11] We have sin α = side opp./hyp. = AC/OC = AC, since we have agreed to set OC = radius = 1. Also, cos α = side adj./hyp. = OA. Now raise a perpendicular to the radius at C and extend it to meet OA extended, at B say; and also to meet the perpendicular to OB, raised at O, at D say. Now we have ∠ODC = ∠ACB = rt.∠ − (rt.∠ −α) = α. In ΔCOB, tan α = side opp./side adj. = BC/OC = BC, and sec α = hyp./side adj. = OB/OC = OB; and in ΔCDO, ctn α = side adj./side opp. = CD/OC = CD, and csc α = hyp./side opp. = OD/OC = OD.

APS 3.2.5. Lewis Carroll had his own odd scheme of representation. In place of the usual abbreviations, he proposed the symbols given in Fig. 3.2.31.[12] Can you guess which symbol goes with which trigonmetrical function? Explain your choice, and Carroll's, by reference to Fig. 3.2.30.

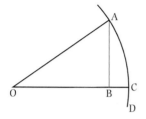

Fig. 3.2.29 The sine and the arrow.

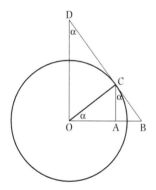

Fig. 3.2.30 Trigonometric functions in a unit circle.

9. Jha, *Aryabhata* (1988), 253–7; the earliest known use of the abbreviation 'sin' occurred in 1634.

10. Berggren, *Episodes* (1986), 132.

11. The scheme developed here comes from François Viète; Hunrath, *Abh. Gesch. Math., 9* (1899), 219.

12. Cf. Eperson, *Math. gaz., 17* (1933), 94.

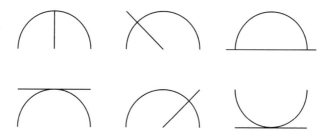

Fig. 3.2.31 Lewis Carroll's symbols for the trigonometric functions.

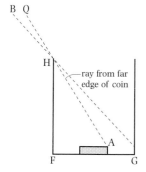

Fig. 3.2.32 A coin out of view in an empty glass.

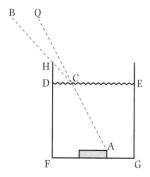

Fig. 3.2.33 The coin raised by filling the glass.

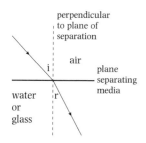

Fig. 3.2.34 The refraction of light.

Snel's law. Trigonometry finds applications in many branches of science. A most important, yet easy, example occurs in optics. When a light ray passes from one medium to another, it bends, or, as the physicists put it with greater drama, is broken or 'refracted'. A simple experiment often used to illustrate the phenomenon is to put a coin in an opaque empty cup and place yourself at B where you cannot see the coin (Fig. 3.2.32). Now have a fellow student fill the cup with water. All of a sudden, without moving your head, you will see the coin; the ray AQ, which missed your eye before, now bends at the surface of the water at C so as to enter your eye at B and disclose the coin (Fig. 3.2.33).

The Greeks knew the effect and, in their usual way, sought to express the phenomenon geometrically. They defined the angle between the incident ray (AC in Fig. 3.2.33) and the perpendicular to the plane (DE) separating the media as the angle of incidence (i) and the angle between the refracted ray (CB) and the same perpendicular as the angle of refraction (r); and they tried to find a relationship between i and r (Fig. 3.2.34). They did not do very well. Nor did the Arabs and the late medieval Christians, despite the importance of the problem for astronomical observation. (Refraction enters into naked-eye astronomy because light rays bend more and more as they penetrate into deeper, and thus denser, layers of the atmosphere.) At the beginning of the seventeenth century, a Dutch geometer named Willebrord Snel proposed a formula connecting i and r that was not only simple but also right:

$$\sin i = n \sin r,$$

where n, the 'index of refraction', is a number that depends upon the natures of the media through which the light passes.

Water has an index of refraction of 1.333 in comparison with air. If your eye (point B) is 21 inches above and 1 foot to the left of the point H in Figs 3.2.32 and 3.2.33, the cup is 4 inches high, and the point A is 2 inches from the line DF, what must the height of the water be in Fig. 3.2.33 to make the coin come into view? We have the situation depicted in Fig. 3.2.35. From the givens $\tan \alpha = 1.75$, from which $\alpha = 60°15'$, and $r = 90° - \alpha = 29°45'$. From

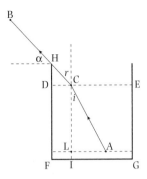

Fig. 3.2.35 The geometry of Fig. 3.2.33.

Snel's law with $n = 1.333$, $\sin i = 0.372$ and $\tan i = 0.401$. Let the unknown height of the water, DF, be x and the distance FI be y. Then

$$\tan \alpha = (\text{FH} - \text{FD})/\text{CD} = (4 - x)/y, \qquad \tan i = (2 - y)/(x - \text{LI}),$$

or

$$1.75y = 4 - x, \qquad 0.4x = 2 - y.$$

(The calculation drops the quantity LI, the height of the coin, which is much smaller than the height of the cup.) At last we can get the unknown x, the height of the water in the glass. It comes out to be 1.67 inches, not quite half way to the top.

3.3. Similarity

We return to the general proposition that the corresponding sides of equiangular triangles are proportional. For Euclid, this is an advanced topic: he does not arrive there until the fourth proposition of his sixth book (VI.4). Figure 3.3.1 shows what is at stake: given that $\angle A = \angle D$, $\angle B = \angle E$, $\angle C = \angle F$, to prove that $AC : DF = BC : EF = AB : DE$. The proof rests on theorems about areas—the spaces enclosed by the sides of plane figures—and on a new concept of equality. The concept is the equality of areas. Figure 3.3.2 rearranges the right triangles made by the sides and diagonals of a rectangle. Since rect. $ABCD = \Delta ABC + \Delta ADC$ and $\Delta ABC + \Delta ADC = \Delta D'CA$, rect. $ABCD = \Delta D'CA$, where '=' signifies 'equal in the sense of area'. Evidently this equality is very different from congruence: no side or angle of the rectangular figure $ABCD$ equals any side or angle of the triangular figure $D'CA$. Yet rect. $ABCD = \Delta D'CA$ in area. This sort of equality is the basis of the Chinese method of demonstration by cutting up figures and rearranging their parts.[13]

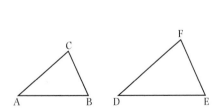

Fig. 3.3.1 Similarity of triangles.

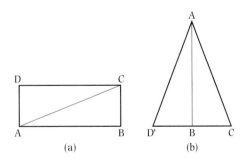

Fig. 3.3.2 Equality of areas does not imply equality or similarity of parts.

13. Supra, §1.2.

Euclid's proof by parallelograms

The path to Euclid VI.4 runs through parallelograms. Figure 3.3.3 presents two parallelograms ABCD and ABEF 'between the same parallels' OO′ and PP′ and standing on the same base AB. (A parallelogram, abbreviated 'par.,' is a four-sided plane figure whose opposite sides are parallel.) As paradoxical as it may appear, par. ABCD = par. ABEF in area; it appears paradoxical because, no matter how far the point F is from C along the line PP′, and consequently no matter how long the sides AF and BE become, the area of ABEF remains the same size. Otherwise stated, knowing the perimeter of a parallelogram, or, for that matter, of any rectilinear figure, tells you nothing about its area. For this reason the Roman orator Quintilian recommended that his colleagues study geometry: 'everything depends on the shape of the figure ... and historians have been taken to task by geometricians for believing the time to circumnavigate an island [a measure of its perimeter] to be a sufficient indication of its size [area].'[14]

Euclid proves the equality of parallelograms between parallels on equal bases in two propositions. The first, I.34, proves that the opposite sides of a parallelogram are equal, as follows (Fig. 3.3.4):

1. Draw the transversal AC   Post. 1
2. ∠1 = ∠3, ∠2 = ∠4   Alt. int. angles (I.29)
3. AC = AC   C.N. 1
4. ∴ ΔADC ≅ ΔABC   *a.s.a.* (I.26)
5. ∴ AD = BC, AB = CD   step 4

Fig. 3.3.3 Parallelograms on the same base between the same parallels are equal (Euclid I.34).

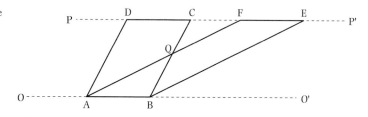

Fig. 3.3.4 Opposite sides of a parallelogram are equal (Euclid I.34).

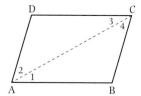

14. Quintilian, *Inst., l*, 10 : 40; ed. Butler (1920), 179.

Proposition I.35 demonstrates that parallelograms on the same base and between the same parallels are equal (Fig. 3.3.3). The proof works by adding and subtracting areas; during the process, a four-sided figure will appear that has only one pair of opposite sides parallel. This figure, with which we shall become very familiar, is called a 'trapezoid'.

1. AB = DC, AB = FE   I.34
2. DC = FE   C.N. 1
3. DF = DC + CF = CF + FE = CE   C.N. 2
4. ∠CDA = ∠FCB   Alt. ext. angles (I.29)
5. AD = BC   I.34
6. ΔFDA ≅ ΔECB (DF = CE, ∠CDA =   s.a.s.
    ∠FCB, AD = BC)
7. trap. AQCD = ΔFDA − ΔQCF =   C.N.3
    ΔECB − ΔQCF = trap. FQBE.
8. par. ABCD = trap. AQCD + ΔAQB =   Q.E.D.
    trap. FQBE + ΔAQB = par. ABEF.

The next step is to generalize the preceding result to parallelograms between the same parallels on equal bases (I.35 refers to the *same* base). Figure 3.3.5, which relates to I.36, has AB = EF = HG = DC, by construction; by adding the lines AH, BG, you can invoke Euclid I.35 to prove that the two parallelograms in Fig. 3.3.5 are equal. You need show only that ABGH is a parallelogram, or, what comes to the same, that ∠HAE = ∠GBF and AH = BG. These relations hold if ΔHAE ≅ ΔGBF. Now AE = AB + BE = BE + EF = BF (since AB = EF by construction); HE = GF (opposite sides of parallelograms); ∠AEH = ∠BFG (alternate interior angles); hence ΔHAE ≅ ΔGBF (s.a.s.).

Euclid I.35 and I.36 have precise analogues for triangles in place of parallelograms. To prove I.37 (the analogue to I.35), you need to show that ΔABC = ΔABD (Fig. 3.3.6). Draw QB∥AC and BR∥AD.

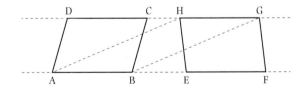

Fig. 3.3.5 Parallelograms on equal bases between the same parallels are equal (Euclid I.35).

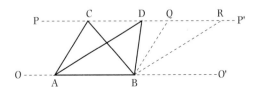

Fig. 3.3.6 Triangles on the same base between the same parallels are equal (Euclid I.37).

Fig. 3.3.7 Triangles on equal bases between the same parallels are equal (Euclid I.38).

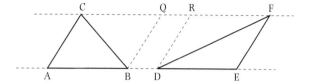

Then (Euclid I.35) par. ABQC = par. ABRD; but par. ABQC = 2∆ABC = par. ABRD = 2∆ABD, whence ∆ABC = ∆ABD, Q.E.D. Euclid I.38, the analog to I.36, is illustrated in Fig. 3.3.7, where, by construction, AB = DE. The proof follows in an obvious way by completing the parallelograms ABQC and DEFR and applying I.36.

APS 3.3.1. We see from Fig. 3.3.4, where ∠1 = ∠3 and ∠2 = ∠4, that the opposite internal angles of a parallelogram are equal. Here is another important but easy property: the diagonals of a parallelogram bisect one another. That means that, in Fig. 3.3.8, DO = OB, AO = OC. These equalities hold if ∆AOB ≅ ∆DOC and ∆AOD ≅ ∆BOC. The congruency follows in both cases from *a.s.a.*, as indicated in the figure. A parallelogram with four equal sides is called a 'rhombus'; a square is a rhombus with right angles. If ABCD were a rhombus, ∆DOC ≅ ∆COB and, therefore, ∠DOC = ∠COB. Since ∠DOB = ∠DOC + ∠COB = 2∠DOC = st.∠, the diagonals of a rhombus (and only, among four-side figures, of a rhombus) meet at right angles.

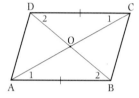

Fig. 3.3.8 The diagonals of a parallelogram bisect one another (APS 3.3.1).

APS 3.3.2. In the trapezoid ABCD (Fig. 3.3.9), E is the midpoint of one of the non-parallel sides AB. Can you show that, if E is connected to the opposite vertices of the trapezoid, the resultant triangle, ∆BEC, has an area half that of the trapezoid? You might consider using I.37 by drawing FG through E parallel to BC. (G is on CD extended). Then (I.37) ∆BEC = ∆BGC = par. FBCG/2. But par. FBCG = trap. ABCD − ∆AEF + ∆GED. Show that ∆AEF ≅ ∆GED and thus that par. FBCG = trap. ABCD.

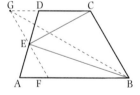

Fig. 3.3.9 Exercise on areas (APS 3.3.2).

APS 3.3.3. E, F, G, and H are the midpoints of the sides of par. ABCD. Join the points to the opposite vertices as shown in Fig. 3.3.10. Can you see that AJCI is a parallelogram too? One way is to recognize that AECG is a parallelogram: AE and GC are parallel

Fig. 3.3.10 Exercise on parallelograms (APS 3.3.3).

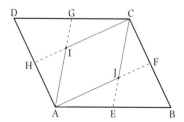

Fig. 3.3.11 The areas of
triangles on equal bases are
proportional to their altitudes
(Euclid VI.1).

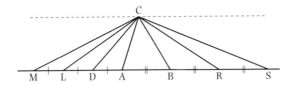

Fig. 3.3.11 The areas of
triangles on equal bases are
proportional to their altitudes
(Euclid VI.1).

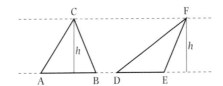

Fig. 3.3.12 Further to Euclid
VI.1.

and equal and therefore so are AG and EC. (You need to prove this
last assertion.) Since AG‖EC, AI‖JC. Similarly, you can see that
IC‖AJ.

Let us now rearrange matters so as to add the concept of proportion. Figure 3.3.11 presents a multitude of triangles between the
same parallels. The bases of the triangles to the left of A are all equal
to AD (AD = DL = LM = MN ...); those of the triangles to the right of
A are all equal to AB (AB = BR = RS = ST ...). All the small triangles
to the left of A are equal to one another (I.38); so are all the triangles to the right of A. Consequently, the larger triangles composed of
the smaller ones are as their bases. For example, $\Delta LAC = \Delta LDC + \Delta DAC = 2\Delta DAC$, $\Delta MAC = 3\Delta DAC$, etc.; but LA = 2DA, MA = 3DA,
etc., and, therefore, $\Delta MAC : \Delta LAC = MA : LA$. Similar results follow
from the triangles to the right of A. Suppose now that the multiplication of triangles has been carried out n times to the left of A and m
times to the right so that, exactly or very closely, $nAD = mAB$. Then
the triangle with base nAD, let us call it ΔCAO, will equal $n\Delta CAD$,
and the triangle with base mAB, call it ΔCAT, will equal $m\Delta CAB$.
Hence $\Delta CAT : \Delta CAO = m\Delta CAB : n\Delta CAD = 1$. ($\Delta CAT = \Delta CAO$ by
Euclid I.38 since they stand between the same parallels and have
equal bases.) It follows that $\Delta CAB : \Delta CAD = n{:}m = AB : AD$; and, in
general, that the areas of triangles contained between the same parallels are as their bases (Euclid VI.1). Alternatively, geometers say
that the areas of triangles of the same height, where height indicates the perpendicular distance between the parallels (Fig. 3.3.12),
are proportional to their bases.

The power of Euclid VI.1 appears from Fig. 3.3.13 (Euclid VI.2),
the first part of which demonstrates that a line parallel to one side of
a triangle cuts the other two sides proportionally. (The second part,
reserved for APS 3.3.4, is the converse of the first.) Given ΔABC,
draw DE‖AB and join D to B and A to E. Imagine a line PQ‖AC; then
VI.1 immediately applies and $\Delta DEA : \Delta DEC = DA : CD$. Similarly,

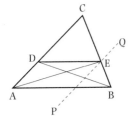

Fig. 3.3.13 A line parallel to
one side of a triangle cuts the
other sides proportionally
(Euclid VI.2).

ΔDEB : ΔDEC = EB : CE. But ΔDEA = ΔDEB (they are triangles between parallels DE and AB). Hence ΔDEA : ΔDEC = ΔDEB : ΔDEC and DA : CD = EB : CE. By adding unity to both sides of this last equation, we have (CD + DA) : CD = (CE + EB) : CE, or, what is the same, CA : CD = CB : CE. Because DE∥AB, ∠CDE = ∠CAB and ∠CED = ∠CBA; therefore, in the circumstances illustrated in Fig. 3.3.13, the sides opposite equal angles in the triangles ABC and DEC are equal. We need only to generalize this result for any pair of triangles whose corresponding angles are equal. Euclid does it in VI.4: 'In equiangular triangles the sides about the equal angles are proportional, and those are corresponding sides which subtend the equal angles.'

Let the two equiangular triangles of Fig. 3.3.1 be joined as in Fig. 3.3.14 by placing D at B and DE along AB extended. Since AC and EF are not parallel, they will meet if extended; let their junction be G. Since ∠CAB = ∠FBE, AG∥BF; and, since ∠FEB = ∠CBA, GE∥CB. The opposite sides of the four-sided figure, or 'quadrilateral', CBFG, are parallel; therefore CBFG is a parallelogram and CG = BF, CB = GF. By applying VI.2 to triangles ABC and AEG, you have AG : AC = AE : AB. Subtract unity from both sides: CG : AC = BE : AB. But CG = BF. Hence BF : AC = BE : AB. Q.E.D.

Euclid gives a further consequence in the useful proposition VI.6: if two triangles have one angle equal and the sides of the angle proportional, the triangles are similar. In Fig. 3.3.15, ∠BAC = ∠EDF and $c : b = f : e$. Draw ΔFDG so that ∠FDG = ∠BAC and ∠DFG = ∠ACB. Then ΔDFG ~ ΔABC and $y : e = c : b$. But $c : b = f : e$, whence $y = f$. We have therefore that ΔFDG ≅ ΔFDE (by s.a.s.); hence they are equiangular, and so are ΔEDF and ΔBAC. Q.E.D.

APS 3.3.4. Remaining with the familiar, take a parallelogram ABCD and bisect its opposite sides at E, G (Fig. 3.3.16). Connect G with A and E with C; can you show that H, F divide the diagonal DB into thirds? A way to proceed is to notice that ΔEFB ~ AHB and that AB = 2EB. It follows from the properties of similar triangles that HB = 2FB or, what is the same, that HF = FB. In exactly the same way, you can show from the similarity of triangles DHG,

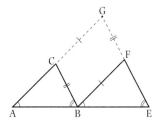

Fig. 3.3.14 Equiangular triangles are similar (Euclid VI.4).

Fig. 3.3.15 Triangles with one angle equal and the sides including it proportional are similar (Euclid VI.6).

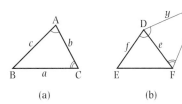

(a)   (b)

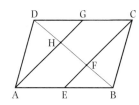

Fig. 3.3.16 Exercise on similarity (APS 3.3.4).

Table **3.3.1** Some words and symbols about plane figures

Word	Meaning
Equality (=)	Refers to equality in area
Ratio (:)	The quotient of one length (or area, or angle) by another; the numerator and denominator must both represent the same sort of quantity
Similar (~)	Refers to triangles the ratios of whose corresponding sides are equal
Isosceles	Refers to a triangle with two equal sides
Hypotenuse	The side opposite the right angle in a right triangle
Parallelogram	A four-sided figure with opposite sides parallel
Perimeter	The sum of the lengths of the sides of a plane figure
Quadrilateral	A plane figure with four sides
Rhombus	A parallelogram with all sides equal
Trapezoid	A quadrilateral with two (and only two) sides parallel

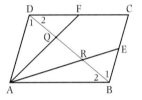

Fig. 3.3.17 Another exercise on similarity (APS 3.3.4).

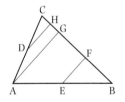

Fig. 3.3.18 A third exercise on similarity (APS 3.3.5).

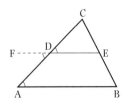

Fig. 3.3.19 A corollary to Euclid VI.2 (APS 3.3.6).

DFC, and from DG = GC, that DH = HF. Altogether, FB = FH = DH. Q.E.D.

While on the subject of splitting the diagonals of parallelograms into thirds, you might be interested in Fig. 3.3.17, in which E, F are the midpoints of the adjacent sides BC, CD. The lines AF, AE intersect the diagonal DB at Q, R. Again DB falls into three equal parts. You have ΔDQF ~ ΔBQA (remember the vertical angle at Q!); therefore, since AB = 2DF, QB = 2DQ and QB + DQ = DB = 3DQ. Proceeding in the same way, using ΔBRE and ΔARD, you have DB = 3RB. The preceding equations make DQ = RB. But DB = DQ + QR + RB = 2DQ + QR = 3DQ, from which QR = DQ. Q.E.D.

APS 3.3.5. In Fig. 3.3.18, D and E are the midpoints of the sides AC, AB, of ΔABC. Drop perpendiculars from D, A, and E to the side BC. It is easy but not very useful to show that DH + EF = AG. From ΔDHC ~ ΔAGC, DH : DC = AG : AC; but DC = AC/2, whence DH = AG/2. From ΔBEF ~ ΔBAG, EF = AG/2. Q.E.D.

APS 3.3.6. The line joining the midpoints of any two sides of a triangle is parallel to the remaining line and equal to half of it. This important truth is a special case of the second part of VI.2 (a line cutting two sides of a triangle proportionally is parallel to the third side). It is more easily demonstrated by VI.6 (Fig. 3.3.19): since AC : CD = BC : CE = 2:1, ΔACB ~ ΔDCE. It follows that ∠CDE = ∠CAB = ∠ADF; and, thence, that DE∥AB.

This special case of VI.6 suggests a pretty theorem. Take any quadrilateral ABCD and join the midpoints of its sides E, F, G, H, in

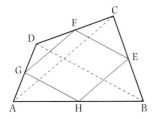

Fig. 3.3.20 The midpoints of the sides of a quadrilateral are the vertices of a parallelogram (APS 3.3.6).

order, as in Fig. 3.3.20. Can you see that EFGH is a parallelogram? No? Draw the diagonals DB, AC. Then, from APS 3.3.6, EF∥DB and HG∥DB, so EF∥HG; and also, EH∥AC∥FG. Q.E.D. And there is more. The sum of the diagonals of the quadrilateral, AC + DB, equals the perimeter of parallelogram EFGH. This follows immediately from the exercise, since EH = FG = AC/2, EF = HG = DB/2.

Sidelines

As important to the triangle as its sides are three sets of three lines each that have the important and interesting property that each set intersects at a single point. These sets are the 'altitudes', that is, the perpendiculars dropped from the vertices onto the opposite sides; the 'medians', the lines connecting the vertices to the midpoints of the sides; and the angle bisectors.

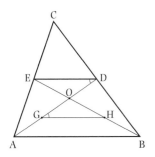

Fig. 3.3.21 The medians of a triangle are coincident.

Medians. It is easiest to begin with medians. In $\triangle$ABC of Fig. 3.3.21, AD and BE are medians intersecting at O. Join E and D; then, as we know, $\triangle$ECD ~ $\triangle$ACB and ED∥AB. This interesting fact allows the inference that AO : AD = BO : BE = 2 : 3 and that, therefore, the third median, from C to AB, passes through O. To obtain the result that the point of intersection of the two medians is two-thirds of the way along from each of their vertices, bisect AO and OB and join the points of bisection G, H. Then (VI.2 again) GH∥AB∥ED. From the similar triangles ECD and ACB, ED : AB = EC : AC = 1 : 2; from the similar triangles GOH, AOB, GH : AB = OG : OA = 1 : 2; hence ED = GH. Triangles EOD and GOH therefore are congruent via *a.s.a* (the base angles are alternate interior angles since ED∥GH), whence OD = OG and OE = OH. But by construction AG = GO and OH = HB. Hence G and O divide AD into thirds and H and O divide BE into thirds: AO : AD = BO : BE = 2 : 3. Q.E.D. But, further, AD and BE are any pair of medians. The third one, proceeding from C, must therefore cut either of the others at a point two-thirds the distance of either from its originating vertex. But that point is O. Hence all the medians meet in a single point.

Altitudes and areas. We rise to altitudes. Several methods of demonstration have been devised, none obvious; the difficulty of the business lies in finding a constraint that will function like the medians' intersecting two-thirds of their way toward the sides they bisect. One classical demonstration makes use of the property that all points on the perpendicular bisector of a line are equidistant from the ends of the line, as shown at Fig. 2.1.28.

The demonstration has two parts: first, the construction of a $\triangle$A′B′C′ (Fig. 3.3.22) the sides of which are parallel to, and twice the

Fig. 3.3.22 The altitudes of a triangle are coincident.

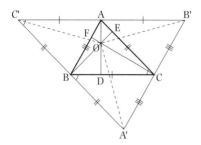

length of, those of △ABC; second, the argument that the perpendicular bisectors raised at A, B, and C intersect at a point. To construct △A′B′C′, draw parallelograms ACBC′ and AB′CB. Extend B′C and C′B to their intersection at A′. If you can prove that A′BAC also is a parallelogram, you will show, as required, that △A′B′C′ and △ABC have their sides mutually parallel. One way to show that A′BAC is a parallelogram is to prove that its opposite sides are equal. That follows from △ABC ≅ △BCA′ (*a.s.a.*), which follows because C′B′∥BC implies that ∠CBA′ = ∠AC′B = ∠BCA (the last step because the opposite angles of a parallelogram are equal); and, by the same reasoning, ∠BCA′ = ∠C′B′C = ∠ABC. Furthermore, A lies at the midpoint of C′B′ by construction; B at that of C′A′ because of parallelograms C′BCA and ABA′C, which require C′B = AC = BA; and C at that of A′B′ because of parallelograms AB′CB and ACA′B.

Now erect the perpendicular bisectors at B and C and call their intersection O. Draw OA, OA′, OB′, and OC′. Then OC′ = OA′ and OB′ = OA′ because of our theorem about the perpendicular bisector. Therefore, OC′ = OB′ (C.N.1). But then, since O lies equidistant from C′ and B′, AO must be the perpendicular bisector of B′C′: the three perpendicular bisectors of the sides of △A′B′C′ meet at a point. We come to the clinch. Continue AO, BO, CO to intersect BC, AC, and AB at D, E, F, respectively. Since AO⊥C′B′ and BC∥C′B′, AD⊥BC; similarly, BE⊥AC, CF⊥AB. Therefore AD, BE, CF are the altitudes of △ABC and they intersect at a single point, O. Q.E.D.

A second classical demonstration uses the constraint that, owing to the right angles at E and F, the four points A, F, O, and E lie on a circle of which AO is the diameter. The relevant theorem is derived in §5.2 at Fig. 5.2.33. Let M be the center of this circle (Fig. 3.3.23). We want to show that ∠FAO, or α, equals ∠FEO, or β, because, using theorems so far developed, we can then show that β = ∠FCB. Thence we shall have △BAD ∼ △BCF, whence ∠BDA = rt.∠, Q.E.D. That $\alpha = \beta$ emerges from the isosceles triangles FMA, AME (FM, MA, and ME are radii): introducing γ for ∠MFE = ∠MEF and using I.32, we have ∠MAE = ∠MEA = rt.∠. − ($\alpha + \gamma$) and β = rt.∠ − γ − ∠MEA = α.

Fig. 3.3.23 An alternative demonstration of the coincidence of altitudes.

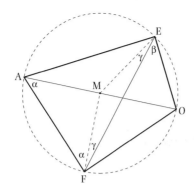

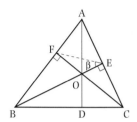

Fig. 3.3.24 Further to the alternative demonstration of the coincidence of altitudes.

Figure 3.3.24 suggests how to show that $\beta = \angle FCB$. Since $\Delta BOF \sim \Delta COE$ (they are right triangles connected by vertical angles), FO : EO = BO : CO. Consequently, via Euclid VI.6, $\Delta FOE \sim \Delta COB$ and, as advertised, $\angle FCB = \angle FEO = \beta = \alpha = \angle FAO = \angle BAD$, making ΔADB and ΔBFC, which have $\angle B$ in common, similar, and $\angle ADB$ a right angle. Q.E.D.

The altitude of a triangle is a very important concept. Considered algebraically, it frees the calculation of the area of a triangle from the constraint under which we have labored so far, of working with proportions. The area of a triangle may be expressed as half the product of any of its altitudes and the side to which it is perpendicular. This formula may be made plausible in a Euclidean manner with the help of Fig. 3.3.25

From VI.1, the areas in Fig. 3.3.25 are as the bases, so

$$\frac{\text{rect.ABCD}}{\text{rect.AEFD}} = \frac{AB}{AE}, \qquad \frac{\text{rect.ABCD}}{\text{rect.GHIJ}} = \frac{AD}{HG}.$$

Let us write, algebraically,

$$\text{rect.ABCD} = AB\,(\text{rect.AEFD}/AE) = AD\,(\text{rect.GHIJ}/HG).$$

These equations would both be satisfied if

$$\text{rect.AEFD} = AE \times AD \quad \text{and} \quad \text{rect.GHIJ} = HG \times AB,$$

that is, if the area of a rectangle were the product of two of its adjacent sides. We shall accept this as our formula. Since, from I.35, parallelogram ABRQ = rect.ABCD, its area is its base AB times the

Fig. 3.3.25 The area of a rectangle.

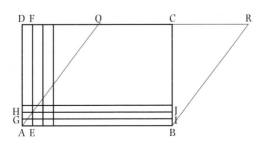

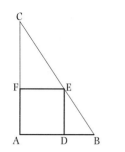

Fig. 3.3.26 Exercise on areas.

perpendicular between the base and its parallel, that is, the 'height'; and, since the area of a triangle is half that of a parallelogram formed on the same base, the area of a triangle is half the product of its base times its altitude.

Say the square ADEF (Fig. 3.3.26) has an area of 100 'area units', which we shall write 'a.u'; examples of area units are square inches, square miles, square meters. The leg AB has a length of 12 linear units 'l.u.' What is the area of $\triangle$ABC? We have $\triangle$EFC ~ $\triangle$BDE (each contains a right angle and $\angle$EBD = $\angle$CEF since FE∥AB); therefore CF : 10 = 10 : 2, or CF = 50 l.u. The area of $\triangle$ABC is half the product of its base AB = 12 l.u. times its height AC = 10 + 50 = 60 l.u., or area = $1/2(12 \times 60)$ = 360 a.u.

The areas of similar triangles are as the squares of their altitudes or bases. Figure 3.3.27 presents two similar triangles ABC, A'B'C'. Choose the corresponding sides AB, A'B', as the bases; the altitudes or heights are then CD = h, C'D' = h'. Now $\triangle$ADC ~ $\triangle$A'D'C' ($\angle$CAD = $\angle$C'A'D' because of the similarity of triangles ABC, A'B'C', and the angles at D, D' are right); therefore $h : h'$ = AC : A'C'. But also, AC : A'C' = AB : A'B'. The heights therefore are as the bases. But the areas are as the products of the heights and the bases. In symbols:

$$\frac{\triangle ABC}{\triangle A'B'C'} = \frac{hAB}{h'A'B'} = \frac{h^2}{h'^2} = \left(\frac{AB}{A'B'}\right)^2, \qquad (3.3.1)$$

Q.E.D.

Angle bisectors. In Fig. 3.3.28, the angles at A and B are supposed to be bisected by lines AI, BJ, meeting at O. Join C to O and extend the line to H. We are to show that $\angle$ACH = $\angle$BCH, that is, that CH bisects the angle at C. Drop perpendiculars from O to the sides AB,

Fig. 3.3.27 Areas of similar triangles are as their altitudes or bases.

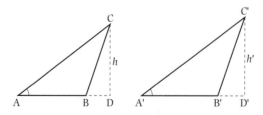

Fig. 3.3.28 The angle bisectors of a triangle are coincident.

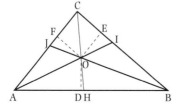

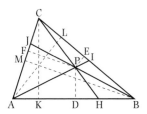

Fig. 3.3.29 A general relationship for coincident lines from the vertices of a triangle.

BC, CA, at D, E, F, respectively. All these perpendiculars are equal: OE = OD and OD = OF because the points on an angle bisector are equidistant from the angle's sides (see Fig. 2.1.31). For the same reason (since OE = OD = OF), O lies on the bisector of ∠ACB, Q.E.D.

Any point within a triangle can of course be joined to each of the vertices. The inquiring mind might ask whether in this general case—where the lines are not necessarily angle bisectors, altitudes, or medians—any relation exists among them. A nice one may easily be deduced via Fig. 3.3.29, in which P is any arbitrary point within the triangle ABC. The nice relation is

$$\frac{PH}{CH} + \frac{PI}{AI} + \frac{PJ}{BJ} = 1. \qquad (3.3.2)$$

To prove it, drop the perpendiculars PD, PE, PF to the sides AB, BC, CA. Now, the area of ΔABC equals the sum of the areas of triangles APB, BPC, and APC, so that

$$\frac{\Delta APB}{\Delta ABC} + \frac{\Delta BPC}{\Delta ABC} + \frac{\Delta APC}{\Delta ABC} = 1.$$

Let AL, BM, CK be the altitudes of ΔABC. Then

$$\frac{\Delta APB}{\Delta ABC} = \frac{AB \times PD}{AB \times CK} = \frac{PD}{CK},$$

$$\frac{\Delta BPC}{\Delta ABC} = \frac{BC \times PE}{BC \times AL} = \frac{PE}{AL},$$

$$\frac{\Delta APC}{\Delta ABC} = \frac{AC \times PF}{AC \times BM} = \frac{PF}{BM}.$$

From the similar right triangles BKC and BDP, PD : CK = PH : CH; from triangles BEP and BLA, PE : AL = PI : AI; from triangles CFP and CMB, PF : BM = PJ : BJ. Hence

$$\frac{\Delta APB}{\Delta ABC} + \frac{\Delta BPC}{\Delta ABC} + \frac{\Delta APC}{\Delta ABC} = \frac{PD}{CK} + \frac{PE}{AL} + \frac{PF}{BM} = \frac{PH}{CH} + \frac{PI}{AI} + \frac{PJ}{BJ} = 1,$$

Q.E.D.

Another way of phrasing the same property is

$$h'_1/h_1 + h'_2/h_2 + h'_3/h_3 = 1, \qquad (3.3.3)$$

where the h's are the altitudes and the h''s are the corresponding perpendiculars to the sides from the arbitrary point P.

APS 3.3.7. Here is a little exercise in medians and similarity. Any two medians mark out a triangle and a quadrilateral, as in the shaded areas in Fig. 3.3.30. These areas are equal, that is ΔAPB = quad.CEPF. Join E and F. We know full well by now that

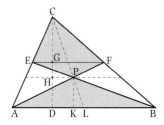

Fig. 3.3.30 Exercise on medians and similarity (APS 3.3.7).

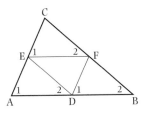

Fig. 3.3.31 Further to
APS 3.3.7.

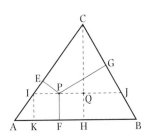

Fig. 3.3.32 Exercise on altitudes
(APS 3.3.8).

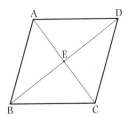

Fig. 3.3.33 Exercise on
rhombus (APS 3.3.9).

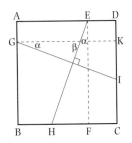

Fig. 3.3.34 Further to
APS 3.3.9.

EF = AB/2. Therefore, according to equation (3.3.1), ΔEFP = ΔAPB/4. Furthermore, ΔEFC = ΔABC/4. (Look at Fig. 3.3.31, where E, F, D are the midpoints of their sides: can you see that the four internal triangles are all equal, and hence that ΔEFC = ΔABC/4?) Now quad. CEPF = ΔEFC + ΔEFP = (ΔAPB + ΔABC)/4. To prove the proposition, you have only to show that ΔABC = 3ΔAPB. Since ΔABC and ΔAPB have the same base, the task reduces to showing that the altitude CD of the larger is three times that (PK) of the smaller. We already know that. Draw the third median CPL; since PL : CL = 1 : 3 and ΔDCL ~ ΔKPL, PK = CD/3.

APS 3.3.8. A simple proposition regarding altitudes may be desirable. Simplicity is at its height in equilateral triangles, like ABC (Fig. 3.3.32). P is any point within the triangle, E, F, and G the feet of perpendiculars dropped from P on the sides. The simple proposition: the sum of the perpendiculars equals the altitude. (All the altitudes of an equilateral triangle are equal.) How would you proceed to show that PE + PF + PG = CH? No doubt by beginning with familiar territory, which can be reached by drawing IJ through P parallel to AB. You have many similar 30°–60°–90° triangles. Take ΔIQC as the reference triangle and, to save writing, set k = CQ/CI. Then PE = kPI, PF = IK = kAI, PG = kPJ, CH = kAC and PE + PF + PG = k(PI + PJ + AI) = k(IJ + AI) = kAC = CH, Q.E.D. (That IJ + AI = AC follows from ΔIJC ~ ΔABC, given as equilateral, which makes IJ = IC.)

APS 3.3.9. A nice theorem about a rhombus and another about that special rhombus, the square, will help crystalize your ideas about similarity and areas. First, the area of a rhombus is half the product of its diagonals. As we know, the diagonals of a rhombus meet at right angles. Hence (Fig. 3.3.33), ΔABD = BD × AE/2 and ΔCBD = BD × EC/2. But rhom. ABCD = ΔABD + ΔCBD = BD(AE + EC)/2 = BD × AC/2. Second, any two lines drawn between opposite sides of a square and intersecting at right angles, as in Fig. 3.3.34, are equal. This amusing property is easily confirmed by drawing EF∥DC and GK∥AD. Then we have ΔGKI ~ ΔEFH (the angles marked α are each complementary to β, and $\angle$GKI = $\angle$EFH = rt.$\angle$), whence EH : EF = GI : GK. But EF = GK = CD, the side of the square. Notice that triangles GKI, EFH are not only similar, they are congruent, which illustrates the proposition that similar triangles any pair of whose corresponding sides are equal are congruent.

APS 3.3.10. Here's a funny one. Extend the sides of any ΔABC by doubling them, as indicated in Fig. 3.3.35. Join the extremities to form ΔXYZ. This big triangle is exactly seven times as large as

Fig. 3.3.35 Exercise on
similarity and areas
(APS 3.3.10).

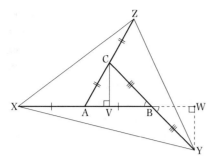

Fig. 3.3.35 Exercise on
similarity and areas
(APS 3.3.10).

ΔABC. How to show it? You might guess that since 7 = (3 × 2) + 1,
it is likely that each of the triangles XBY, ZCY, and ZAX is twice
ABC. And so it is. Take ΔXBY as an example. Drop a perpendicular from Y onto XB extended at W. Also draw the altitude CV of
ΔABC to the base AB. You have ΔXBY = (XA + AB)(WY)/2 = AB
× WY, and ΔABC = AB × CV/2. But ΔCVB ≅ ΔYWB (they are
similar and have a pair of equal sides, CB = BY), so CV = WY. You
have confirmed your guess that ΔXBY = 2ΔABC. The same
demonstration using the other altitudes and bases and addition—
ΔXYZ = ΔXBY + ΔZCY + ΔZAX + ΔABC—confirms that ΔXYZ =
7ΔABC. You are now in a position to draw a triangle 49 times as
large as ΔABC, should you wish.

Extraordinary lines

You have proceeded far enough that you may enjoy the following
three theorems for the pleasure they afford. They come from very different eras in the history of geometry: Euler's line from the eigth-
eenth century, Ceva's theorem from the seventeenth, and Menealaus'
from the first. We shall take them up in antichronological order.

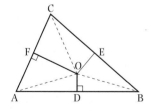

Fig. 3.3.36 Perpendicular
bisectors of the sides of a triangle
are coincident.

Euler's line. Like the medians, the altitudes, and the angle bisectors,
the perpendicular bisectors of the sides of a triangle intersect in a
point. This last coincidence is very easily demonstrated. Let O in Fig.
3.3.36 be the point of intersection of the perpendicular bisectors of
the sides AB and AC. From O draw DE to the midpoint E of BC. We
are to prove that ∠OEC = ∠OEB. Join the vertices to O. Then, since
ΔAOF ≅ ΔCOF and ΔAOD ≅ ΔBOD (both by *s.a.s.*), OA = OB = OC.
Consequently, O lies equidistant from B and C and OE is the perpendicular bisector of BC. Q.E.D.

A remarkable connection exists between the points of intersection
of the medians (M), the altitudes (L), and the perpendicular bisectors
(O) of a triangle. It is that M lies on the line joining L and O two-
thirds of the way from L to O. This truth was revealed in 1765 by
Leonhard Euler, one of the greatest mathematicians of all time, and

also one of the most productive. Generations of learned editors have been laboring at publishing his collected works since 1911; they have now taken more time to edit them than it took him, working alone, to compose them. The edition has reached many dozens of volumes, in French, German, and Latin, with an end just barely in sight. Euler contributed to all branches of the mathematics and mathematical physics of his time: sublime contributions, like setting the basis of analytical mechanics, and trivial ones, like popularizing the abbreviations for the trigonometric functions with which you are familiar. His work on plane geometry falls somewhere in between.

One reason that Euler's work occupies so much space is that he did not always find the shortest paths or clearest notation in his proofs. Although he warned at the outset of his discussion of the points L, M, and O that success lay in choosing suitable quantities for analysis, 'ne in calculos taediosissimos et omnino inextricabiles delabamur', 'lest we bog down in very tedious and altogether inextricable calculations', his method, though straightforward, would be tiresome to follow.[15] Fortunately, the 'theorem of Euler's line' admits an astonishingly simple demonstration.[16]

In Fig. 3.3.37, CD is the median to AB and O and M are the points of intersection of the perpendicular bisectors and the medians, respectively. Draw OM and extend it to a point L (we do not yet know that L is the intersection of the altitudes) so that LM = 2MO. (This apparently arbitrary assignment is recommended by the fact that M divides CD in the ratio 2 : 1.) Join C and L and extend CL to intersect AB at E. We are to prove that CE is an altitude, that is, that CE⊥AB. The method is to show that CE (or what for this purpose is the same, CL) is parallel to OD⊥AB. That CL∥OD follows immediately from the similarity of triangles LCM and MDO, which may be proved as follows. We have CM : MD = 2 : 1 and, by insightful construction, LM : MO = 2 : 1; furthermore, the angles between the proportional sides, ∠CML and ∠DMO, are equal (vertical angles); therefore, by Euclid VI.6, ΔLCM ~ ΔMDO. From the similarity, ∠CLO = ∠DOL, which are alternate interior angles under the transversal LO to the lines CL, DO. Therefore CL∥OD (Euclid I.29) and CE is an altitude, Q.E.D. It remains only to show that L is the point of intersection of all three altitudes. The demonstration follows immediately from Fig. 3.3.38, which presents the situation of Fig. 3.3.37 for a second altitude, ALG.

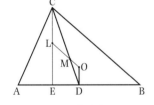

Fig. 3.3.37 Euler's line.

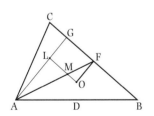

Fig. 3.3.38 Further to Euler's line.

15. Euler, *Opera omnia*, ser. 1, vol. 26 (1953), 140, 149.
16. Dörrie, *100 great problems* (1965), 141–2.

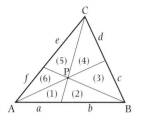

Fig. 3.3.39 Ceva's theorem.

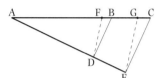

Fig. 3.3.40 A theorem on ratios, re Ceva's theorem (Euclid V.17).

Ceva's line. Giovanni Ceva, a mathematician and engineer, and also a marquis, hit upon the following remarkable property.[17] Take any point P within any triangle ABC; draw lines from the vertices of the triangle through the point to the opposite sides; the product of the ratios of the segments into which the process splits the sides equals unity. This opaque utterance may be clarified from Fig. 3.3.39. We are to show that

$$\frac{a}{b} \times \frac{c}{d} \times \frac{e}{f} = 1. \tag{3.3.4}$$

The relation is proved most easily by invoking ratios of areas, as in the analysis of Fig. 3.3.39.

Let P be the point and number the areas as indicated. Then, since areas with the same altitudes are as their bases,

$$a/b = (1 + 5 + 6)/(2 + 3 + 4) = (1)/(2) = (5 + 6)/(3 + 4).$$

The last step makes use of the proposition that, if $x = p/q = r/s$, then also (Euclid V.17) $x = (p - q)/(r - s)$. The inference can be demonstrated geometrically via Fig. 3.3.40, where $AC = p$, $AB = q$, $AE = r$, $AD = s$, and, because CE∥DB, $p : q = r : s$. Cut off $AF = s$ from AB and $AG = r$ from AC. Then triangles FBD and GCE are similar ($\angle B = \angle C$ by alternate interior angles and $\angle DFB = \angle EGC$ as supplements of the equal base angles in the isosceles triangles ADF, AEG); hence $(p - r):(q - s) = GE : FD = r : s = x$. Q.E.D.

To return to Ceva's theorem:

$$c/d = (1 + 2 + 3)/(4 + 5 + 6) = (3)/(4) = (1 + 2)/(5 + 6),$$

$$e/f = (3 + 4 + 5)/(1 + 2 + 6) = (5)/(6) = (3 + 4)/(1 + 2).$$

Hence $(a/b) \times (c/d) \times (e/f) = (5 + 6)/(3 + 4) \times (1 + 2)/(5 + 6) \times (3 + 4)/(1 + 2) = 1$. Q.E.D. Ceva's theorem admits its converse, that is, if the segments a, b, c, d, e, f satisfy equation (3.3.4), the 'cevians' (as lines from any vertex of a triangle to the opposite side might be called) making them are coincident. For say that one of the cevians does not pass through the point P at which the other two intersect. Then draw the cevian that does go through P and let it make segments a', b' on the line AB. Ceva's theorem requires that $(a'/b')(ce/df) = 1$. But by hypothesis, $(a/b)(ce/df) = 1$. Therefore $a'/b' = a/b$, which, since $a + b = a' + b'$, is only possible if $a' = a$, $b' = b$.[18]

Since medians, altitudes, and angle bisectors are cevians, the coincidences demonstrated earlier are all immediate consequences of Ceva's theorem. The case of medians requires no thought: since they

17. Chasles, *Aperçu* (1837), 294–6.
18. Cf. Coxeter and Greitzer, *Geometry revisited* (1967), 4–9.

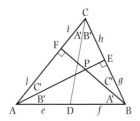

Fig. 3.3.41 The coincidence of altitudes proved via Ceva's theorem.

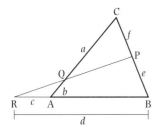

Fig. 3.3.42 Menelaus' theorem.

create equal segments in the lines they strike, $a/b = c/d = e/f = 1$ and they are coincident by Ceva's converse. The case of the altitudes also follows easily. In Fig. 3.3.41, AE and BF are altitudes meeting at P; CD is a cevian from C perpendicular to AB at D; and the primes indicate complementary angles, that is, $\angle A' = \mathrm{rt.}\angle - \angle A$, etc. We want to show that CD also passes through P by demonstrating that equation (3.3.4) holds for the several segments. We have from the similar triangles FBC, EAC, that $i/h = BC/AC$; from $\triangle AEB$, $\triangle BDC$, that $g/f = AB/BC$; and from $\triangle AFB$, $\triangle ADC$, that $e/j = AC/AB$. Multiplying the ratios together, you have $(i/f) \times (g/h) \times (i/j) = 1$, as required.

Menelaus' line. Menelaus' theorem resembles Ceva's, or, rather, the reverse, since Ceva's came a millenium and a half later. Outside any triangle ABC and on any one of its sides extended, pick any point R; from R draw any line intersecting the other two sides (Fig. 3.3.42). The product of the ratio of the resulting segments is again unity, taking the 'segments' of the line on which R lies as the side extended and the extension. All will be clear from the figure. We are to show that, again,

$$\frac{a}{b} \times \frac{c}{d} \times \frac{e}{f} = 1. \tag{3.3.4}$$

As you might expect, every thing follows nicely from similar triangles. But the figure presents none. You must make them, by drawing the parallel to AC through B and extending it to meet RP extended at S (Fig. 3.3.43). That makes $\triangle RBS \sim \triangle RAQ$ and $\triangle QCP \sim \triangle SPB$. From the first pair, $c:d = b:BS$; from the second, $BS:a = e:f$. Therefore $c:d = b:ae/f$, or $1 = (a/c)(b/d)(e/f)$. Thinking to draw a line like BS makes the challenge, and the difficulty, of geometry.

Some practical problems

Problems of an Egyptian surveyor. The theorems about similarity and area allow you to set up as an old Egyptian tax advisor. (Since there is no evidence that the old Egyptians knew these theorems, which we have from the Greeks, you would have a great advantage over

Fig. 3.3.43 Proof of Menelaus' theorem.

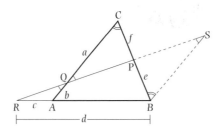

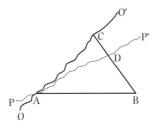

Fig. 3.3.44 Problem for an Egyptian tax advisor.

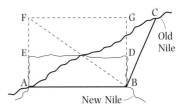

Fig. 3.3.45 Another Egyptian tax problem.

your competitors.) The triangular plot of land ABC (Fig. 3.3.44) lying along the Nile in its bed OO′ brings its owner a tax burden of 1000 sesterces. A flood shifts the Nile to PP′. What should the new tax be? Fortunately the papyrus containing the survey of the area survived the flood: it shows that BC = 1000 paces. To calculate the new tax you need only measure BD. Let it be 800 paces. Then, in accordance with Euclid VI.1, ΔABD : ΔABC = BD : BC; and, in accordance with the rules of good King Sesostris, new tax : old tax = ΔABD : ΔABC = BD : BC = 800 : 1000. The new tax is 800 sesterces.

Here are a few more problems that might have come to you as an Egyptian surveyor and tax calculator. Your client's original triangular field ABC (Fig. 3.3.45) became the rectangle ABDE after the flood. Measurement discloses that the height of the original field AF is twice that of the new, so that AF = 2EA. This case allows no tax relief, since rect. ABDE = ΔABC: we have ΔABC = ΔABF (by Euclid I.37) = ($\frac{1}{2}$) rect. ABGF (since ΔBAF ≅ ΔFGB); and ($\frac{1}{2}$) rect. ABGF = rect. ABDE. The case provides an instance of the general theorem (Euclid I.41) that a parallelogram has twice the area of a triangle on the same base and between the same parallels. Figure 3.3.46 will remind you of the truth of this theorem. You have ΔABC = ΔABE (Euclid I.37) = ($\frac{1}{2}$) par. ABDE.

Your discontented client now wants to know if he has gained or lost river frontage in the rearrangement of his land. In the old plot, the river ran along AC = FA/sin ∠CAB. In the new plot it bathes a length AE + ED + DB = FA + ED = FA + AB. The ratio of the river frontages is therefore

$$\frac{\text{new}}{\text{old}} = \frac{\text{FA} + \text{AB}}{\text{FA}/\sin\angle\text{CAB}} = \left(1 + \frac{\text{AB}}{\text{FA}}\right)\sin\angle\text{CAB}.$$

From old papyrus records, you know that AB = 800 paces, FA = 1000 paces. Hence new frontage/old frontage = (1.8) sin ∠CAB. Is this number larger or smaller than the old? Only measurement of ∠CAB will tell. You go to the site, take out your angular measurer,

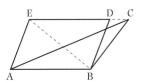

Fig. 3.3.46 Further to Fig. 3.3.45.

place its base along AB, line up its alidade with the old river bed, and find ∠CAB = 34°. Did your client gain or lose river frontage? What would be the answer if ∠CAB < 34°? And if ∠CAB > 34°?

The plane table. In the field, surveyors do not use hand-made quadrants but nicely graduated scales and telescopic sights. For many purposes, the 'plane table' used by the French army engineers shown at work in Fig. 3.3.47 is quite satisfactory. It was the ordinary surveying technique during the seventeenth and eighteenth centuries. It works by similar triangles. Begin (Fig. 3.3.48) with the table, covered with a fresh sheet of paper, in situation (1). Aim the alidade AC, which rotates around A, at the distant point E. Draw a line on the paper along AC. Then move the table to situation (2) along the line AB extended through a measured distance AA′. Move the alidade to B′ and sight E along B′D. Let the intersection of B′D and A′C be E′; it is the representative or projection of E on the plane

Fig. 3.3.47 Napoleon's surveyors at work. From Berthaut, *Ingénieurs géographes* (1902), frontispiece.

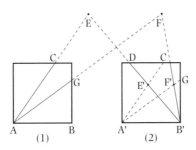

Fig. 3.3.48 Operation of the surveyor's plane table.

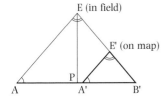

Fig. 3.3.49 Further to the plane table.

table. The same process performed on a number of distant points F′, G′, H′, ..., produces images F′, G′, H′, ... that make the outline of a scaled map. The scale is determined by the width of the table AB = A′B′ and the distance AA′ = BB′. As appears from Fig. 3.3.49, AB′/A′B′ = AE/A′E′; the unknown distance AE to the distant point E may be calculated from the measured distances AB′, A′B′, and A′E′: AE = A′E′(AB′/A′B′). Here A′B′/AB′ indicates the scale of the map: if AB′ = 100A′B′, the scale is 1 to 100. Common practical scales for country maps are an inch to a mile, that is, 1 to 63 360, and a centimeter to a kilometer, or 1 to 100 000. A city map might be drawn at 1 : 10 000. From the value of AE and the formula sin A = EP/AE, you have the perpendicular distance EP from E to the baseline AB′.

Another Egyptian problem. The formula for the area of a triangle has been known for 4000 years. The mathematical papyrus written in Egypt in the sixteenth century BC by the copyist Ahmose records, according to him, wisdom that was at least 400 years old in his time. The original from which he copied therefore antedated Euclid by a millenium and a half. Ahmose gave his work the modest sort of title fancied by mathematicians: 'The correct method of reckoning, for grasping the meaning of things and knowing everything that is.'[19] In fact, he did not know much about geometry; but he delivered the areas of triangles and circles in special cases that imply that he possessed a general rule. To give the example we already know (see Fig. 1.0.2), he asks for the area of a triangle with a base of 4 l.u. and a height of 10 l.u. and directs, as an answer, that you divide the 4 by 2 and multiply the result by 10 to obtain 20 a.u. In a more exciting exercise, he works out the area of a trapezoid. His recipe: take half the sum of the base and the 'cut-off line' and multiply that by the distance between them.[20] A proof of his formula is given in Exercise 3.5.7.

19. Robins and Shute, *Rhind math, papyrus* (1987), 11.
20. Chace, *Rhind math. papyrus* (1927), *l.* 92–93 (problems 51,52, resp.); cf. Gillings, *Mathematics* (1982), 138–9.

Fig. 3.3.50 Another problem for
the Egyptian tax advisor.

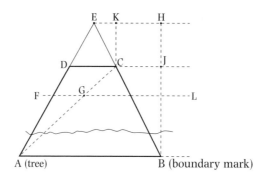

Let us return to the no longer entirely hypothetical case book of
the old Egyptian surveyor. This time your client has a trapezoidal
field ABCD through which the river flows as illustrated in Fig.
3.3.50. He wants to know how to find the size of the field without
crossing the river. You point out to him that the area of the field
ABCD is the difference between the areas of the similar triangles AEB
and DEC. Since the areas of similar triangles are as the squares of
their bases,

$$\frac{\Delta ABE}{\Delta EDC} = \frac{AB^2}{CD^2} = \frac{AE^2}{ED^2}, \quad \text{or} \quad \frac{\text{trap. ABCD}}{\Delta EDC} = \frac{AE^2 - ED^2}{ED^2}. \quad (3.3.5)$$

To solve the problem, you need only to find AE without crossing the
river.

From point D walk toward the tree at A a convenient distance DF,
say 100 l.u. Now turn to your left through an angle equal to ∠EDC
and walk parallel to DC, laying out a line along FL. Then go to C and
walk toward A until you encounter FL; mark the point of the
encounter G. Measure FG; let it be 60 l.u. Now ΔDAC ~ ΔFAG, since
∠CDA = ∠GFA by parallels and ∠DAC is common; therefore AD/AF
= CD/FG. But AD = AF + FD; hence

$$\frac{AF + FD}{AF} = \frac{CD}{FG}, \quad \text{or} \quad 1 + \frac{FD}{AF} = \frac{CD}{FG}, \quad \text{or} \quad AF = \frac{FD \times FG}{CD - FG}.$$

Measure CD; let it be 120 l.u.

$$AF = 100 \times \frac{60}{120 - 60} = 100 \text{ l.u.}$$

Then AD = AF + FD = 200 l.u. Further measurement shows
DE = 40 l.u., whence AE = 240 l.u. At last, therefore, according to
equation (3.3.5), the trapezoid ABCD equals $\Delta EDC \times (240^2 - 40^2)/40^2$
$= 35 \Delta EDC$, that is, 35 times the area of ΔEDC. What is the area of
ΔEDC in a.u.? Well, ΔEDC = DC × CK/2. You know DC = 120 l.u.
You have two ways to obtain CK: either measure directly the

perpendicular distance between the parallels DJ and EH or measure ∠EDC. If you choose the second method, CK = DE sin ∠EDC = 40 sin ∠DEC:

$$\text{trap. ABCD} = 35\left(\frac{120 \times 40}{2}\right) \sin \angle \text{DEC} = 84\,000 \sin \angle \text{DEC a.u.}$$

Say ∠DEC = 45°. Then trap. ABCD = 59 397 a.u.

Geometry and gravity. Among Galileo's most important discoveries was the law of free fall, that is, the relationship between the distance through which a body falls without impediment under gravity and the length of time elapsed since it began its descent. After many false starts, he discovered that the fundamental characteristic of free fall is constant acceleration, which means that the velocity with which the body moves at any time is proportional to the elapsed time. The algebraist would express this characteristic as $v = gt$, where g is a constant number that measures the pull of gravity and v is the body's velocity at any time t. The geometer would write $v_A : v_B = t_A : t_B$, where A, B indicate two different points on the body's trajectory.

Galileo derived his rule relating distance and time using geometry. In Fig. 3.3.51, the line OE represents time and the parallels AC, BD, the velocity of the body at the instants A, B respectively. The angle EOF is drawn so that AC = gOA, that is, so that tan ∠AOC = g. In a moment of great insight, he identified the *areas* of the triangles AOC, BOD, with the *distances* covered by the falling body in the times t_A, t_B. To obtain the relationship he sought between distance fallen, d, and time elapsed, t, he had only to resort to elementary geometry:

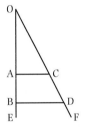

Fig. 3.3.51 Galileo's geometrical demonstration of the law of falling bodies.

$$\frac{d_A}{d_B} = \frac{\Delta \text{AOC}}{\Delta \text{BOD}} = \frac{\text{OA} \times \text{AC}}{\text{OB} \times \text{BD}} = \frac{t_A^2}{t_B^2}.$$

The algebraist (and today's physicist) would write $d_A = \Delta \text{AOC} = gt_A^2/2$, or, since A is an arbitrary instant, $d = gt^2/2$.

Galileo made many applications of this rule of geometrical interest. For example, in his last book, published while he was under house arrest for teaching that the earth goes around the sun, he showed that the length of time required for a ball to slide down inclined planes that are chords of a vertical circle is the same for all the planes provided the ball always reaches the lowest point on the circle. (The Inquisition considered this curious theorem harmless.) The theorem and its proof will be clear from Fig. 3.3.52. Let O be the center of the vertical circle ACB of diameter AB = $2r$. Galileo showed that t_{AB} (the time of free fall through the vertical diameter) equals t_{CB} (the time required to slide without friction along the chord CB) regardless of where C lies on the circle. The proof requires the

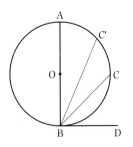

Fig. 3.3.52 A Galilean theorem about descent along inclined planes.

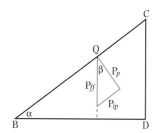

Fig. 3.3.53 The physics of Galileo's theorem.

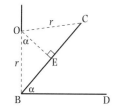

Fig. 3.3.54 The geometry of Galileo's theorem.

assumption that the acceleration under gravity down an inclined plane is to the acceleration in free fall as the sine of the angle of inclination of the plane (∠CBD in Fig. 3.3.52) to the sine of inclination of the vertical (∠ABD = rt.∠).

The reasonableness of this assumption is indicated in Fig. 3.3.53. Let p_{ff} represent the pull of gravity in free fall, p_{ip} the pull of gravity along the incline BC: that is, suppose that the portion p_p of p_{ff} acting perpendicular to the plane does not cause the body to slide, since p_p pulls the body against the plane, not along it. Now $p_{ip} = p_{ff} \sin \beta = p_{ff} \sin \alpha$, since α and β are each the complement of ∠BCD. The time-distance rule for sliding down an inclined plane without friction is therefore taken to be $d = (g \sin \alpha) \, t^2/2$.

To return to Galileo's problem, connect O and C in Fig. 3.3.52 to form the isosceles triangle BOC (Fig. 3.3.54). Drop the perpendicular bisector OE from O to BC. Then ∠BOE = α (it and ∠EBD = α are each the complement of ∠OBE) and BC = $2r \sin \alpha$. We have therefore

$$\frac{t_{AB}^2}{t_{CB}^2} = \frac{2r/g}{BC/g \sin \alpha} = \frac{2r \sin \alpha}{2r \sin \alpha} = 1.$$

Since α can be any acute angle, it follows that the time of descent along all chords through B is the same. The book from which this demonstration comes, Galileo's *Discourse concerning two new sciences*, set the basis of the theory of the motion of bodies on which Newton built his theory of the world.

3.4. Deception

One of the charms and challenges of geometry is the need to stay awake during demonstrations. If you nod, you can be swindled easily, primarily by yourself. There follow several examples of deception, in which correct reasoning results in a nonsensical conclusion. They repay careful study.

A. Take any acute-angled triangle ABC (Fig. 3.4.1); bisect its base BC at D and erect the perpendicular bisector; bisect the angle opposite BC and continue the bisector until it intersects the perpendicular bisector of BC. Let their point of intersection be O. Drop perpendiculars from O to the sides AB, AC, at F and E, respectively, and join O to the vertices B and C. You can prove easily that AF = AE, and that FB = EC (ΔAOF ≅ ΔAOE, ΔBOD ≅ ΔCOD, ΔFOB ≅ ΔEOC). From this proof it follows, by C.N. 1, that AB = AC, that is, that ΔABC is isosceles. But ΔABC was taken to be *any* acute-angled triangle. You seem to have proved that all such triangles are isosceles, which is nonsense.

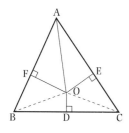

Fig. 3.4.1 All acute triangles seem to be isosceles.

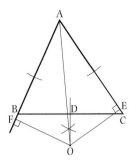

Fig. 3.4.2 Corrective to Fig. 3.4.1.

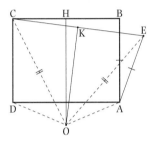

Fig. 3.4.3 All right angles seem to be obtuse.

Your reasoning was impeccable. The nonsense arose from the diagram. Now follow the instructions for construction faithfully, using your compass to bisect BC and ∠BAC, as in Fig. 3.4.2. You will find that O falls outside ΔABC and that one of the perpendiculars from O to the sides AB and AC comes inside, and the other outside, the triangle. Show that AF = AE, and, consequently, that AC = AE + EC > AB = AF − BF. All therefore is right with the world. Notice that the apparent paradox of this problem can be rephrased as a theorem—'the perpendicular bisector of one side of a triangle does not meet the bisector of the opposite angle within the triangle'—and proved by *reductio and absurdum*. The theorem admits exceptions if the triangle is equilateral or isosceles. What occurs in these cases?

B. Draw the rectangle ABCD (Fig. 3.4.3) and, from A, at any convenient angle, the line AE = AB = CD. Join C and E. Find the midpoints H of BC and K of CE and erect the perpendicular bisectors. Let them meet at O. Connect O with A, E, C, and D. Show that ∠CDO = ∠EAO and ∠ADO = ∠DAO. Therefore ∠CDA = ∠CDO − ∠ADO = ∠EAO − ∠DAO = ∠EAD. But ∠CDA = rt.∠ and ∠EAD > rt.∠. You have proved that a right angle equals an angle larger than a right angle.

This paradox can be exploded, like the previous one, by a drawing. It will look like Fig. 3.4.4. Show again that ∠EAO = ∠CDO, ∠DAO = ∠ADO. From the obvious relations ∠CDO = ∠ADO + rt.∠ and ∠EAD = 2st.∠ − ∠EAO − ∠DAO, show that ∠EAD = rt.∠ + ∠DOA, that is, that ∠EAD > rt.∠. Again, all is right in a well-drawn world.

Fig. 3.4.4 Corrective to Fig. 3.4.2.

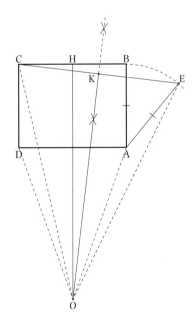

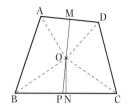

Fig. 3.4.5 All quadrilaterals seem to be parallelograms.

And, again, the apparent paradox can be rephrased as a theorem. Try your hand at it.

C. Draw the quadrilateral ABCD having no pair of sides parallel and one pair of opposite sides, AB and CD, equal (Fig. 3.4.5). Find the midpoints M, N of AD and BC and raise the perpendicular bisectors. Let them meet at O and extend MO to P. Join O to the vertices of the quadrilateral. Show that ∠MOA = ∠MOD and that ∠AOB = ∠DOC. Then, by C.N.1, ∠MOB = ∠MOC. Since ∠BOP = st.∠ − ∠MOB and ∠COP = st.∠ − ∠MOC, ∠BOP = ∠COP. But by construction ∠BON = ∠CON. These two equalities can be satisfied simultaneously only if P coincides with N. But that makes MON perpendicular to both AD and BC; AD must therefore be parallel to BC, which contradicts the premise. You have arrived at another absurdity.

By now you know the escape. Make a proper drawing, like Fig. 3.4.6. O falls outside the quadrilateral. Prove that P does not coincide with N by showing that ∠NOP > 0. (You might want to argue that for ∠NOP = 0, ∠BAD + ∠ABC = st. ∠, that is, that AD∥BC.)

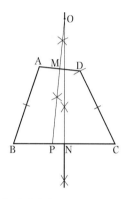

Fig. 3.4.6 Corrective to Fig. 3.4.5.

D. ABCD is a rectangle 5 units in width and 13 units long. Draw diagonal AC and raise perpendiculars at E and G, 8 units from A and from C, respectively, to meet the diagonal at F and H (Fig. 3.4.7). Now imagine the pieces cut out and reassembled into a square as in Fig. 3.4.8. The area of the square is $8 \times 8 = 64$; that of the rectangle, $5 \times 13 = 65$. What has happened to the extra unit? There must be a swindle somewhere in the assembly. Can you spot it? It is easiest to see with the pieces colored as in plate IV. The green and yellow triangles go together as proposed; but the red and blue trapezoids do not quite fit. Calculate EF from similar triangles AEF, ABC in Fig. 3.4.7: $EF = 40/13$. But $AD = 5$. Therefore $AD + EF \neq DE$ as suggested by Fig. 3.4.8, although the error is less than one percent: $5 + 40/13 = 105/13 = 8.08$ rather than 8.

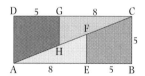

Fig. 3.4.7 Area apparently lost in converting a rectangle into a square.

This paradox, first published by a German mathematician in 1868, fell into the hands of that great lover of paradox, Lewis Carroll, who generalized it.[21] Carroll made the sides of the rectangle $n − a$ (width) and $2n − a$ (length), and the side of the square n. The points E and G therefore come a distance n from A and C, respectively. In the case illustrated in Fig. 3.4.7, $n = 8$, $a = 3$. In the general case, the area of the rectangle, $(n − a)(2n − a)$, exceeds that of the square, n^2, by one unit:

$$2n^2 − 3na + a^2 − n^2 = 1,$$

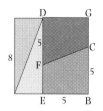

Fig. 3.4.8 Further to the conversion of a rectangle into a square.

21. Weaver, *Am. math. monthly:* 45 (1938). 234–6.

or

$$n^2 - 3na + a^2 - 1 = 0. \qquad (3.4.1)$$

The equation is indeterminate unless a (or n) is specified.

Let us insist on integral solutions to equation (3.4.1). Since the general solution to the quadratic runs

$$n = \left[3a \pm \sqrt{9a^2 - 4(a^2 - 1)}\right]/2,$$

our insistence means that $5a^2 + 4$ must be a perfect square. We already know one such value of a : $a = 3$, $5a^2 + 4 = 49$. The easy way to find the next acceptable value of a is by trial and error. If you go in order, you will succeed first with $a = 8$, $5a^2 + 4 = 324 = 18^2$. If $a = 8$, $n = 21$, according to equation (3.4.1). You have that when $a = 3$, $n = 8$, and when $a = 8$, $n = 21$; does it occur to you that 21 might be the next acceptable value of a? If so, you are right: $5(21)^2 + 4 = 2209 = 47^2$. The corresponding value of n is 55. Shall we try $a = 55$? It works: $5(55)^2 + 4 = 15{,}129 = 123^2$, and $n = 144$. The error in DE (Fig. 3.4.8) in the rearrangement of a rectangle of width $n - a = 89$ and length $2n - a = 233$ into a square of side 144 is far too small to notice. Instead of DE = 144, you have FE + DA = $55.004 + 89 = 144.004$.

The Euclidean propositions introduced in this chapter are listed in Table 3.4.1.

Table 3.4.1 Euclidean propositions in Chapter 3

I.5	In isoscles triangles the angles at the base are equal to one another.
I.6	If the base angles of a triangle are equal, it is isosceles.
I.7	Two given lines drawn from the end points of a third line and on the same side of it can meet in only one point.
I.8	Congruence by *s.s.s.*
I.9	To bisect a given angle.
I.10	To bisect a given finite straight line.
I.11	To erect a perpendicular at a given point on a given line.
I.26	Congruence by *a.s.a.*
I.35	Parallelograms on the same base in the same parallels are equal.
I.36	Parallelograms on equal bases in the same parallels are equal.
I.37	Triangles on the same base in the same parallels are equal.
I.38	Triangles on equal bases in the same parallels are equal.

3.5. Exercises

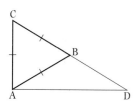

Fig. 3.5.1 Exercise 3.5.1.

3.5.1. If an external angle of a triangle, and the opposite angle on the same base, each be bisected, and the bisectors extended until they meet, the angle they form is half the other angle opposite the external angle. Show, therefore, that in Fig. 3.5.1, ∠BEC = ∠BAC/2.

[Let ACD be the external angle, ∠ABC the opposite angle on the same base; then ∠ACE = ∠ECD = β (say) and ∠ABE = ∠EBC = γ (say). We have 2β = ∠BAC + 2γ, ∠BEC = st.∠ − γ − (st.∠ − β) = β − γ, whence ∠BAC = 2(β − γ) = 2∠BEC, as required.]

3.5.2. ΔABC (Fig. 3.5.2) is equilateral, ∠CAD = rt. ∠. Show that BD = BC.

[If BD = BC, then AB = BD since ΔABC is equilateral; we should therefore try to prove that ΔABD is isosceles. That follows easily: since ∠ACB = ∠CAB, ∠BAD and ∠BDA are complements of equal angles and therefore equal.]

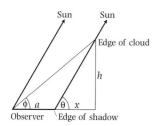

Fig. 3.5.2 Exercise 3.5.2.

3.5.3. You observe the edge of a cloud at an altitude of 20° and the sun above it at an altitude of 35°. The shadow cast by the edge of the cloud falls on an object that you know is 2300 yards from your position. How high is the cloud? [From *Ladies diary* of 1721, in Leybourne, *l*, 108.]

[Let θ be the altitude of the sun, φ that of the cloud's edge (Fig. 3.5.3). We have tan φ = h/(a + x), tan θ = h/x, whence

$$h = a \frac{\tan \phi \tan \theta}{\tan \theta - \tan \phi}.$$

Putting in φ = 20°, θ = 35°, a = 2300 yds, we have h = 1743 yds; the cloud is a mile high, to within 1 percent.]

Fig. 3.5.4 Exercise 3.5.4. From Chace *et al.*, *The Rhind papyrus* (1927), problem 52, vol. 2, plate 74.

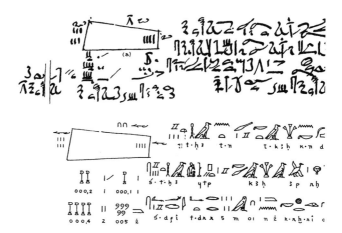

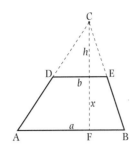

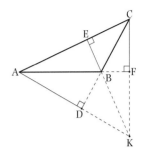

Fig. 3.5.5 Further to Exercise 3.5.4.

Fig. 3.5.6 Exercise 3.5.5.

3.5.4. Prove that the area of a truncated triangle (that is, a trapezoid) is half the sum of the parallels times the perpendicular distance between them. You will thus supply a proof of the procedure used by Ahmose, in the 52nd problem of the Rhind Mathematical Papyrus (Fig. 3.5.4), which gives 100 a.u. for the area of a trapezoid with parallel sides of 4 l.u. and 6 l.u. (In fact the scribe made a computational error and got 200 a.u. instead of the correct answer according to his own rule; the human race has been making mistakes in arithmetic for over 5000 years.)

[In Fig. 3.5.5, where $DE = b$, $AB = a$, $CF \perp AB$, $DE \| AB$, $2\triangle ABC = a(h + x)$, $2\triangle CDE = bh$, 2trap. $ABDE = a(h + x) - bh = ax + h(a - b)$. But from the similar triangles ABC and DEC, $h : b = (h + x) : a$, or $bx = h(a - b)$; hence trap. $ABDE = (a + b)x/2$, Q.E.D.]

3.5.5. Prove that the altitudes in an obtuse triangle coincide in a point when extended.

[The altitudes in $\triangle ABC$ (Fig. 3.5.6) are AF, BE, CD, which, when extended outside the triangle, meet at K. The proof runs parallel to that for an acute triangle as given in the text.]

3.5.6. Find a point X on $AC \perp AB$ (Fig. 3.5.7) equidistant between two non-parallel straight lines AB, CD, that intersect at an inaccessible point. (If the point, call it Q, were accessible, the problem would be solved by bisecting $\angle AQC$ and extending the bisector until it intersected with AC; the intersection would be X, the desired point.)

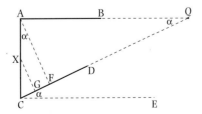

Fig. 3.5.7 Exercise 3.5.6.

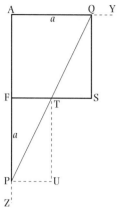

Fig. 3.5.8 Exercise 3.5.7.

[Construction: Draw CE∥AB and drop the perpendicular AF from A onto CD; label ∠DCE, α, and the base AC, a. The problem requires the point X such that AX = XG. Now AX = a – CX, XG = CX cos α, cos α = AF/a, whence a – CX = CX(AF/a) or CX = a^2/(a + AF), Q.E.D.]

3.5.7. To complete the preceding problem, the length CX should be constructed by straight edge and compass. The construction is shown in Fig. 3.5.8: draw AZ and AY at right angles; along AZ lay out AP = AF + a; along AY, AQ = a; join P and Q. Draw FS = a perpendicular to AF. Prove that FT, where T is the intersection of PQ and FS, equals the required length CX.

Complete the rectangle FTUP. Then ∠FPT = ∠PTU and ∆PAQ ~ ∆TUP, whence PU : TU = a : (a + AF). But PU = FT, TU = a, so FT : a = a : (a + AF), or FT = a^2/(a + AF) = CX.]

3.5.8. The circumstances being the same as those in Exercise 3.5.6 above, and Y being any point between the lines AB, CD, find the line that goes through Y and the inaccessible point Q of intersection of AB and CD. You will find it convenient to define the position of Y by its distance p along, and its distance q beneath, AB. Then, in Fig. 3.5.9, you are to find an analytical expression for the length AZ, which fixes the point Z and hence the required line ZYQ.

Suppose that QYZ can be drawn, and draw it. Then, from the similar triangles QHY, QAZ, AZ : q = (p + x)/x; and, from ∆QHI, tan α = HI/x = (a – p tan α)/x. According, AZ = aq/(a – p tan α), which is the analytic solution required. The length AZ can be constructed in the manner indicated in Fig. 3.5.10.]

3.5.9. Bisect a triangle (that is, divide it into two equal areas) by a line parallel to one side.

[Let DE be the desired line (Fig. 3.5.11); if you find the altitude AP of ∆ADE, you have solved the problem. You have ∆ADE/∆ABC = (AP/AQ)^2. Since ∆ABC is divided into equal parts by DE, ∆ADE/∆ABC = 1/2 = (AP/AQ)^2, or AP = AQ/$\sqrt{2}$. Hence AP is the side of a square of which AQ is the diagonal.]

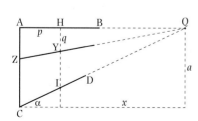

Fig. 3.5.9 Exercise 3.5.8.

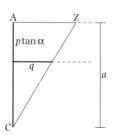

Fig. 3.5.10 Further to Exercise 3.5.8.

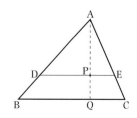

Fig. 3.5.11 Exercise 3.5.9.

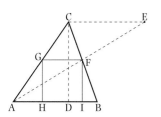

Fig. 3.5.12 Exercise 3.5.10.

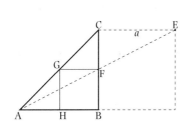

Fig. 3.5.13 Further to Exercise 3.5.10.

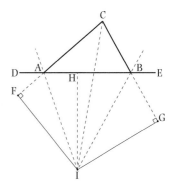

Fig. 3.5.14 Exercise 3.5.11.

3.5.10. In Fig. 3.5.12, CD is the altitude of ΔABC perpendicular to the base AB. Draw CE parallel to AB and equal to CD. Connect A and E intersecting BC at F. Prove that F is the corner of the square circumscribed by ΔABC, that is, that if FI⊥AB and GF⊥FI, then GF = FI.

[To find the method for the general case, it might be useful here, as it often is, to take the simplest case first, in which ΔABC is a right triangle. In Figure 3.5.13, CB⊥AB and CE = CB = a. Since ∠CEF = ∠FAB (alternate interior angles) and ∠AFB = ∠CFE (vertical angles), ΔAFB ~ ΔCFE, whence FB : AB = (a − FB) : a. Furthermore, ΔGFC ~ ΔABC, whence GF : AB = (a − FB) : a. It follows that FB = GF. In the general case, Fig. 3.5.12, we need three pairs of similar triangles. From ΔAFB ~ ΔCFE, CF : FB = a : AB. From ΔCDB ~ ΔFIB, FI : FB = a : CB. From ΔCGF ~ ΔCAB, CF : GF = CB : AB. Hence FI = a(CB/FB) = (AB/CB)CF = GF, Q.E.D.]

3.5.11. The bisectors of external angles CAD and CBE of ΔABC (Fig. 3.5.14) intersect at I. From I, the perpendiculars IF and IG are let fall on CA extended and CB extended, respectively. Prove that CF = CG = s, half the perimeter of ΔABC.

[Draw the perpendicular IH from I to AB. Then, from the analysis of Fig. 2.1.33, IF = IH = IG, and CI bisects ∠FCG; therefore, ΔCFI ≅ ΔCGI (h.s.) and CF = CG, which demonstrates the first part of the problem. Now, ΔAFI ≅ ΔAHI (h.s.) and ΔIHB ≅ ΔIGB. Hence FA = AH and HB = BG. We have finally, CG = CB + BG, CF = CA + AF, CG + CF = CA + CB + AF + BG = CA + CB + AH + HB = 2s. But CG = CF. Hence CG = CF = s, Q.E.D.]

3.5.12. Let us suppose that when Marcellus besieged Syracuse, Archimedes stood at the corner of the city wall a distance a from a ditch that ran parallel to it. To his left, a distance b along the wall, he had placed one of his catapults. The distance from the catapult to a line perpendicular to the ditch was c. If Archimedes' line of sight toward the center of the camp ran perpendicular to the wall and

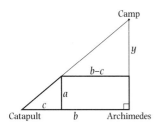

Fig. 3.5.15 Exercise 3.5.12.

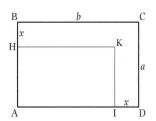

Fig. 3.5.16 Exercise 3.5.13.

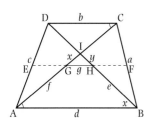

Fig. 3.5.17 Exercise 3.5.14.

ditch, show that he stood a distance ab/c from the center of the camp. [After *Ladies diary* (1725), in Leybourn, *l*, 150.]

[In Fig. 3.5.15, $(y + a)/b = a/c$, whence the distance from Archimedes to the camp $= ab/c$, Q.E.D.]

3.5.13. 'A gentleman has a garden of a rectangular form and wants to make a walk of equal width half way round to take up half the garden. What must be the width of the walk?' [*Ladies diary* (1707), in Leybourn, *l*,2.]

[In Fig. 3.5.16, let x be the width of the path, a and b ($b > a$) the sides of the rectangle ABCD. Then the area of the path is $xb + x(a - x)$, and that of the remaining garden, AIKH, $(b - x)(a - x)$. Equating the two areas, you have

$$x^2 - x(a + b) + ab/2 = 0$$

whence $x = [a + b \pm (a^2 + b^2)^{1/2}]/2$. You must choose the minus sign, since the path's width cannot exceed the semi-perimeter of the garden. Hence

$$x = [a + b - (a^2 + b^2)^{1/2}]/2.$$

Although the path may appear excessive in occupying half the garden, the gentleman was following the standard medieval proportions of a cloister, which made the central garden equal in area to the surrounding path.[22]

3.5.14. Draw the diagonals of a trapezoid; bisect the diagonals; join the midpoints. Prove that the line joining the midpoints is half the difference of the parallel sides of the trapezoid.

[In figure 3.5.17, ABCD is the trapezoid, $AC = f$ and $BD = e$, the diagonals, $GH = g$ the line connecting the midpoints. The proof follows from showing that GH∥AB and that DE = EA, CF = FB. We know that if E and F are the midpoints of AD, BC, then EF∥AB. We need only show therefore that EF intersects AC and BD at their midpoints, which follows immediately from the similar triangles ADC and AEG.

Now, from the similar triangles GHI, ABI, DCI,

$$\frac{g}{b} = \frac{x}{f/2 - x} = \frac{y}{e/2 - y},$$

$$\frac{g}{d} = \frac{x}{f/2 + x} = \frac{y}{e/2 + y}.$$

---

22. Foster, *Patterns* (1991), 117–21, after Villard de Honnecourt.

(To save writing, $a = BC$, $b = DC$, $c = AD$, $d = AB$, $e = BD$, $f = AC$.) We have, therefore,

$$g = bx/(f/2 - x) = dx/(f/2 + x) = by/(e/2 + y) = dy/(e/2 + y).$$

From these equations, $x = f(d - b)/2(d + b)$, $y = e(d - b)/2(d + b)$. With these values for x and y,

$$g = bx/(f/2 - x) = by/(e/2 - y) = (d - b)/2, \quad \text{Q.E.D.}]$$

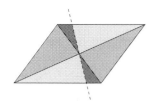

Fig. 3.5.18 Exercise 3.5.15.

3.5.15. Any line drawn through the intersection of the diagonals of a parallelogram so as to intersect opposite sides divides the parallelogram into equal quadrilaterals.

[The Chinese method of coloring the parts (Fig. 3.5.18) would make the proposition evident: use vertical and alternate interior angles for a proof.]

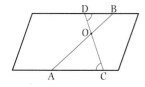

Fig. 3.5.19 Exercise 3.5.16.

3.5.16. Let O be any point within a parallelogram. Draw two lines AB, CD through O so as to cut the same pair of opposite sides (Fig. 3.5.19). Can you prove that $OA \times OD = OB \times OC$?

[If you rewrite the proposed equality as $OA : OB = OC : OD$, you can guess that it follows from symmetrical triangles. Indeed, by vertical and alternate interior angles, $\Delta DOB \sim COA$, whence $OA : OB = OC : OD$.]

3.5.17. ABCD is any parallogram, E, F, G, H the midpoints of its sides (Fig. 3.5.20). Show that the figure AQCP made by joining the midpoints to the opposite vertices is a parallelogram equal in area to par. ABCD/3.

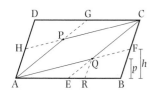

Fig. 3.5.20 Exercise 3.5.17.

[You will find that things go well if you draw QR∥CB and set $AB = CD = 2a$, $BC = AD = 2b$. From $\Delta AQR \sim \Delta AFB$, $QR:(a + ER) = b:2a$; from $\Delta ERQ \sim \Delta EBC$, $QR : ER = 2b : a$, whence $ER = a/3$, $QR = 2b/3$. Now par. AQCP = par. ABCD $- 2\Delta EBC - 2\Delta AEQ$. But par. $ABCD = 2a \times 2h = 4ah$; $\Delta EBC = ah$; $\Delta AEQ = ap/2$. Since $p/QR = h/b$, $\Delta AEQ = (a/2)QRh/b = (a/2)(h/b)(2b/3) = ah/3$. Therefore par. AQCP $= 4ah - 2ah - 2ah/3 = (4/3)ah = \text{par.ABCD}/3$. You forgot to prove that AQCP is a parallelogram. A winning strategy is to show that $AG = EC$ and $PG = EQ$, whence, by subtraction, $AP = QC$; and, similarly, that $PC = AQ$.]

3.5.18. Show that the bisectors of the angles of a rectangle intersect so as to form a square.

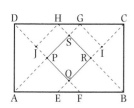

Fig. 3.5.21 Exercise 3.1.18.

[In Fig. 3.5.21, the bisectors AG, CE, BH, DF form the quadrilateral PQRS, which you are to prove is a square. It is certainly a rectangle, since the angle bisectors meet at right angles ($\angle RCB = \angle CBR$ = rt.$\angle 2$, which makes $\angle CRB = $ rt.$\angle$, etc.). Draw GI parallel to BH (and, therefore, perpendicular to CE); then $GI = RS$. Also, draw HJ∥AG; then $HJ = PS$. To prove PQRS is a square, that is, that

PS = RS, you need to show that GI = HJ. Prove it by showing ΔDJH $\cong$ ΔGIC: they both are isosceles and similar, and their bases DH and GC are equal.]

3.5.19. Show that the bisectors of the angles of a quadrilateral intersect to form a quadrilateral whose opposite angles are supplementary.

[In Fig. 3.5.22, the dotted lines indicate the angle bisectors; we must show that $\angle$AGD + $\angle$BEC = st.$\angle$. We have

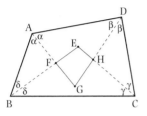

$$\angle AGD = st.\angle - (\alpha + \beta), \quad \angle BEC = st.\angle - (\gamma + \delta),$$
$$\delta = st.\angle - (\alpha + \beta + \gamma),$$

whence $\angle$BEC = st.$\angle$ − δ − γ = α + β = st.$\angle$ − $\angle$AGD, Q.E.D. It follows that if ABCD is a parallelogram, EFGH is a rectangle; the adjacent angles of a parallelogram are supplementary, so α + β = rt.$\angle$ = $\angle$BEC = $\angle$AGD. We already know that if ABCD is a rectangle, EFGH is a square.]

3.5.20. An old Chinese general led his army to a river with a steep bank (Fig. 3.5.23). Standing on the bank, he held a stick 6 feet long perpendicular to himself. When its near end was 0.5 feet below his eye, he sighted the opposite side of the river over the far end of the stick. Without moving anything but his eye, he spied the near side of the river in line with a mark on the stick 2.6 feet from the near end. Having finished gazing, he lowered a rope down the bank and found the length to be 30 feet. What was the width of the river if his eye was 5 feet above the ground? [From Libbrecht, *Chinese mathematics* (1973), 130–1.]

Fig. 3.5.22 Exercise 3.5.19.

Fig. 3.5.23 Exercise 3.5.20.
From Needham, *Science and civilization*, 3 (1970), 30.

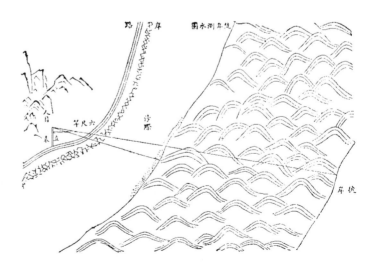

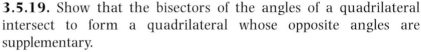

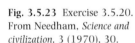

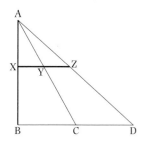

Fig. 3.5.24 Further to Exercise 3.5.20.

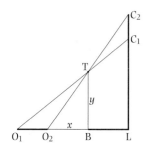

Fig. 3.5.25 Exercise 3.5.21.

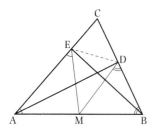

Fig. 3.5.26 Exercise 3.5.22.

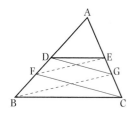

Fig. 3.5.27 Exercise 3.5.23 (Pascal's theorem).

[The situation is represented in Fig. 3.5.24, where A is the eye, XB the height of the bank, XZ the stick, CD the river. Similar triangles provide all the information necessary: AX : AB = XY : BC and AX : AB = XZ : BC + CD). We know everything but BC, the distance from the foot of the bank to the river, and CD, the width of the river sought. From the first equation, BC = 182 feet. From the second, CD = 238 feet. A good-size river.]

3.5.21. In the *Ladies diary* for 1790, a schoolmaster from Liverpool set the following problem. 'Being in a room opposite to the side of a window, the bottom of which was just the height of my eye, I observed that up the edge of the window I could see 42 courses of bricks in a wall on the opposite side of a street; but walking in a direct line towards the window 5 yards, I found that I could see 72 courses. Required the height of the window, supposing the breadth of the street to be 12 yards, and 4 courses of brick work to the foot in height?' The question was answered by a Miss Betty Claxon, from a village near Newcastle, by finding first the distance x between the second position of the observer and the bottom of the edge of the window.

[Betty's solution (Fig. 3.5.25): Let O_1, O_2, be the two positions of the observer, B the bottom and T the top of the window, C_1 and C_2 the heights of the courses observed from O_1 and O_2, L the lowest bricks seen. We are given $O_1O_2 = 5$ yd, BL = 12 yd, $LC_2 = (72/4)$ ft = 6 yd, $LC_1 = 3.5$ yd. Set $O_2B = x$ and the unknown sought, the height of the window BT, $= y$. From similar triangles O_1TB, O_1C_1L, $y : (x + 5) = 3.5 : (x + 17)$; from similar triangles O_2TB, O_2C_2L, $y : x = 6 : (x + 12)$. Eliminate y between these two equations: you have $x^2 + 17 - 84 = 0$, or $(x - 4)(x + 21) = 0$, or $x = 4$. Hence $y = 24/16 = 1.5$ yd. The problem is solved in the Chinese manner at Fig. 4.4.22.]

3.5.22. BE and AD are altitudes, M the midpoint of the base AB, in Fig. 3.5.26. Can you show that EM = DM and that $\angle MED = \angle MDE = \angle C$?

[This is a problem related to the second derivation of the coincidence of the three altitudes of a triangle: A, B, D, and E lie on a circle centered on M, hence EM and DM, being radii, are equal (see the analyses at Fig. 3.2.23 and 5.2.33). To show that the base angles of $\triangle MED = \angle C$, note that $\angle EMB = 2\angle A$ and $\angle DMB = \text{st.}\angle - 2\angle B$, and that, therefore, $\angle EMD = \angle EMB - \angle DMB = \text{st.}\angle - 2\angle C$ (Euclid I.32). But $\angle MED = \angle MDE = (\text{st.}\angle - \angle EMD)/2 = \angle C$. Q.E.D.

3.5.23. Take a triangle ABC (Fig. 3.5.27) and draw a line DE across it parallel to BC. Now pick any point F on BD and join it to E; draw BG parallel to FE; join F and C. According to Pascal (for this is Pascal's theorem), FC is parallel to DG.

[You need to show that AD : AG = AF : AC. You know from the construction that AF : AB = AE : AG ($\triangle$AFE $\sim$ $\triangle$ABG) and that AD : AB = AE : AC ($\triangle$ADE $\sim$ $\triangle$ABC). Eliminate AE between the two equations and you will arrive at what was to be demonstrated. The great mathematician David Hilbert made much use of Pascal's theorem in placing geometry on a new axiomatic footing; Hilbert, *Foundations* (1971), 46–51.]

4 Many cheerful facts about the square of the hypotenuse

4.1. The theorem of Pythagoras

The easy way: by similarity

According to the ancient tradition, the proposition that the square made with the hypotenuse as its side equals the sum of the squares made with the legs as sides was first established by Pythagoras, a geometer who lived a century or so before Euclid. Tradition also ascribes to Pythagoras, whose writings have perished, the invention of geometrical proof and, as an exercise of the method, the theorem that the sum of the angles in a triangle equals a straight angle. It is not known how Pythagoras found the proposition of the squares or how he proved it; but the interest he is said to have entertained in the ratios of line segments suggests a possible route. In Fig. 4.1.1, ΔBCA is the given right triangle; note the convention of labelling the angles with capital letters and the sides opposite them with the same letters in lower case. Drop the perpendicular from C onto AB; let the intersection be D. Then the three triangles BCA, ADC, and BDC are all similar.

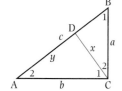

Fig. 4.1.1 The Pythagorean theorem proved by symmetry.

To help get the proportions right, you can label the equal angles as in the figure and make explicit the corresponding sides. Thus, from ΔBCA and ΔADC, with AD = y,

$$\frac{\text{side opp.}\angle 1}{\text{side opp. rt.}\angle} = \frac{b}{c} = \frac{y}{b}, \quad \text{or} \quad b^2 = cy;$$

From the first and third,

$$\frac{\text{side opp.}\angle 2}{\text{side opp. rt.}\angle} = \frac{a}{c} = \frac{c-y}{a}, \quad \text{or} \quad a^2 = c(c-y) = c^2 - cy;$$

By addition, therefore,

$$a^2 + b^2 = c^2.$$

The similarity of the triangles, on which the argument depends, requires ∠BCA to be a right angle.

Another way, which requires a little more algebra, makes clearer the relation between the Pythagorean theorem and squares considered as areas. The areas of the triangles BCA, ADC, and BDC are, respectively, $ab/2$, $xy/2$, and $x(c-y)/2$. From the similar triangles BCA and BDC, $x = ab/c$; we already know that $y = b^2/c$ and $c-y = a^2/c$. Therefore

$$\Delta ADC = xy/2 = (ab/2)(b^2/c^2)$$

$$\Delta BDC = x(c - y)/2 = (ab/2)(a^2/c^2).$$

But $\Delta BCA = \Delta ADC + \Delta BDC$. Therefore

$$(ab/2) = (ab/2)(b^2/c^2) + (ab/2)(a^2/c^2), \text{ or}$$

$$c^2 = a^2 + b^2.$$

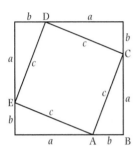

Fig. 4.1.2 The same proved by squaring the triangle.

Euclid proceeds similarly in an easy derivation he gives long after he has perplexed the student with the infamous windmill, which will bother us too momentarily. Euclid's easy demonstration (VI.31) makes use of VI.19, which proved that the areas of similar triangles are as the squares of their corresponding sides. We have $\Delta ABC/\Delta ADC = c^2/b^2$, $\Delta ABC/\Delta BDC = c^2/a^2$. But $\Delta ABC = \Delta ADC + \Delta BDC$. Dividing by ΔABC, we have $1 = b^2/c^2 + a^2/c^2$, which recovers the theorem of Pythagoras.

A demonstration that operates directly on the given triangle is illustrated in Fig. 4.1.2. Complete a square of side $a + b$ and draw within it the square ACDE, each side of which equals the hypotenuse. The large square of area $(a + b)^2$ is made up of four triangles each equal in area to $ab/2$ plus $\square$ACDE. Therefore

$$(a + b)^2 = 2ab + c^2$$

from which, again, $a^2 + b^2 = c^2$.

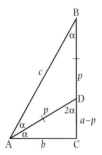

Fig. 4.1.3 Exercise on the Pythagoran theorem (APS 4.1.1).

APS 4.1.1. If one angle of a right triangle, A say, is twice the other, show that $c = 2b$. Figure 4.1.3 shows what is involved. Let's set the equal angles of the isosceles triangle ADB = α. Then, from the similar triangles ABC, DAC,

$$\frac{\text{side opp. } \alpha}{\text{side opp. rt.}\angle} = \frac{b}{c} = \frac{a - p}{p}, \quad \frac{\text{side opp. } 2\alpha}{\text{side opp. rt.}\angle} = \frac{a}{c} = \frac{b}{p}.$$

Eliminate p; you have

$$2b^2 + bc - c^2 = (2b - c)(b + c) = 0,$$

whence $2b = c$.

APS 4.1.2. Here are two related problems. In Figs. 4.1.4 and 4.1.5, ΔABC is isoceles; in Fig. 4.1.4, P is any point on AB; in Fig. 4.1.5, P is any point on AB extended. Can you show that, in the first case, $AC^2 - CP^2 = AP \times PB$ and, in the second, $CP^2 - CB^2 = AP \times PB$? You need only drop the altitudes from C onto AB at D

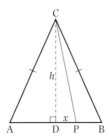

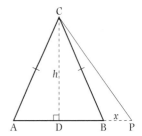

Fig. 4.1.4 Another exercise on the Pythagoran theorem (APS 4.1.2).

Fig. 4.1.5 Further to APS 4.1.2.

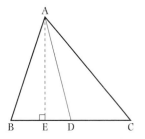

Fig. 4.1.6 Another Pythagorean exercise, in a theorem of Pappus (APS 4.1.3).

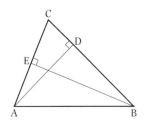

Fig. 4.1.7 A variant of Pappus' theorem (APS 4.1.4).

and apply the Pythagorean theorem. Since the altitude to the base of an isosceles triangle bisects it, $AD = DB = c/2$; then $AP = c/2 + x$, $PB = c/2 - x$, and $AP \times PB = c^2/4 - x^2 = AC^2 - h^2 - (CP^2 - h^2) = AC^2 - CP^2$. In the second case, $AP \times PB = (c + x)x = (c/2 + x)^2 - (c/2)^2 = CP^2 - h^2 - (CB^2 - h^2) = CP^2 - CB^2$.

APS 4.1.3. The Greek geometer Pappus, who lived 500 years after Euclid, played with relationships of the kind just presented. In his play (Fig. 4.1.6), D is the midpoint of BC and the relationship he found is $AB^2 + AC^2 = 2(AD^2 + CD^2)$. Can you prove it? Hint: drop the altitude onto BC at E and use the Pythagorean theorem on triangles AEB, AEC, and AED.

APS 4.1.4. A nice relationship may be obtained from a similar game on altitudes. Let AD, BE (Fig. 4.1.7) be a pair of altitudes. Then $AB^2 = AC \times AE + BC \times BD$. To show it, write the Pythagorean theorem for the four triangles ADC, ADB, BEC, BEA. You will need to write $CE = AC - AE$, $CD = BC - BD$.

The hard way: on the windmill

Euclid's proof (I.47), called 'the windmill' by ages of students because of the resemblance of the classical figure (4.1.8) to the sails of a mill, is an ingenious and purely geometrical demonstration using literal squares drawn on the sides of the right triangle. The proof, which is worth exhibiting formally, proceeds by dividing up the square on the hypotenuse ACDE by a perpendicular (BK) dropped from the right angle and showing that each of the resultant rectangles (KLCD and KLAE) is equal to the square on one of the legs. Propositions adduced in the proof but not previously demonstrated are marked with an asterisk.

1. Draw the squares on the sides   I.46*
2. ABH, CBG are straight lines   I.14*
3. $\angle DCA = \angle BCI$ Post. 4

Fig. 4.1.8 Pythagoras' theorem demonstrated by the 'windmill' (Euclid I.47).

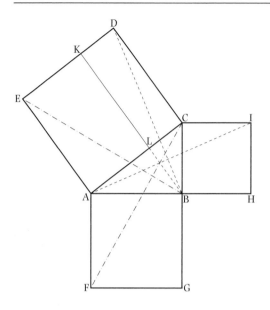

4. ∠DCB = ∠DCA + ∠ACB =   C.N.1
    ∠BCI + ∠ACB = ∠ACI
5. BC = CI, AC = DC   Sides of squares
6. ∴ ΔDCB ≅ ΔACI   *s.a.s.*
7. Draw BK‖DC   Post. 5
8. rect. KLCD = 2ΔDCB (a parallelogram   I.41
    and a triangle between the same parallels)
9. ☐CBHI = 2ΔACI   I.41
10. ∴ rect. KLCD = ☐CBHI   Step 6, C.N.1
11. ∠EAB = ∠FAC   As in Steps 3,4
12. AF = AB, AE = AC   As in step 5
13. ΔEAB ≅ ΔFAC   *s.a.s.*
14. rect. KLAE = 2ΔEAB   As in step 8 (I.41)
15. ☐ABFG = 2ΔFAC   I.41
16. ∴ rect. KLAE = ☐ABFG   C.N.1
17. rect. KLCD + rect.KLAE = ☐ACDE   C.N.4
18. ∴ ☐ACDE = ☐ABFG + ☐CBHI   steps 10, 16, C.N.1

   Propositions I.14 and I.46, cited at steps 1 and 2, are evident enough, but deserve demonstration for completeness. Proposition I.14 asserts that if from a point B on a line AB two others, CB and BD, diverge so that the angles CBA and ABD sum to a straight angle, CBD is a straight line (Fig. 4.1.9). The proof, as in many obvious propositions, runs by reduction to absurdity. If BD does not lie along CB extended, let CBE be a straight line. Then ∠CBA + ∠ABE = st.∠. (Euclid had bothered to prove, in I.13, that the adjacent angles made by two straight lines sum to a st.∠.) But (returning to I.14) ∠CBA + ∠ABD are supposed to sum to a straight angle; we have therefore

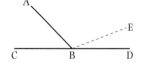

Fig. 4.1.9 A proposition ancillary to Euclid I.47 (Euclid I.14).

∠CBA + ∠ABE = ∠CBA + ∠ABD, or ∠ABE = ∠ABD. That is to set the part equal to the whole, which violates CN 5.

Proposition I.46 tells how to erect a square on a given straight line. It is not worth rehearsing.

The windmill proof may strike you as a conjuror's trick. If so, you are not alone. A textbook writer of the early seventeenth century broke his head trying to figure out how Euclid, or Pythagoras, or whoever it was, had found so obscure a demonstration. 'I've thought a lot about it, but have never managed the proof by any other way than by proportions [similarity, the easy way].' And the philosopher Schopenhauer objected that the way of the windmill more amazed than instructed the student. 'Lines are drawn, we know not why, and it afterwards appears that they were traps which close unexpectedly and take prisoner the assent of the astonished reader.'[1] That is the charm and frustration of geometry. Many different approaches to the same problem can succeed. Many of them would never occur to even the most experienced geometers. But the connoisseur can appreciate the elegant wherever found, and the connoisseur of geometry esteems the windmill, which needs no algebra or additional figures, as the most beautiful deomonstration of the theorem of Pythagoras.

Let us do some exercises on the windmill. Join the corners of the sails as shown in Fig. 4.1.10. The triangles thus formed—EAF, DCI, GBH—are equal in area. Prove it. You might extend IC and AF, and drop perpendiculars on them from D and E, respectively. Then area

Fig. 4.1.10 An exercise on the windmill.

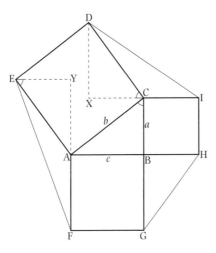

1. Tournes, *Euclid* (1611), 85; Schopenhauer. *The world as will and idea* (1891), as quoted by Cajori, *Mathematics* (1928), 81.

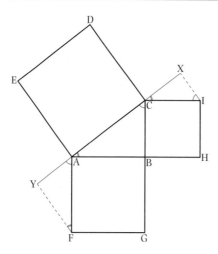

Fig. 4.1.11 Another exercise on the windmill.

$\Delta DCI = CI \times DX/2 = ac/2$. $DX = c$ because $\Delta DXC \cong \Delta ABC$ via *a.s.a*: $\angle DCX = \angle ACB$ (both are the complement of $\angle XCA$) and $CA = DC = b$. Similarly, $\Delta AEF = AF \times EY/2 = ca/2$, and, by inspection, $\Delta GBH = ac/2$. Q.E.D.

In Fig. 4.1.11, drop perpendiculars from the corners of the smaller sides onto the hypotenuse extended. The construction makes $CX = AY$. By considering complementary angles, you can show easily that $\Delta AFY \sim \Delta CAB \sim \Delta ICX$. Hence $CX : a = c : b$, $AY : c = a : b$, whence $AY = CX$. There are many other simple relationships among the lines of Fig 4.1.10 and 4.1.11, for example, in the first figure, that $EF^2 + GH^2 = 5BC^2$.

APS 4.1.5. Here's one last swing on the windmill. Extend parallel sides of the square on the hypotenuse to intersect the side of an opposite square and a side extended, as at N, M (Fig. 4.1.12). Then $\Delta AFM \sim \Delta CIN$ and $\Delta AFM + \Delta CIN = 2\Delta ABC$. The similarity is easily shown by marking complementary angles, as in the figure. In fact, the triangles are congruent. We have $\Delta AFM \cong \Delta ABC$ (*a.s.a.*), and, likewise, $\Delta CIN \cong \Delta ABC$ (*a.s.a.*), whence $\Delta AFM \cong \Delta CIN$ (C.N. 1). Since each is equal in area to ΔABC, their sum is twice ΔABC, as promised.

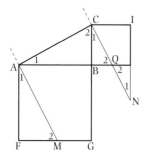

Fig. 4.1.12 A third exercise on the windmill (APS 4.1.5).

Some wider thoughts

Right angles and right thinking. The Pythagorean theorem and its converse (Euclid I.48) bring the first book of the *Elements* to a close. They therefore mark a climax, the high point toward which all the apparatus of definitions, postulates, and propositions was aimed. The diagram of the windmill may be taken as a symbol of geometry, and hence of Greek thought, which generations of Europeans have regarded as the bedrock of their culture. The French poet and critic Paul Valéry, writing just after World War I, during which Europe did its best to destroy itself, asked what it meant (or had meant) to be a European. His answer: to live under the legacies of Rome (civil administration and military strength), Christianity (religion and morality), and Greece (thought and culture). The emblem and message of this last legacy was a certain indestructible pattern or model of intellectual inquiry and demonstration. 'Greek geometry was this incorruptible model; not only the model offered to every science that aims at its perfection, but also an incomparable model of the most typical qualities of the European intellect. I never think of classical art without inevitably choosing the monument of Greek geometry as its best example.' These sentences, which expressed a hope in the rebirth of a European civilization and the practice of its ideals, were reprinted as a preface to an artistic American edition of Euclid published toward the end of World War II. The edition contained only the first book. Its implicit, powerful message ran: here, in this beautifully presented chain of clear, controlled, dispassionate reasoning, beginning with the definition of a point and ending with the theorem of Pythagoras, is one of the most precious legacies of the civilization we are fighting to preserve.[2]

A curious example of the belief that the Pythagorean theorem symbolizes European civilization—indeed, all human culture—occurred during the nineteenth century. In 1821 a German physician and astronomer named Franz Paula von Gruithuisen, who thought that intelligent creatures inhabited the planet Mars, suggested that we might open a conversation with them by displaying an indication of our own claims to intelligence. This display would have to be discernible at a great distance and recognizable immediately as the work of an advanced civilization. Gruithuisen's proposal: plow and plant huge fields in such a way that, from Mars, they would have the appearance of the windmill diagram or, as others suggested, dig canals suggestive of the Pythagorean theorem in Siberia or in the Sahara, fill them with oil, and set them afire.[3] Later in the century, an English astronomer who had managed to spot a moon of Mars through a telescope referred to the old German proposals and judged them to be practicable, or almost so. Perhaps with this encouragement, a rich lady left the Académie des sciences of Paris a large legacy in 1891 as a reward to anyone who could find a way to communicate with another planet, 'and to receive a response'.[4] The experiment has not yet been tried, leaving undecided

2. Valéry. *Variétés* (1924), 42–7 (quote); *Euclid: Book I* (1944). 'Introduction'.
3. Naber, *Theorem* (1908), 3; Crowe, *Debate* (1986), 203–8.
4. Hall, Roy. Astr. Soc., *Month. not. 38* (1878), 207; Rebière, *Mathématiques* (1926), 456.

whether the inhabitants of Mars have no telescopes, no intelligence, or no existence.

The irrational. 'Reason' and 'ratio' come from the same Latin word, *ratio*, the root meaning of which is calculation or plan. Calculations involve the proportion of numbers, hence ratio in the specialized usage of arithmetic and geometry. Calculation also implies the operation of an intelligence, of a faculty of mind able to calculate; this useful power, for which the Romans used their portmanteau word *ratio*, we call the rational faculty, or the reason. For us, 'irrational' means 'unreasonable' or, perhaps, 'crazy'; to an ancient geometer, it could also mean 'not expressible as the ratio of numbers'. The combination of meanings gave to a quantity that could not be expressed as a ratio of integers the reputation of being both crazy and incalculable.

That there are such quantities came as a very disagreeable surprise to the Pythagoreans who discovered them. Once detected, irrationals turned up everywhere. Right triangles might be considered almost human: few are perfectly rational. You need only try to write down the length of the diagonal of a square to appreciate this truth. Let the side of the square be any whole number a. Let us be rational and suppose that the diagonal's length can be written $(p/q)a$, where p and q are whole numbers. Further, let us require that p and q contain no common factors, that we have cancelled out all factors that occurred in both numerator and denominator. Therefore, p and q cannot both be even numbers. By Euclid I.47, the square of the diagonal of a square of side a is $2a^2$, so that, if the diagonal can be written as $(p/q)a$,

$$(p^2/q^2)a^2 = 2a^2, \text{ or } p^2 = 2q^2.$$

This last equality requires that p be an even number, since the squares of odd numbers are odd. Therefore, let us write $p = 2r$, where r is a whole number. Substituting $2r$ for p in the preceding equation, we have $4r^2 = 2q^2$, or $q^2 = 2r^2$. Now q must be even. But we agreed at the outset that p and q had no common factors; and yet our analysis shows that if the diagonal of a square is expressed as a rational fraction of its side, p and q must both contain a factor of 2. We have ended in absurdity and contradiction. Very irrational. It follows that the assumption that the diagonal can be written as the quotient of whole numbers where the side is taken to be a whole number is false. Likewise, if the diagonal is assigned a whole number, the sides cannot be expressed as rational fractions.

The 'Pythagorean triplets' (3, 4, 5), (5, 12, 13), etc., are exceptional in being rational. (If you expressed them as whole numbers and rational fractions, you could write them as 1, 4/3, 5/3, etc.)

Geometers from Pythagoras to Euclid worried about whether the existence of irrationals menaced their reasoning about lines and triangles. For example, the demonstrations about similarity we considered earlier (see Fig. 3.3.11) depend upon the division of two lines into many equal parts by finite numbers m, n, so that m parts of the one equalled n parts of the other. But what if, as in the case of the diagonal and side of a square, it is not possible to make the division by any numbers whatsoever? You will doubtless be glad to hear that after a minute investigation of the nature of proportion in Book V, Euclid concluded that the propositions of his geometry held (and hold!) irrespective of the existence of irrationals.

The question how to find Pythagorean triplets naturally arises. Mathematicians have identified several different sets, which have been given the names of heroes of ancient geometry. The 'Pythagorean' triplet *par excellence* is n, $(n^2 - 1)/2$, $(n^2 + 1)/2$, when n is any odd number. The 'Platonic' set concerns even values of n; the 'Euclidean' any two numbers whatsoever.[5] It may be useful to display them in a table (Table 4.1.1). Check the algebra to ensure that the formulas work. In what set(s) do the triplets used by the Vedic priests, (5, 12, 13), (8, 15, 17), fall?

Table 4.1.1 Pythagorean triplets

Set	Generator(s)	Triplet
Pythagorean	n, any odd number	n, $(n^2 - 1)/2$, $(n^2 + 1)/2$
Platonic	n, any even number	n, $n^2/4 - 1$, $n^2/4 + 1$
Euclidean	x, y, any unequal numbers	$x^2 - y^2$, $2xy$, $x^2 + y^2$

If you do not have the luxury of dealing with Pythagorean triplets, how can you solve arithmetic problems involving right angles without consulting tables or using a calculator? Neither might be available and, what is worse, relying upon them makes you dependent on other peoples' work. There is a very good method of obtaining close approximations to the value of square roots that, in principle, you could have made use of in your old Egyptian surveying business.

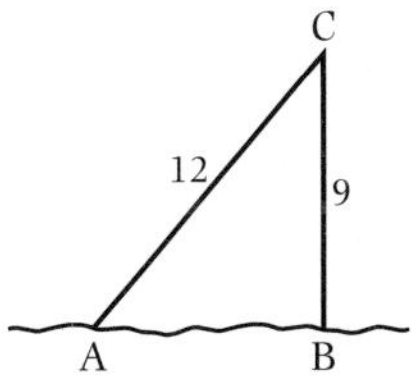

Say you have a client who wants to know the river frontage of the triangular parcel pictured in Fig. 4.1.13. The Pythagorean theorem gives the square of the frontage, AB^2, equal to $(144 - 81) = 63$ a.u. What is AB? Since 64 is a square number ($8 \times 8 = 64$), AB must be

Fig. 4.1.13 A Pythagorean problem for the Egyptian tax collector.

5. Halsted, *Mensuration* (1881), 146, attributing the 'Euclidean' form to Francis Maseres, an eighteenth century mathematician.

only a little less than 8. As an approximation, let the square root of 63 be 8 − x. We know that x is smaller than 1. Then

$$(8 - x)^2 = 63, \quad \text{or} \quad 16x - x^2 = 1.$$

Since x<1, x^2<< ('much smaller than') 16x; therefore, a good approximation to x is 1/16, obtained from the preceding equation by ignoring (that is, throwing out) the very small x^2. You can confidently tell your client that his river frontage amounts to 8 − 1/16 = 7.9375 l.u. A more exact calculation gives $\sqrt{63}$ = 7.9373. The client will not discern the difference. Let us take our length unit to be a kilometer, or 1000 meters, which equals about 5/8 mile. Then the difference between the approximate and the exact value of $\sqrt{63}$ translates to 0.2 m, or 20 cm. Now 20 cm is under 8 inches, so that an error of 20 cm in almost 8 kilometers amounts to 8 inches in 5 miles. The technique of approximation you have used is very powerful.

Another client wants to know the height of the great pyramid at Gizeh (Fig. 4.1.14). You tell him to measure the length of one side of the base and also the shortest line along the sloped face from top to

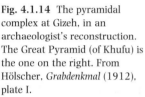

Fig. 4.1.14 The pyramidal complex at Gizeh, in an archaeologist's reconstruction. The Great Pyramid (of Khufu) is the one on the right. From Hölscher, *Grabdenkmal* (1912), plate I.

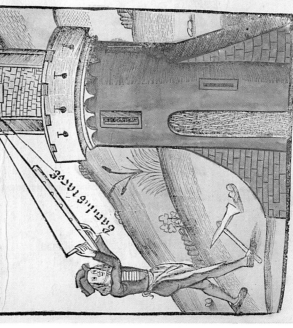

Plate I. 'Geometria' and 'Surveying'. From Reisch, *Margarita philosophica* (1512).

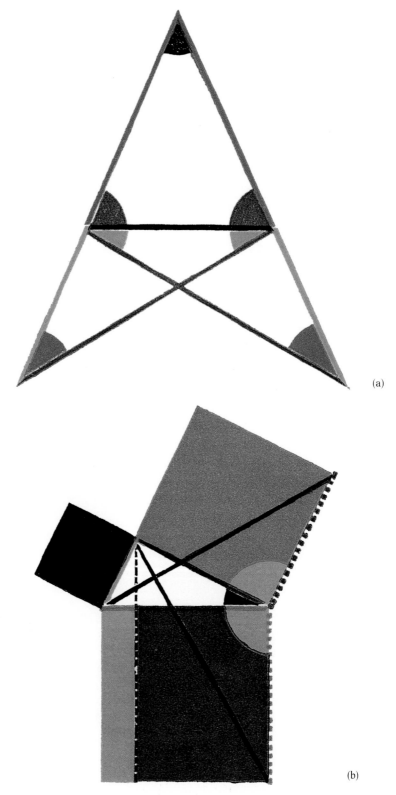

Plate II. Demonstrations in color: (a) 'The Bridge of Asses,' Euclid I. 5, discussed in text on pp. 83–6, (b) 'The Windmill,' Euclid I. 47, discussed on pp. 145–8. From Byrne, *First six books* (1847). Reproduced with permission from the Bodleian Library, University of Oxford, shelfmark 2 Δ 576.

(a)

(b)

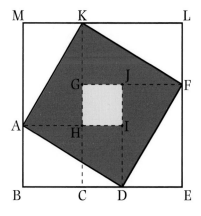

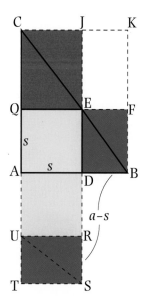

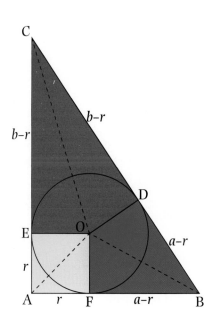

Fig. III. The Chinese method of color and cut. The figures are discussed in the text on pp. 154, 157, 213.

Plate IV.   School of Athens, 1510–11 (fresco) by Sanzio of Urbino Raphael (1483–1520)
Vatican Museums and Galleries, Vatican City, Italy/Bridgeman Art Library, London/New York.

(a)

Plate V. Islamic geometrical mosaics:
(a) Detail from a wall in the 'four
hands' pattern, Madrasat al
Būʻnāniyyah, Fez, Morocco, 1350–55
(b) Panel in a classical pattern,
Zlayjiyyah Cooperative, Fez. From
Hedgecoe and Damluji, *Zillij* (1992),
233, 247.

(b)

(a)

Plate VI. Islamic geometrical mosaics:
(a) Interior wall in the '16-pointed star'
pattern, Al Saʻdiyyin Tombs, Marrakesh,
Morroco, 1550s
(b) A simpler star from the decoration of
the Madrasat al Būʻnāniyyah, Fez,
Morocco, 1350–55. From Hedgecoe and
Damluji, *Zillij* (1992), 151, 202.

(b)

Plate VII. Santa Maria del Fiore, Florence. From Bartoli, *Disegno* (1994), plate 1.

Plate VIII. The death of Archimedes. From Winter, 'Der Tod' (1924), [plate 1]. Courtesy of Caltech Archives.

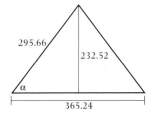

Fig. 4.1.15 Exercise on the Great Pyramid.

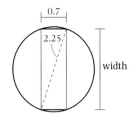

Fig. 4.1.16 A Chinese Pythagorean problem (APS 4.1.6).

bottom. He returns with the numbers indicated in Fig. 4.1.15, where the l.u. is the so-called 'sacred cubit' of 25.025 inches. Not knowing the theorem of Pythagoras, the client can go no further. You will charge a good fee for your superior knowledge, especially since the numbers are large and require a little computation. From the data presented, you calculate the height at 232.52 sacred cubits, or 484.9 feet. Since you will want to approximate a square root at the end of the calculation, you might as well make some approximations along the way. The square of the slant height, 295.66, is, very closely, $(295 + 2/3)^2 \sim 87\ 419$; the square of half the side of the base, 182.62, is, also very closely, $(182 + 5/8)^2 \sim 33\ 352$; hence the square of the unknown height h is 54 066, or, as near as we shall figure, 54 070. Now $54\ 070 = (5.407)(10\ 000)$, so $h = 100\sqrt{5.407}$. You can see that the root is a little larger than 2.3, whose square is 5.29. Setting $(2.3 + x)^2 = 5.407$, we have, approximately, $4.6x = 0.117$, or $x = .025$, $h = 232.5$ sacred cubits. Try the same calculation on your calculator. You should be pleased with the accuracy of your approximation. Your client too would have been amazed at your prowess. He could not have measured h directly, since the pyramid is almost entirely solid within.

APS 4.1.6. An old Chinese problem asks for the maximum width of a plank 0.7 units thick that can be cut from a cylindrical log 2.25 units in diameter. This comes to requiring the leg of a right triangle whose other leg is 0.7 l.u. and hypotenuse 2.25 l.u. (Fig. 4.1.16). An obvious guess leads to the approximate solution, width = 2.14 l.u. Can you show how to obtain the approximation?

APS 4.1.7. The Babylonians approximated $\sqrt{2}$ by a method that can be paraphrased as follows.[6] Set $xy = 2$ and make a guess at the root, say x_1. Then $y_1 = 2/x_1$. If you chose x_1 too large, then y_1 will be too small, and the average, $(x_1 + y_1)/2$, will be closer than either. Take then $x_2 = (x_1 + y_1)/2$ as your second guess and calculate $y_2 = 2/x_2$. Again, either x_2 or y_1 is too large, the other too small, and $x_3 = (x_2 + y_2)/2$ better than either. Show that the reasonable choice of $x_1 = 1.5$ gives a value for $x_3 = 1.4142$, a value accurate to about one part in a thousand.

APS 4.1.8. Among the various Vedic prescriptions for building later altars on the site of earlier ones is that each should have an area one unit larger than its predecessor.[7] To be precise, the first altar must have an area of 7.5 units, the second an area of 8.5

6. Neugebauer and Sachs. *Texts* (1945), 43.
7. Kulkarni, *Layout* (1987), 124–5.

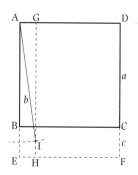

Fig. 4.1.17 A Vedic Pythagorean problem (APS 4.1.7).

units; and, to make things easy, let us have our altars square. What is the length of the side of the second altar if that of the first is $a = \sqrt{15/2}$ units? How could you lay out the side of the second from that of the first geometrically?

Let the side of the second altar be b. Then according to the rule of succession, $b^2 = 8.5$ and b may be found by the Babylonian method to be 2.944 to the third approximation with the obvious choice of $x_1 = 3$. To obtain b geometrically, begin with the altar ABCD of side a (Fig. 4.1.17). Add a strip BEFC of width $c = (2/15)a$; its area is therefore $ac = (2/15)a^2 = 1$, as required. By definition, $b^2 = a(a + c)$. Now draw GH∥AE with AG = EH = $c/2$. Take a rope of length $a + c/2$ and attach one end at A; draw it tight and sweep it until its free end meets GH at I. Then $b = $ GI, since from Pythagoras' theorem, which the Vedic priests used freely, even if they did not know it,

$$GI^2 = AI^2 - AG^2 = (a + c/2)^2 - (c/2)^2 = a(a + c) = b^2.$$

4.2. The Chinese Pythagoras

As we know, the ancient Chinese handled problems involving right triangles without invoking the Greek concept of symmetry. And they had a full (and justified) confidence in what the Western world calls the Pythagorean theorem without the benefit of the Greek notion of proof. The Chinese operated under the principle of cut and place: they would dissect (or, when necessary, duplicate) parts of a problematic diagram and rearrange the parts into other figures they could solve. To keep track of the process, they identified the parts with different colors.

Figure 4.2.1 presents the oldest recorded proof of the Pythagorean theorem. (It is illustrated in its original colors on Plate III.) The Chinese Pythagoras has copied the given triangle ABD four times and fitted the four pieces—ΔAID, ΔDJF, ΔFGK, and ΔKHA together to form the square ADFK. He then reasoned

ADFK = (red) triangles + (yellow) square = 4ΔABD + □GHIJ.

But 4ΔABD = rect. ABDI + rect. DEFJ = (□ABCH + rect. CDIH) + (□CEFG − rect. CDIH − □GHIJ). Replacing 4ΔABD with its equal he had

$$\square ADFK = \square ABCH + \square CEFG,$$

which is Pythagoras' theorem.[8]

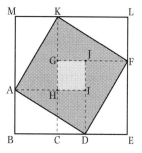

Fig. 4.2.1 Chinese form of the Pythagorean theorem.

8. Swetz and Kao, *Was Pythagorus Chinese?* (1977), 66; Lam and Shen, *AHES, 30* (1984), 88–9, 95.

Figure 4.2.1 expresses many other relationships, for example,

$$\square GHIJ = \square ADFK - 4\triangle AID,$$

$$\square BELM = \square ADFK + 4\triangle AID,$$

and, by their addition,

$$2\square ADFK = \square GHIJ + \square BELM. \qquad (4.2.1)$$

These relationships can be understood as forms of the Pythagorean theorem. For let AD = c, BD = b, AB = a; then the equations (4.2.1) become

$$(b - a)^2 = c^2 - 2ab,$$

$$(b + a)^2 = c^2 + 2ab, \qquad (4.2.2)$$

$$2c^2 = (b - a)^2 + (b + a)^2.$$

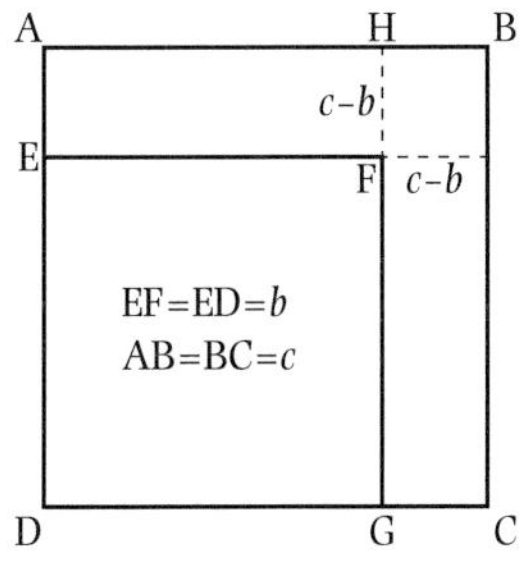

We would say that all of these equations are derivative from $c^2 = a^2 + b^2$ and we would not distinguish any of them as special or distinct cases. Not so the Chinese. Each of equations (4.2.2) represented to them a solution to a right-triangle problem; which solution they chose depended on which elements of the problem were given. For example, if they knew the sum and difference of the legs, they would use the last of equations (4.2.2) to obtain the hypotenuse.

The practical working of the scheme may be seen from Fig. 4.2.2, in which $\square ABCD = c^2$ and $\square DGFE = b^2$. You can see immediately that

$$\square ABCD - \square DGEF = \text{rect. AEFH} + \text{rect. BCGH},$$

or

$$c^2 - b^2 = b(c - b) + c(c - b),$$

or

$$a^2 = (c + b)(c - b). \qquad (4.2.3)$$

Let us suppose that we know one leg, a, and the difference between the hypotenuse and the other leg, $c - b$, and want to know b. Then, to do it in the Chinese way, which explicitly expresses the unknown in terms of the givens, we must manipulate equation 4.2.3, as follows:

$$c + b = a^2/(c - b), \quad c - b + 2b = a^2/(c - b),$$

$$b = \tfrac{1}{2}[a^2/(c - b) - (c - b)]. \qquad (4.2.4)$$

Equation (4.2.4) is the form desired. It allowed the Chinese to solve a problem with which we are familiar: the floating lotus.[9]

A _____ H ___ B

$c{-}b$

E _____ F $c{-}b$

EF=ED=b
AB=BC=c

D _____ G ___ C

Fig. 4.2.2 A Chinese Pythagorean problem.

9. The following material comes from Lam and Shen, *AHES, 30* (1984), 91–2, 97, 100, and Swetz and Kao, *Was Pythagorus Chinese?* (1977), 30–5, 44–5.

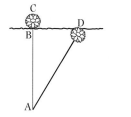

Fig. 4.2.3 The lotus problem.

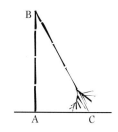

Fig. 4.2.4 The bamboo version of the lotus problem.

Recall that in the floating lotus, Fig. 4.2.3, we are told the maximum height of the flower above a lake when the plant is upright and the distance along the horizontal the flower has moved when it just touches the water. We have BC = 1 chih = AC − AB = $c - b$, and BD = 5 chih = a; therefore, from equation (4.2.4) we have for the unknown depth of the lake BA = b,

$$b - (25 - 1)/2 = 12 \text{ chih.}$$

The Western-style solution to this problem would not invoke the special formula (4.2.4), but set up an equation directly from the diagram. Call the unknown AB, x, $AD^2 = (x + 1)^2 = AB^2 + BD^2 = x^2 + 25$, whence $x = 12$.

Because they dealt with many special cases rather than with a single, supple equation, it took long study with a master for a Chinese to learn how to solve right-triangle problems. Consider the problem of the broken bamboo. Here (Fig. 4.2.4) we are told the distance of the top of the broken shaft from the base (3 chih) and the original length of the tree (10 chih) and asked for the height of the break. The Westerner would write AB = x and $BC^2 = (10 - x)^2 = AB^2 + AC^2 = x^2 + 9$, whence $x = 4 \ 11/20$. The Chinese apprentice could not proceed with the equations so far developed, because they do not express the unknown in terms of the given knowns. He would have to turn to the master, who would observe that the problem of the broken bamboo required for its solution a return to the relationship (4.2.3),

$$a^2 = (c + b)(c - b)$$

This time we know $(c + b)$, not $(c - b)$, and again want b. So we write

$$c - b = a^2/(c + b), \quad c + b - 2b = a^2/(c + b),$$

$$b = \frac{1}{2}[(c + b) - a^2/(c + b)]. \tag{4.2.5}$$

The apprentice plugs in the numbers:

$$b = (10 - 9/10)/2 = 4 \ 11/20,$$

as before.

APS 4.2.1. In Fig. 4.2.5, CD is a wall 10 chih high, and you know the difference AB = 1 chih between a ladder's length BD and the distance BC of its far end from the wall. Find the length of the ladder Chinese style. You need a formula expressing BD = c in terms of a and $c - b$. Obtain one from equation (4.2.3) from which you can derive $c = (100 + 1)/2 = 50.5$ chih.

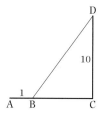

Fig. 4.2.5 The problem of the leaning ladder (APS 4.2.1).

The problem of the leaning ladder, which recurs through the ages, may be associated with the exercise of taking heights at a distance in order to prepare equipment for storming a tower. We can infer as much from a lament by Hero of Alexandria, voiced some 2000 years ago: 'How many times in an attack on a stronghold have we arrived at the foot of the ramparts and found that we had made our ladders ... too short!'[10]

APS 4.2.2. A rope hangs limply from a pole so that 3 chih of it lies on the ground. When pulled taut to the ground, its end is 8 chih from the base of the pole. Find the length of the rope in the Chinese manner.

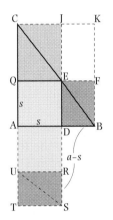

Fig. 4.2.6 A square inscribed in a right triangle, the Chinese way.

The method of cut and place may be used advantageously to find the length of a side of a square inscribed in a right triangle. In Fig. 4.2.6, ΔABC is the triangle, AD = DE = s, the side sought; AC = b, AB = a. (The figure is shown in Chinese color on plate III). Complete the rectangles CADJ and ABKC. Extend CA to T, making AU = AQ, UT = DB, so that AT = AB = a. Now we have

rect. ABKC = 2ΔABC = two (red) Δs, two (blue) Δs, two (yellow) $\Box$s.

(That rect. EFJK = $\Box$ADEQ is evident from the figure, since the diagonal CB divides rect. ABKC into two equal parts and the triangles on either side of it are equal in pairs.)

But rect. CJST also equals two red triangles (ΔCJE = ΔCQE), two blue triangles (rect. RSTU = 2ΔEBD since UR = ED and UT = DB), and two yellow squares. Therefore,

$$\text{rect. ABKC} = \text{rect. CJST,}$$

or

$$ab = s(a + b),$$

whence $s = ab/(a + b)$. This method comes from Liu Hui's commentary on the *Jin zhang suanshu*.[11] A more direct way would have noticed that, since $\Box$QEDA = rect. JKEF, $s^2 = (a - s)(b - s)$; but that would have produced s only implicitly, whereas the Chinese favored explicit expressions for the unknown in terms of the givens.

APS 4.2.3. The Greek way (Fig. 4.2.7) may appear easier. From the similarity of triangles ABC and DBE,

$$b : a = s : (a - s),$$

or $s = ab/(a + b)$, as before.

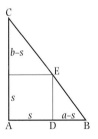

Fig. 4.2.7 The same, the Greek way (APS 4.2.3).

10. Skyring Walters, Newcomen Society, *Trans.*, 2 (1921), 46.
11. Cf. Lam and Shen, *AHES, 30* (1984), 103.

4.3. More trigonometry

Functions of sums and differences of angles

For many purposes, it is useful to know the trigonometrical functions of the sum and difference of angles, for example $\sin(\alpha + \beta)$ ('the sine of alpha plus beta'), or $\cos(\alpha - \beta)$ ('the cosine of alpha minus beta'). Finding these functions has the further attraction of being in itself an instructive exercise in trigonometry.

Functions of sums. Let ABC (Fig. 4.3.1) be a 'unit circle', which, you will recall, is a circle whose radius we arbitrarily set equal to 1 l.u. Then, by definition, $\cos(\alpha + \beta) \equiv$ side adjacent/hypotenuse = OD/OC = OD, since we have set OC equal to 1. Similarly, OE = $\cos\alpha$, BE = $\sin\alpha$, and, if we drop a perpendicular from C on OB, meeting OB at F, CF = $\sin\beta$, OF = $\cos\beta$. Our first exercise will be to find $\cos(\alpha + \beta)$ in terms of functions of the constituent angles α and β.

From Fig. 4.3.1, OD = $\cos(\alpha + \beta)$ = OG − DG. Here G is the foot of the perpendicular dropped from F onto OA. Draw HJ from H, the intersection of OB and CD, parallel to OA until it cuts FG at J. DGJH is a rectangle; therefore DG = HJ and $\angle$FHJ = α (by alternate interior and vertical angles). Now

$$OG = OF \cos\alpha = \cos\alpha\cos\beta,$$

and

$$DG = HJ = HF \cos\alpha.$$

But in ΔHCF, $\tan\alpha$ (tangent = side opp./side adj.) = HF/CF, or HF = CF $\tan\alpha$ = $\sin\beta\tan\alpha$. In all,

$$DG = HJ = \sin\beta\tan\alpha\cos\alpha = \sin\alpha\sin\beta,$$

and

$$\boxed{\cos(\alpha + \beta) = OG - DG = \cos\alpha\cos\beta - \sin\alpha\sin\beta}$$

We may obtain $\sin(\alpha + \beta)$ in the same fashion from the same figure:

$$\sin(\alpha + \beta) = CD = CH + HD,$$

$$CH = CF/\cos\alpha = \sin\beta/\cos\alpha,$$

$$HD = OH \sin\alpha = (OF - HF) \sin\alpha,$$

$$OF = \cos\beta, \quad HF = \sin\beta\sin\alpha/\cos\alpha.$$

Altogether,

$$\sin(\alpha + \beta) = \frac{\sin\beta}{\cos\alpha} + \sin\alpha\left(\cos\beta - \frac{\sin\beta\sin\alpha}{\cos\alpha}\right)$$

$$= \sin\alpha\cos\beta + \frac{\sin\beta}{\cos\alpha}(1 - \sin^2\alpha)$$

$$\boxed{\sin(\alpha + \beta) = \sin\alpha\cos\beta + \cos\alpha\sin\beta}$$

In the last step, the equation $\sin^2\alpha + \cos^2\alpha = 1$, which states the Pythagorean theorem, was exploited. We might have obtained $\sin(\alpha + \beta)$ another way, by making use of the formula already established for $\cos(\alpha + \beta)$. In Fig. 4.3.1, HD = OD tan α = tan α cos $(\alpha + \beta)$. Hence

$$\begin{aligned}
\sin(\alpha + \beta) = \text{CH} + \text{HD} &= \sin\beta/\cos\alpha + \tan\alpha\cos(\alpha + \beta)\\
&= \sin\beta/\cos\alpha + (\sin\alpha/\cos\alpha)(\cos\alpha\cos\beta - \sin\alpha\sin\beta)\\
&= \sin\beta/\cos\alpha + \sin\alpha\cos\beta - (\sin\beta/\cos\alpha)(\sin^2\alpha)\\
&= \sin\alpha\cos\beta + \cos\alpha\sin\beta.
\end{aligned}$$

Some numerical values. These formulas allow calculations of the values of some of the trigonometrical formulas. Notice first that sin 30° = 0.5. This follows from the fact that the bisector of any angle of an equilateral triangle is also the perpendicular bisector of the opposite side (Fig. 4.3.2). Hence sin 30° = sin∠ACD = (AD/2)/AC = 1/2. From the general relation, $\cos^2\alpha + \sin^2\alpha = 1$, cos 30° = $\sqrt{3/4}$ = 0.866. Values for the sine and cosine of 15° follow readily. Take the expression for $\cos(\alpha + \beta)$ for the case $\alpha = \beta$:

$$\cos(\alpha + \alpha) = \cos2\alpha = \cos^2\alpha - \sin^2\alpha = 1 - 2\sin^2\alpha,$$

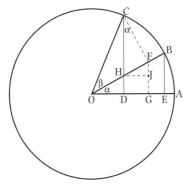

Fig. 4.3.1 Trigonometric functions of the sum of angles.

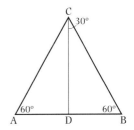

Fig. 4.3.2 An equilateral triangle, for calculating trigonometric functions.

whence

$$\sin^2\alpha = (1 - \cos 2\alpha)/2,$$

$$\sin^2 15° = (1 - \cos 30°)/2 = 0.067,$$

or

$$\sin 15° = 0.259, \qquad \cos 15° = \sqrt{1 - \sin^2 15°} = 0.966.$$

From the definitions of the functions, $\sin 15° = \cos 75°$, $\cos 15° = \sin 75°$. Furthermore, we know that $\sin 0° = 0$ (the side opposite is zero), whence $\cos 90° = 0$; also, $\cos 0° = 1$, $\sin 90° = 1$. Finally, in an isosceles right triangle, the hypotenuse is $\sqrt{2}$ times either leg, so that $\sin 45° = \cos 45° = 1/\sqrt{2} = 0.707$. These values are summarized in Table 4.3.1. Note that the sine and cosine cannot exceed 1 and that the tangent can take any value, including ∞ ('infinity').

Table 4.3.1 Some values of trigonometric functions

Angle	Sine	Cosine	Tangent
0°	0	1	0
15	0.259	0.966	0.268
30	0.500	0.866	0.577
45	0.707	0.707	1
60	0.866	0.500	1.732
75	0.966	0.259	3.732
90	1	0	∞

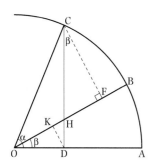

Fig. 4.3.3 Trigonometric functions of the difference of angles.

Functions of differences. In Fig. 4.3.3, which reproduces the unit circle, draw DK parallel to FC. We have $\angle KDH = \angle DCF = \beta$ (by alternate interior angles) and

$$OF = \cos(\alpha - \beta) = OK + KH + HF.$$

Since, however, $OK = OD \cos \beta = \cos \alpha \cos \beta$, $KH = DH \sin \beta$, and $HF = HC \sin \beta$, $KH + HF = \sin \beta(DH + HC) = DC \sin \beta = \sin \alpha \sin \beta$, whence

$$\boxed{\cos(\alpha - \beta) = OF = \cos \alpha \cos \beta + \sin \alpha \sin \beta}$$

This formula permits a definition of the cosine of a negative angle. Let $\alpha = 0$. Then

$$\cos(-\beta) = \cos \beta.$$

Furthermore, from the formula for $\cos(\alpha + \beta)$ we can find the cosine of a straight angle:

$$\cos 180° = \cos(90° + 90°) = -1.$$

Evidently, $\sin 180° = 0$.

The sine of the difference of two angles also may be read from Fig. 4.3.3:

$$\sin(\alpha - \beta) = CF = CH \cos \beta,$$

$$CH = CD - HD = \sin \alpha - \cos \alpha \tan \beta,$$

$$\sin(\alpha - \beta) = \cos \beta(\sin \alpha - \cos \alpha \tan \beta) = \sin \alpha \cos \beta - \sin \beta \cos \alpha.$$

By setting $\alpha = 0$, we learn that $\sin(-\beta) = -\sin \beta$.

One further point is needed. We have been reasoning as if α and β are always acute angles. The preceding formulas for the functions of differences of angles allow us to extend trigonometry to obtuse angles (angles greater than 90°) also. Let β be obtuse, then $(180° - \beta)$ is acute:

$$\sin(180° - \beta) = \sin 180°\cos \beta - \cos 180°\sin \beta = \sin \beta.$$

The value of the sine of an obtuse angle is the value of its supplement. Similarly,

$$\cos(180° - \beta) = \cos 180°\cos \beta - \sin 180°\sin \beta = -\cos \beta.$$

The value of the cosine of an obtuse angle is the negative of the value of its supplement.

François Viète, one of the first Europeans to develop trigonometry (this some 400 years ago), observed that if you know the sines and cosines of all angles up to 30°, you know the rest.[12] Write an angle greater than 30° and less than 60° as 60° − x. We have the identity,

$$\sin x = \sin(60 + x) - \sin (60 - x), \tag{4.3.1}$$

which you should check, remembering that $\cos 60° = 1/2$. Say we wish the sine of 35°. Then $x = 25°$, and

$$\sin 35° = \sin 85° - \sin 25° = \cos 5° - \sin 25°.$$

Equation (4.3.1) also enables you to find the functions of angles between 60° and 90°. For let us take an interest in $\sin 75°$:

$$\sin 15° = \sin 75° - \sin 45°,$$

which gives $\sin 75°$ since, on the basis of Viète's formula, we have assumed knowledge of the functions of the angles up to 60°.

---

12. Hunrath, *Abh. Ges. Math.*, 9 (1899), 228.

Ptolemy's theorem. A theorem known by the name of the Greek astronomer Claudius Ptolemy, who was to astronomy what Euclid was to geometry, allows another way to derive the trigonometric functions of the sums and differences of angles. The theorem: the rectangle made by the diagonals *e, f* of any quadrilateral inscribed in a circle equals the sum of the rectangles made by the pairs of opposite sides. In the symbols of Fig. 4.3.4, *ef = ac + bd*. The proof, one of the most elegant in all plane geometry, is accomplished by drawing a line from one of the vertices, say C, onto a diagonal so as to make two sets of similar triangles. One such magical line is CE in Fig. 4.3.4, drawn so as to make ∠DCE = ∠ACB = ∠1. Now, as will be proved in §5.2, angles contained within the circumference of a given circle, like the angles at A, B, C, and D in Fig. 4.3.4, are equal if they intercept equal arcs on the circle. Since circumferential angles CDB and CAB intercept the same arc, BC, they also are equal; hence ΔDCE ∼ ΔACB. The second set of similar triangles is DCA and ECB. Their similarity follows from ∠EBC = ∠CAD = ∠3 (both subtend arc DC) and ∠DCA = ∠ECB (each equals ∠1 + ∠ECQ).

From ΔDCE ∼ ΔACB, *x : a = c : f*, where *x* = DE and *f* = AC. From ΔDCA ∼ ΔECB, *(e − x) : d = b : f*. We have therefore

$$e - ac/f = db/f, \quad \text{or} \quad ef = ac + db, \quad \text{QED}$$

The use of Ptolemy's theorem in trigonometry appears from taking one of the diagonals as the diameter of the circle. That presents the situation of Fig. 4.3.5. The theorem gives

$$AC \times BD = AD \times BC + DC \times AB. \tag{4.3.2}$$

Let ∠AOD = *α*, ∠DOC = *β*. We may then write (with *r* = radius AO):

$$AD = 2r \sin \alpha/2, \quad DC = 2r \sin \beta/2, \quad AC = 2r \sin(\alpha + \beta)/2,$$

$$DB = 2r \sin(\text{st.}\angle - \alpha)/2 = \cos \alpha/2,$$

$$BC = 2r \sin(\text{st.}\angle - \alpha - \beta)/2 = 2r \cos(\alpha + \beta)/2.$$

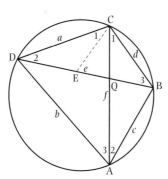

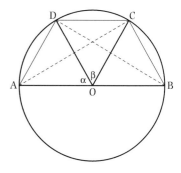

Fig. 4.3.4 Ptolemy's theorem.

Fig. 4.3.5 Application of Ptolemy's theorem to the trigonometric functions.

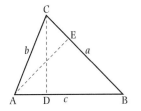

Fig. 4.3.6 The law of sines in an acute triangle.

After substitution in equation (4.3.2),

$$\sin(\alpha + \beta)/2 \cos \alpha/2 = \sin \alpha/2 \cos(\alpha + \beta)/2 + \sin \beta/2.$$

This last equation takes on a familiar form with the substitutions $(\alpha + \beta)/2 = \gamma$, $\alpha/2 = \lambda$:

$$\sin \gamma \cos \lambda - \cos \gamma \sin \lambda = \sin(\gamma - \lambda).$$

The laws of sines and cosines. In any acute triangle ABC, such as Fig. 4.3.6, there rules a 'law of sines',

$$\boxed{\frac{\sin A}{a} = \frac{\sin B}{b} = \frac{\sin C}{c}}$$

It follows directly from Fig. 4.3.6, where CD, AE are altitudes: $\sin A = CD/b$, $\sin B = CD/a$, so $b \sin A = a \sin B$. Similarly, $c \sin B = b \sin C$.

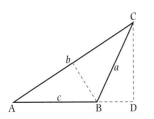

Fig. 4.3.7 The law of sines in an obtuse triangle.

The same relation holds for an obtuse angle. In Fig. 4.3.7, the perpendicular dropped from B onto AC gives $\sin A/a = \sin C/c$. That from C onto AB extended gives $\sin A = CD/b$, $\sin\angle CBD = CD/a$. But $\sin\angle CBD = \sin(180° - B) = \sin B$. Therefore, again,

$$\frac{\sin A}{a} = \frac{\sin B}{b} = \frac{\sin C}{c}.$$

The 'law of cosines' for an acute triangle may be derived from Fig. 4.3.8, where CD⊥AB. The Pythegorean theorem applied to ΔADC and to ΔBDC yields

$$b^2 = CD^2 + x^2, \quad a^2 = CD^2 + (c - x)^2.$$

Subtracting the second of these equations from the first gives

$$a^2 - b^2 = c^2 - 2cx.$$

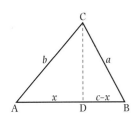

Fig. 4.3.8 The law of cosines in an acute triangle.

But $x = b \cos A$, whence 'the law of cosines':

$$\boxed{a^2 = b^2 + c^2 - 2bc \cos A}$$

We also have $b^2 = c^2 + a^2 - 2ac \cos B$, etc.

If A were a right angle, the law of cosines would express the Pythagorean theorem, since $\cos 90° = 0$. If A is obtuse, Fig. 4.3.9 holds:

$$b^2 = CD^2 + x^2, \quad a^2 = CD^2 + (c + x)^2,$$

$$a^2 - b^2 = c^2 + 2cx = c^2 + 2bc \cos\angle CAD.$$

But $\cos\angle CAD = \cos(180° - A) = -\cos A$, so that, again,

$$a^2 = b^2 + c^2 - 2bc \cos A.$$

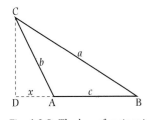

Fig. 4.3.9 The law of cosines in an obtuse triangle.

Fig. 4.3.10 Easy exercise on the law of sines.

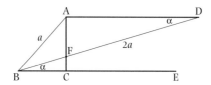

In words: the square of any side of a triangle is the sum of the squares of the other two sides less twice the product of those sides with the cosine of the angle included between them.

Here is an easy exercise on the law of sines. In Fig. 4.3.10, AD∥BE, AC⊥BE, and FD = 2AB. Show that ∠DBC = ∠ABC/3, or, what is the same, that ∠ABD = 2α. We are given that FD = 2a; therefore, AD = 2a cos α. The law of sines applied to ΔABD gives

$$\frac{\sin \angle ABD}{2a \cos \alpha} = \frac{\sin \alpha}{a}, \quad \text{or} \quad \sin \angle ABD = 2 \sin \alpha \cos \alpha = \sin 2\alpha,$$

whence ∠ABD = 2α, as required.

And here is an easy one on the law of cosines: show that, for any triangle (Fig. 4.3.11),

$$ab \cos C + ac \cos B + bc \cos A = (a^2 + b^2 + c^2)/2. \quad (4.3.3)$$

The law of cosines delivers

$$2ab \cos C = a^2 + b^2 - c^2, \quad 2ac \cos B = a^2 + c^2 - b^2,$$
$$2bc \cos A = b^2 + c^2 - a^2,$$

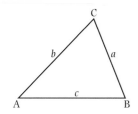

Fig. 4.3.11 Easy exercise on the law of cosines.

so the left side of equation (4.3.3) equals $(a^2 + b^2 + c^2)/2$, as required. Equation (4.3.3) is occasionally useful.

APS 4.3.1. Take *any* triangle and draw squares upon it (Fig. 4.3.12). The relationship $3(a^2 + b^2 + c^2) = EF^2 + GH^2 + ID^2$ follows easily from the law of cosines, which gives $EF^2 = b^2 + c^2 - 2bc \cos\angle EAF$, etc. But, since ∠EAF + 2 rt.∠ + ∠CAB = 2st.∠, we have ∠EAF = st.∠ − ∠CAB; and therefore, since the cosine of the supplement of an angle is the negative of the cosine of the angle, cos∠EAF = −cos∠CAB, or, by an obvious change in notation, −cos A. In sum,

$$EF^2 + FG^2 + GH^2 = 2(a^2 + b^2 + c^2) + (bc \cos A + ac \cos B + ab \cos C).$$

The right side of the equation is just $3(a^2 + b^2 + c^2)$, as you can see by reference to equation (4.3.3), which thus, as alleged, does prove occasionally to be useful.

Trigonometrical identities. Relationships involving the trigonometric functions only, like sin 2α = 2 sin αcos α, are called identities. They

Fig. 4.3.12 A more difficult exercise on cosines (APS 4.3.1).

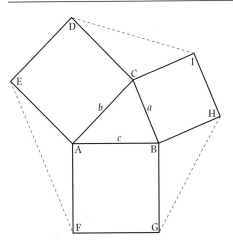

are true irrespective of the values of the angles that appear in them. Although most of these identities are not useful in themselves, concocting and demonstrating them have very considerable value as exercises. Facility in the algebra of trigonometry is a great desideratum for a practicing geometrician. The invention of trigonometric identities became a competitive sport among mathematicians of the Renaissance after they had thoroughly absorbed both Greek geometry and the commentaries on it by Arab experts who had translated and developed the ancient heritage. Our new acquaintance François Viète, who, among much else, introduced the practice of using letters to represent algebraical quantities, is credited with a few identities that are pleasing in form if not very useful in practice. They are

$$\sin \alpha - \sin \beta = 2 \cos(\alpha + \beta)/2 \sin(\alpha - \beta)/2,$$

and

$$\frac{a - b}{a + b} = \frac{\tan(\alpha - \beta)/2}{\tan(\alpha + \beta)/2}.$$

In the second identity, a and b represent the sides of a triangle opposite the angles α and β, respectively.

The first identity can be demonstrated readily by writing the left side in terms of functions of half angles, that is, as $2 \sin \alpha/2 \cos \alpha/2 - 2 \sin \beta/2 \cos \beta/2$. We then must show that

$$\sin \alpha/2 \cos \alpha/2 - \sin \beta/2 \cos \beta/2 = \cos(\alpha + \beta)/2 \sin(\alpha - \beta)/2.$$

Expanding the right side of the previous equation and rearranging terms, we have

$$\sin \alpha/2 \cos \alpha/2(\sin^2\beta/2 + \cos^2\beta/2)$$
$$- \sin \beta/2 \cos \beta/2(\sin^2\alpha/2 + \cos^2\alpha/2)$$
$$= \sin \alpha/2 \cos \alpha/2 - \sin \beta/2 \cos \beta/2,$$

since, in consequence of the Pythagorean theorem, $\sin^2\phi + \cos^2\phi = 1$, no matter what the value of ϕ. The second of Viète's identities follows immediately from the first by replacing a and b via the law of sines ($\sin\alpha/a = \sin\beta/b$). We are to show that

$$\frac{\sin\alpha - \sin\beta}{\sin\alpha + \sin\beta} = \frac{\tan(\alpha-\beta)/2}{\tan(\alpha+\beta)/2}$$

$$= \frac{2\sin(\alpha-\beta)/2\cos(\alpha+\beta)/2}{2\cos(\alpha-\beta)/\sin(\alpha+\beta)/2}.$$

The factors of 2, which of course cancel, are placed in the last equation to allow immediate application of Viète's first identity, which makes the numerators equal. It also makes the denominators equal; just write it with $-\beta$ for β.

Here is an improbable one generated from deriving equation (3.3.2) trigonometrically. Show that

$$\frac{\sin\beta}{\sin 2\beta}\frac{\sin(\alpha+2\beta)}{\sin(\alpha+\beta)} + \frac{\sin\alpha}{\sin(\alpha+\beta)}\frac{\sin(2\alpha+\beta)}{\sin 2\alpha}$$

$$+ \frac{\sin\beta}{\cos\alpha}\frac{\cos(\beta-\alpha)}{\sin 2\beta} = 2. \qquad (4.3.6)$$

Let $Q = R + S$, where

$$R = \frac{1}{\sin(\alpha+\beta)}\left(\frac{\sin(\alpha+2\beta)}{2\cos\beta} + \frac{\sin(2\alpha+\beta)}{2\cos\alpha}\right)$$

and

$$S = \frac{\cos(\beta+\alpha)}{2\cos\alpha\cos\beta}.$$

(Note the use of $\sin 2x = 2\sin x\cos x$.) Place R over a common denominator,

$$R = \frac{\cos\alpha\sin(\alpha+2\beta) + \cos\beta\sin(2\alpha+\beta)}{2\sin(\alpha+\beta)\cos\alpha\cos\beta} = \frac{T}{2\sin(\alpha+\beta)\cos\alpha\cos\beta},$$

$$T = \cos\alpha(\sin(\alpha+\beta)\cos\beta + \cos(\alpha+\beta)\sin\beta)$$

$$+ \cos\beta(\sin(\alpha+\beta)\cos\alpha + \cos(\alpha+\beta)\sin\alpha)$$

$$= \sin(\alpha+\beta)(\cos(\alpha+\beta) + 2\cos\alpha\cos\beta).$$

With this last expression for T,

$$R + S = \frac{3\cos\alpha\cos\beta - \sin\alpha\sin\beta + \cos\beta\cos\alpha + \sin\beta\sin\alpha}{2\cos\alpha\cos\beta} = 2.$$

Therefore, $Q = R + S = 2$, Q.E.D.

APS 4.3.2. A generalized law of sines holds among the lines and angles of a quadrilateral. For a triangle, $a \sin B = b \sin A$. For a quadrilateral, $ab \sin A + cd \sin C = bc \sin B + ad \sin D$. Note that each product in the equation involves an angle and the sides that include it and that the equation relates the sum of these products for opposite angles.

To demonstrate the equality, make some triangles as in Fig. 4.3.13 and apply the law of sines. You have

$$\frac{\sin E}{d} = \frac{\sin C}{x} = \frac{\sin D}{y}, \quad \text{and} \quad \frac{\sin E}{b} = \frac{\sin B}{a + x} = \frac{\sin A}{c + y}.$$

If you perform the indicated algebra, you will find that

$$x = \frac{ad \sin C}{b \sin B - d \sin C}, \qquad y = \frac{cd \sin D}{b \sin A - d \sin D};$$

which, when substituted back, gives

$$ab \sin A + cd \sin C = bc \sin B + ad \sin D, \qquad (4.3.7)$$

Q.E.D.

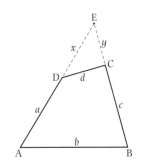

Fig. 4.3.13 A more difficult exercise on sines (APS 4.3.2).

Semi-practical surveying problems

Egyptian–Greek type. Back at the old Egyptian survey bureau, a client appears with a problem to stump the experts. His lot is triangular, one side of which is river frontage. He wants to know immediately how long it is; the river is rapidly rising and he needs the information for a claim for losses. Unfortunately you cannot pace out the frontage because a thick hedge runs right down to the water; and you cannot survey by boat because the river is at a dangerous stage. Fortunately you know the law of cosines.

You run to your client's plot before the waters have swept away the boundary markers A and C (Fig. 4.3.14). You cut through the hedge to the markers and raise poles at each visible from the marker B. At B you measure the angle ABC with your quadrant; you find it to be 37.5°. The distances AB and BC are known from previous plots as 1000 l.u. and 300 l.u., respectively. The law of cosines gives

$$AC^2 = (1000)^2 + (300)^2 - 2(300)(1000)\cos 37.5°.$$

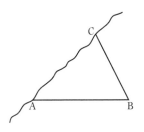

Fig. 4.3.14 Surveying via the law of cosines.

You promise to report the value of AC as soon as you have a chance to check your trig tables at the office. Your importunate client is too worried to let you out of his sight. He wants his answer immediately.

Fortunately, you have a few tricks. All you need is cos 37.5°. You know that sin 30° = 1/2, and you know the formulas for the functions of the sums and differences of angles. You obtain the values of cos 15° and sin 15° in Table 4.3.1 by the process we have already been through. You need the functions for 7.5°. Nothing easier:

$$\sin^2 7.5° = (1 - \cos 15°)/2 = 0.017,$$

$$\sin 7.5° = 0.131, \quad \cos 7.5° = 0.991.$$

The formula for $\cos(\alpha + \beta)$ gives

$$\cos 37.5° = \cos 30° \cos 7.5° - \sin 30° \sin 7.5°$$

$$= (0.866)(0.991) - (0.5)(0.131) = 0.793.$$

You can now calculate,

$$AC^2 = 1\ 000\ 000 + 90\ 000 - (600\ 000)(0.793) = 614\ 200.$$

The table of square roots is back in the office. Your client grows impatient. You reason as follows: 614 200 = (61.42)(10 000); the square root of 10 000 is 100; the square root of 61.42 is something less than 8. Using the trick exploited earlier, you write for x, the unknown square root of 61.42, 8 − y, supposing y to be considerably smaller than 8. Thus

$$(8 - y)^2 = 61.42,$$

or, to first approximation,

$$y = 0.16, x = 7.84.$$

You report AC = 784 l.u.

Your client reviews the calculations. He is not satisfied. He observes that $x^2 = 61.47$, not 61.42. He would be over-credited with river frontage and probably, therefore, also overtaxed, if he accepted the number. He asks you to do better. You reply that your answer is as good as needs be and a closer approximation is unlikely to change the results by as much as 1 l.u. He insists. You make no further objection (you are paid by the hour) and reason as before. Now you set $x = 7.84 - y$ and write

$$(7.84 - y)^2 = 61.42, \quad y = 0.003.$$

As you supposed, the difference between your first and second approximation to AC amounts to less than 1 l.u.—0.3 l.u. to be exact. Your method of approximation again proved very effective.

The tower of Pisa. Galileo is said to have dropped a ball of lead and another of wood from the top of the leaning tower in the university town of Pisa, where he was a professor of mathematics 400 years ago, in order to refute the views of the followers of Aristotle, who

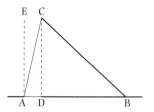

Fig. 4.3.15 Galileo's legendary experiment from the Tower.

claimed that the heavier ball would hit the ground first. Alas! it appears that some of his students invented the story in order to add drama to an otherwise dull problem in physics.[13] But nothing in the true history prevents our making a drama of the geometry of the operation. Fig. 4.3.16 shows the scene of action. Say that the balls landed x feet from the base of the tower. Suppose that at the time of the drop a colleague of Galileo's stood watching at the point B (Fig. 4.3.15); she caught Galileo's eye (at C, say), and, whipping out her quadrant, measured angle ABC to be 65°. Intrigued, Galileo countered with his quadrant, and found ∠ACB to be 29.5°. After this romantic interchange, the observer at B paced out the distance to the base of the tower: it was 97 feet. Galileo and the lady then sat down over a bottle of wine to calculate the height of the tower CD, its length AC, the angle by which it declined from the vertical ∠EAC, and the horizontal distance AD by which it leans over its base.

They knew ∠ACB = 29.5°, ∠ABC = 65°, AB = 97 feet. From the law of sines,

$$\frac{AC}{\sin 65°} = \frac{AB}{\sin 29.5°}, \qquad AC = 178.62 \text{ feet.}$$

Now ∠CAD = st.∠ − ∠ACB − ∠ABC = 85.5°. Therefore the declination from the vertical, ∠CAE, is 4.5°. The height, CD, follows from CD = 178.62 × sin 85.5° = 178 feet. The overhang, AD, equals 178.62 cos 85.5° = 14 feet. At the time of her observation, the lady stood 83 feet from the pile of dropped rocks. Had she measured this distance, instead of her separation from the tower, the preceding calculations could have been accomplished without using the law of sines. It is not always clear what is best to measure.

The Chinese way. The applications to surveying that we have made rested on the theory of similar triangles. That was the Greek way. It was by no means the only way. The ancient Chinese, whose interests in geometry also arose in part from a need to resurvey land after floods, and who, like the Egyptians, raised one of their kings into a geometer, made do solely with linear measurements. The basis of their approach will be clear from Fig. 4.3.17, which represents a theorem about the areas of rectangles. In any rectangle ABCD, pick a point E on a diagonal and pass lines through E parallel to the sides of the rectangle until they cut the sides as at G, H, I, J. Then—this is the theorem—the rectangle under the diagonal, GAHE, equals the rectangle above the diagonal EICJ. We saw a special case of this theorem in the Chinese derivation of the side of a square inscribed in a right triangle (Fig. 4.2.6).

13. Cooper, *Tower of Pisa* (1935), 27–33, 53–5.

Fig. 4.3.16 The Leaning Tower of Pisa. From Carli, *Piazza* (1956), 33.

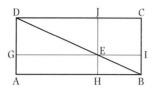

Fig. 4.3.17 Surveying in the Chinese manner.

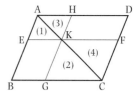

Fig. 4.3.18 A Greek demonstration of the principle behind **Fig.** 4.3.17 (Euclid I.43).

So nice a proposition did not escape Euclid. It appears, generalized, as I.43, applied to parallelograms of which, as we know, a rectangle is only a special case. The Chinese text does not prove the theorem on which its examples rest. Euclid's proof runs as follows. The areas to be proved equal (Fig. 4.3.18) are the parallelograms EBGK and HKFD, which Euclid calls 'the complements about the diameter'. The demonstration uses only addition, subtraction, and the proposition that the diagonal of a parallelogram splits it into two equal areas (I.34). For

$$\text{par. EBGK} = \Delta ABC - \Delta 1 - \Delta 2 = \Delta ACD - \Delta 3 - \Delta 4 = \text{par. HKFD}.$$

The Chinese use of the theorem in practice is illustrated in Fig. 4.3.19, the so-called 'sea-island problem'. It comes from a text composed about 500 years after Euclid. The problem is to measure the height of the mountain on the island (on the left), and its distance from the flat mainland, without crossing the sea. The geometry is laid out in Fig. 4.3.20, where AI is the height of the mountain, and AB the distance across the sea; LB and KD are vertical sighting poles of height a; BC and DE, the known distances at which the observer stands to sight the top of the mountain above the poles; and BD, the known distance between the poles. We want IA and AB in terms of BC, BD, DE, and a.

For ease in bookkeeping, let us write x for IA and y for AB. The general theorem expressed in Fig. 4.3.17—that the complementary rectangles, below and to the left, and above and to the right, of the diagonal are equal—may be applied to rect. AEFI to yield rect. ADKM = rect. KJFG. But rect. ADKM = rect. ABLM + rect. BDKL = $ay + a$BD, where a, the height of the sticks, and BD, the distance between them, are known. Also, rect. KJFG = GK × KJ = IM × DE = $(x - a)$DE, where DE is the measureable distance at which the eye must be placed to see the peak of the island in line with the top of the second stick. The geometrical relation, rect. ADKM = rect. KJFG, therefore yields the algebraic equation

$$ay + a\text{BD} = (x - a)\text{DE}. \qquad (4.3.8)$$

Fig. 4.3.19 The traditional 'sea-island problem.' From Needham, *Science and civilization*, 3 (1970), 572.

Fig. 4.3.20 Geometry of the sea-island problem, Chinese style.

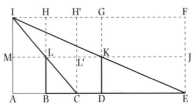

A second equation is necessary to solve for the unknowns x and y. Another touch of the general theorem will bring the second equation. In rect. ACH'I, rect. ABLM (the rectangle below and to the left of the diagonal IC), equals rect. LL'H'H (the rectangle above and to the right). Expressed algebraically,

$$ay = HL \times BC = (x - a)BC, \tag{4.3.9}$$

where BC is the measurable distance at which the eye must be placed to see the peak of the island in line with the top of the first stick. To solve the simultaneous equations (4.3.8) and (4.3.9), substitute $ay = (x - a)BC$ from the second into the first, and obtain

$$(x - a)BC + aBD = (x - a)DE,$$

$$x = \frac{aBD}{DE - BC} + a, \qquad y = \frac{BD \times BC}{DE \times BC}.$$

The Greeks would have solved the sky-island problem by using the pairs of similar triangles IML, LBC and IMK, KDE. From the first pair,

$$\frac{IM}{ML} = \frac{LB}{BC}, \quad \text{or} \quad \frac{x-a}{y} = \frac{a}{BC}.$$

This produces $ay = (x - a)BC$, that is, equation (4.3.9). From the second pair,

$$\frac{IM}{ML + LK} = \frac{KD}{DE}, \quad \text{or} \quad \frac{x-a}{y + BD} = \frac{a}{DE}.$$

This last yields $a(y + BD) = (x - a)DE$, that is, equation (4.3.8). Although the Chinese and the Greek methods differ fundamentally in principle, they give the same results in numbers. Neither is better for the sea-island problem, although the Greek method of proportions had the advantage of leading to trigonometry, which is superior for calculation to similar triangles and to rectangles.

The trigonometer would use Fig. 4.3.21 and write $\tan \alpha = (x - a)/y$, $\tan \beta = (x - a)/(y + BD)$. Putting $(x - a)$ from the first equation into the second, he or she would have

$$y = \frac{BD \tan \beta}{\tan \alpha - \tan \beta}. \tag{4.3.10}$$

To check that this answer agrees with the earlier, express the tangents as ratios of measurable lines:

$$\tan \alpha = a/BC, \quad \tan \beta = a/DE,$$

which, when substituted into equation (4.3.10), gives

$$y = \frac{aBD/DE}{(a/BC) - (a/DE)} = \frac{BD \times BC}{DE - BC}.$$

As appears from equation (4.3.10) the trigonometer would measure the angles α and β and the distance between the sticks BD, and not bother with the shadow distances BC and DE. The Chinese and the Greek surveyors ignored the angles and took the shadow distances. The Chinese relied on relationships derived from the equality of areas of rectangles, which rested on the principle that the

Fig. 4.3.21 The sea-island problem, Greek style.

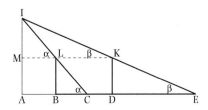

Table 4.3.2 Euclidean propositions in Chapter 4

I.14	If two angles joined at their vertices add to a straight angle, their bases make a straight line
I.34	The opposite sides of a rectangle are equal, and its diagonals bisect it
I.43	The complements of a parallelogram about its diameter are equal
I.47	The Pythagorean theorem
I.48	Converse of the Pythagorean theorem
VI.1	Triangles and rectangles under the same height are to one another as their bases
VI.2	A straight line drawn parallel to one side of a triangle will cut its sides proportionally, and conversely
VI.4	The sides opposite the equal angles of similar triangles are proportional
VI.5	Converse of VI.4
VI.6	Triangles with one angle equal and corresponding sides of the angle proportional are similar

whole equals the sum of its parts. The Greeks used relationships deriving from the theory of similar triangles, which rested on the elaborate principles set forth in the previous chapter.

The Euclidean propositions introduced in this chapter are enumerated in Table 4.3.2.

4.4. Exercises

4.4.1. A fish loiters at the NE corner of a rectangular body of water. A heron standing at the NW corner spies the fish. When the fish sees the heron looking at him, he swims toward the south. When he reaches the south shore he has the unwelcome surprise of meeting the heron, which has calmly walked due south along the shore, turned at the SW corner of the water, and proceeded due east, to arrive simultaneously with the fish on the south shore. Given that the water measures 12 units by 6 units, and that the heron walks as quickly as the fish swims, find the distance the fish swam. [Bag, *Mathematics* (1979), 158–9; Jha, *Aryabhata* (1988), 210–11.]

[Let the fish start at C and swim along CE (Fig. 4.4.1); the heron begins at D and walks the path DA + AE. Since the two actors move at the same speed, CE = DA + AE. A solution is easily obtained from the theorem of Pythagoras. Set AB = CD = a = 12, BC = AD = b = 6, AE = x, CE = y. We have the pair of equations $y^2 = b^2 + (a - x)^2$ and $y = b + x$. Hence

$$b^2 + 2bx + x^2 = b^2 + a^2 - 2ax + x^2,$$

or $x = a^2/2(a + b) = 4$. Hence y, the fish's path, is 10 units.

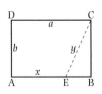

Fig. 4.4.1 Exercise 4.4.1.

This problem comes from the Hindu tradition (Aryabhata and Bhaskara). Their solution invokes not Pythagoras but a theorem that depends on the properties of a circle. We shall do the problem again, Indian-style, in Chapter 5.]

4.4.2. Show that in Fig. 4.4.2, $EF^2 + GH^2 + DI^2 = 6AC^2$
[$EF^2 = EY^2 + (FA + AY)^2$, $DI^2 = DX^2 + (XC + CI)^2$, $GH^2 = BG^2 + BH^2$. But $FA = BG = AB = c$, $CI = BH = BC = a$; furthermore, from the congruent triangles AYE, CXD, ABC (a.s.a.), $AY = DX = c$, $EY = CX = a$. On substitution, $EF^2 + GH^2 + DI^2 = a^2 + (2c)^2 + (2a)^2 + c^2 + a^2 = 6(a^2 + c^2) = 6AC^2$.]

4.4.3. An oak grows in an open field. From it, the distances to the three nearest corners of a square garden are 78, 59.16, and 78 yards. Find the length of the side of the square.
[In Fig. 4.4.3, $FD = FB = a = 78$ yds, $FC = b = 59.16$ yds, F the oak, and ABCD the garden. Since $FD = FB$, $\triangle FDB$ is isosceles, and FE is the perpendicular bisector of the diagonal of the garden. Let the half diagonal be y. Then, from Euclid I.47,

$$a^2 = (b + y)^2 + y^2, \text{ or } 2y^2 + 2by + (b^2 - a^2) = 0,$$

and

$$y = \left(-b + \sqrt{2a^2 - b^2}\right)/2$$

from which we find the side of the square, $y\sqrt{2}$, about 24 yds.]

4.4.4. A hawk sits on a pole 18 cubits high. A rat has his hole at the foot of the pole. The rat is 81 cubits from his hole when they see one another. The rat runs for his home, the hawk sweeps down and catches him. How far does the capture occur from the base of the

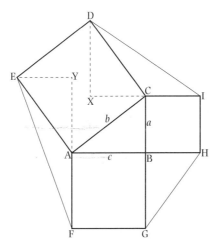

Fig. 4.4.2 Exercise 4.4.2.

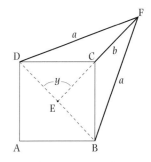

Fig. 4.4.3 Exercise 4.4.3.

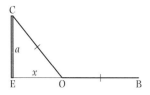

Fig. 4.4.4 Exercise 4.4.4.

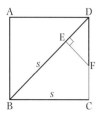

Fig. 4.4.5 Exercise 4.4.5.

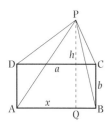

Fig. 4.4.6 Exercise 4.4.6.

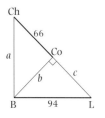

Fig. 4.4.7 Exercise 4.4.7.

pole on the assumption that the hawk and the rat both move in straight lines and that (what is scarcely credible) they move at the same speed. [Jha, *Aryabhata* (1988), 208–9.]

[The improbable assumption allows us to set CO = OB in Fig. 4.4.4; here CE is the pole, C and B the original positions of hawk and rat, respectively. Let EO = x, CE = a, EB = b; then CO = OB = $b - x$. We have from $\triangle$CEO, $(b - x)^2 = a^2 + x^2$, whence $x = (b^2 - a^2)/26 = 77/2$ cubits. The Hindu mathematician Bhāskara, from whom this problem comes, uses the Indian method based on circles, rather than the Pythagorean theorem, as explained in Chapter 5.]

4.4.5. In Fig. 4.4.5, ABCD is a square of side s, BE cut off along a diagonal also equals s, and $\angle$DEF = rt.$\angle$. Show that EF = FC.

[FC = s − DF = $s - \sqrt{2}$ED = $s - \sqrt{2}(\sqrt{2}s - s) = s(\sqrt{2} - 1) =$ ED = EF, Q.E.D.]

4.4.6. On rectangle ABCD (Fig. 4.4.6), draw any triangle DPC. Show that PA^2 + PC^2 = PD^2 + PB^2.

[Let h be the altitude of the triangle, a and b the rectangle's sides, PQ the perpendicular from P on AB, x = AQ. Then PA^2 = x^2 + $(b + h)^2$, PC^2 = h^2 + $(a - x)^2$, PB^2 = $(a - x)^2$ + $(h + b)^2$, PD^2 = x^2 + h^2, whence, with a little algebra, we have the relationship sought.]

4.4.7. A man hired a horse in London at 3 pence a mile. He rode 94 miles due West to Bristol then due north to Chester, whence he returned toward London for 66 miles, which put him in Coventry. From Coventry he made a round trip to Bristol and then continued on to London. How much did he owe for the hire of the horse if the road from Coventry to Bristol meets the road from Chester to London at right angles? [*Ladies diary* (1712), in Leybourn, *1*, 25. No doubt the puzzle was concocted to pass the time on the London–Chester road, which, according to a contemporary guide book, 'had the hard misfortune of being continually execrated for its dulness.' Pennant, *Journey* (1782), 'Advt.']

[In Fig. 4.4.7, L is London, B Bristol, Ch Chester, Co Coventry; the trip amounted to the sum of the perimeters of $\triangle$ChBCo and $\triangle$LBCo. We must therefore find the missing sides of the triangles, a, b, c. From $\triangle$ChBL, $(66 + c)^2 = 94^2 + a^2$; from the pair of triangles ChBCo and LBCo, $b^2 = a^2 - 66^2 = 94^2 - c^2$; from these two equations, a can eliminated and c = 66.62 found. The values b = 66.31 and a = 93.56 follow immediately. The total journey equaled 452.81 miles, which, at 3 pence per mile, amounted to 5ℓ. 13s. 2^1/$_4$d. (Until 1971, English money had 12 pence (12d) to the shilling (s) and 20s to the pound (ℓ).)

4.4.8. Given that the side of a rhomboidal garden (remember, a rhombus is a parallelogram with all sides equal) is 8.75 chains and

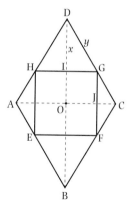

Fig. 4.4.8 Exercise 4.4.8.

the side of the square inscribed in it is 6 chains, determine the area of the garden. [*Ladies diary* (1738), in Leybourne, *1*, 266.]

[In Fig. 4.4.8, ABCD is the rhombus of side a, EFGH the square of side $2c$. Let I be the midpoint of HG; set ID $= x$, DG $= y$. Then from ΔDIG and the similar triangles GJC, DOC,

$$y^2 = x^2 + c^2,$$
$$c : (a - y) = (x + c) : a,$$

resulting in

$$(x^2 + c^2)(x + c)^2 = a^2 x^2. \tag{4.4.1}$$

This vexatious equation can be solved more easily than appears. The left side expanded and rearranged gives

$$(x^2 + c^2)(x + c)^2 = (x^2 + c^2)^2 + 2cx(x^2 + c^2) = [(x^2 + c^2) + cx]^2 - c^2 x^2.$$

Substituting in equation (4.4.1), we have

$$x^2 + c^2 + cx = x(a^2 + c^2)^{1/2}.$$

We now have x as the solution to a quadratic, which will give two values, one for DI and the other for CJ. Since Γ = area of ABCD = 4ΔCOB, we have

$$\Gamma = 4(\tfrac{1}{2})(x_1 + c)(x_2 + c),$$

where x_1, x_2 are the solutions to

$$x^2 + x(c - \sqrt{a^2 + c^2}) + c^2 = 0.$$

With $a = 8.75$, $2c = 6$, $x_1 = 4$, $x_2 = 2.25$, and $\Gamma = 73.5$ chains^2.]

4.4.9. Viète gives the following trigonometric identity in connection with his demonstration that a knowledge of the sines and cosines of angles up to 30° allows computation of all the rest:

$$(2\sin(\alpha - \beta)/2)^2 = (\sin \alpha - \sin \beta)^2 + (\cos \alpha - \cos \beta)^2.$$

Prove it.

[From the formula for the cosine of the sum of any two angles x you have $\cos 2x = \cos^2 x - \sin^2 x = 1 - 2\sin^2 x$. Hence the left side of the proposed identity equals $2(1 - \cos(\alpha - \beta))$, which is the same as the right side, as you will see when you expand them both.]

4.4.10. Show that the sum of the squares of the diagonals e, f plus the square of twice the distance between their midpoints g equals the sum of the squares of the sides of the trapezoid, i.e.

$$e^2 + f^2 + (2g)^2 = a^2 + b^2 + c^2 + d^2.$$

Fig. 4.4.9 Exercise 4.4.10.

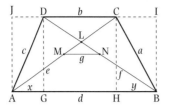

[Repeated application of the law of cosines to Fig. 4.4.9 gives

$$e^2 = a^2 + d^2 - 2ad \cos B = b^2 + c^2 - 2bc \cos D,$$
$$f^2 = c^2 + d^2 - 2cd \cos A = a^2 + b^2 - 2ab \cos C.$$

Add all four equations together:

$$2(e^2 + f^2) = 2(a^2 + b^2 + c^2 + d^2)$$
$$- 2(cd \cos A + ad \cos B + ab \cos C + bc \cos D).$$

To prove the proposition, we must show that

$$(2g)^2 = cd \cos A + ad \cos B + ab \cos C + bc \cos D.$$

Drop the perpendiculars DG, CH onto AB; extend DC to intersect at I the perpendicular raised at B; and extend CD to meet the perpendicular raised at A at J. Set AG = x, CI = HB = y. We have therefore

$$\cos A = x/c, \quad \cos B = y/a, \quad \cos C = -\cos\angle ICB = -y/a,$$
$$\cos D = -\cos\angle JDA = -x/c.$$

We may now write

$$cd \cos A + ad \cos B + ab \cos C + bc \cos D$$
$$= x(d - b) + y(d - b)$$
$$= (x + y)(d - b) = (d - b)^2 = (2g)^2,$$

Q.E.D. (We have directly from Fig. 4.4.9 that $x + y = d - b$; and, from Exercise 3.5.14, that $g = (d - b)/2$.

4.4.11. A lazy surveyor found the height of a tower without measuring any angles as follows. Some distance from the tower he sighted its top at an angle he called α. He then drew the complement of α on a piece of paper and walked forward until he sighted the tower at rt.$\angle$ – α. Next he drew an angle twice α and marched ahead on the same straight line until he saw the tower's top at 2α. If the distance between the first and second stations was a and that between the second and third stations b, show that y, the height of the tower, is

$$y^2 = (b + 3a/2)/(b + a/2).$$

[*Ladies diary* (1761), in Leybourn, *3*, 196.]

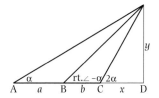

Fig. 4.4.10 Exercise 4.4.11.

[Let the stations of the lazy surveyor be A, B, C, and the bottom of the tower D (Fig. 4.4.10); put x for the unknown distance CD. We have

$$\tan \alpha = y/(a + b + x), \quad \text{ctn } \alpha = \tan (\text{rt.}\angle - \alpha) = y/(b + x),$$
$$\tan 2\alpha = y/x.$$

Since ctn = 1/tan, the first two equations multiplied together give $y^2 = (a + b + x)(b + x)$. It remains to show that $x = a/2$. By writing $\tan 2\alpha = \sin 2\alpha/\cos 2\alpha$ and expanding, you obtain $\tan 2\alpha = 2 \tan \alpha/(1 - \tan^2 \alpha)$. But $\tan^2 \alpha = \tan \alpha/\text{ctn } \alpha = (b + x)/(a + b + x)$, and, therefore, $1 - \tan^2 \alpha = a/(a + b + x)$. Then

$$\tan 2\alpha = y/x = \frac{2 \tan \alpha}{1 - \tan^2 \alpha} = \frac{2y}{a + b + x} \times \frac{a + b + x}{a},$$

from which $x = a/2$, Q.E.D.]

4.4.12. Prove that if the sums of the squares of opposite sides of any quadrilateral are equal, its diagonals intersect at right angles. [*Ladies diary* (1757), in Leybourn, *3*, 141–2.]

[In Fig. 4.4.11, let α be $\angle$DOC and write the law of cosines for all four of the interior triangles into which the diagonals divide the quadrilaterals:

$$a^2 = f^2 + h^2 + 2fh \cos \alpha, \quad b^2 = e^2 + h^2 - 2eh \cos \alpha,$$
$$c^2 = e^2 + g^2 + 2eg \cos \alpha, \quad d^2 = f^2 + g^2 - 2fg \cos \alpha.$$

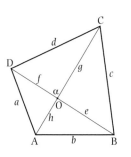

Fig. 4.4.11 Exercise 4.4.12.

We are given $a^2 + c^2 = b^2 + d^2$. In that case, the foregoing equations give

$$\cos \alpha(eh + gf + hf + eg) = 0.$$

Since the expression in parentheses cannot be zero, $\cos \alpha = 0$ and $\alpha = \text{rt.}\angle$, Q.E.D.]

4.4.13. Let the quadrilateral in the preceding problem have its diagonals equal. Show that its area Γ is

$$\Gamma = (\text{diagonal})^2 / 2 = \frac{1}{4} [a^2 + c^2 + \sqrt{2(b^2d^2 + a^2c^2)}].$$

[We have

$$a^2 = f^2 + h^2, \quad b^2 = e^2 + h^2, \quad c^2 = e^2 + g^2, \quad d^2 = f^2 + g^2,$$

and the equal diagonals $p = e + f = g + h$. These equations, combined in pairs, yield

$$2pf = a^2 - b^2 + p^2,$$
$$2ph = b^2 - c^2 + p^2.$$

Inserting the values of f and h from these expressions into $a^2 = f^2 + h^2$, one gets

$$4a^2p^2 = (p^2 + a^2 - b^2)^2 + (p^2 + b^2 - c^2)^2,$$

which, when solved for $p^2/2$, gives the stated form for Γ, Q.E.D.]

4.4.14. In Fig. 4.4.12, CD = 16 l.u. bisects $\angle$ACB, which is 60°, and divides AB such that AD : DB = 4 : 5. Find the area of the triangle. The problem yields to brute force and also, if you think before you calculate, to an easier, gentler method. [Leybourn, *3*, 343.]

[Via brute force: the law of sines applied to $\triangle$ADC and $\triangle$DBC gives $\sin \beta = (4/5)\sin \alpha$, $\sin(\alpha + \beta) = \sin(\text{st.}\angle - \angle ACB) = \sin 60° = \mu$. From these equations,

$$\sin^2\alpha = \mu^2 + \lambda^2\sin^2\alpha - 2\lambda\mu \sin \alpha \, (1 - \sin^2\alpha)^{1/2},$$

where $\lambda = 4/5$. Rearranging, squaring, and putting in the numbers produce $2.05 \sin^4 \alpha - 2.46 \sin^2 \alpha + 0.563 = 0$; which, when solved, gives $\alpha = 70.88°, 33.66°$. Call these last solutions α_1, α_2. Repeat the calculation for the case that $\lambda = 5/4$, that is, that α and β are interchanged. This changes the equation to $5 \sin^4 \alpha - 3.844 \sin^2\alpha + 0.563 = 0$ and the solutions to $\alpha_3 = 49.12°$, $\alpha_4 = 26.34°$. This seems to be a cornucopia of results: we seek but one value for α. But notice that $\alpha_1 + \alpha_3 = 120°$, $\alpha_2 + \alpha_4 = 60°$. We must therefore take $\alpha_1 = \alpha$, $\alpha_3 = \beta$ to make the sum of the angles of $\triangle$ABC 180°.

We are asked to find the area = (AB/2)CE = (AB/2)(AC sin α). AB and AC may be obtained by application of the law of sines to $\triangle$ACD and $\triangle$DCB. You will find that Γ = area $\triangle$ABC = 4(csc 70.88° + csc 49.12°) 16 sin 100.88° = 149.64 a.u.]

4.4.15. The same problem can be solved more easily and directly by noticing that AC : BC = AD : DB = 4 : 5. That being assumed, make CF = CD/4 = 4 l.u., and draw FG∥AB (Fig. 4.4.13); since $\triangle$FCG ~ $\triangle$ACB, CG = 5. The law of cosines applied to $\triangle$FCG gives FG = 4.5826; the law of sines, sin α = (5/4.5826)$\sqrt{3}$/2 = 0.94491, α = 70.88°. It remains to prove that AC : BC = AD : DB in any triangle in which CD, the line from the vertex C to the base AB, bisects the angle at the vertex $\angle$ACB.

[The relationship follows from the law of sines. Call each half of the bisected angle, ACD and BCD, γ. Since $\sin\angle$CDA = $\sin\angle$CDB, (AC/AD) sin γ = (BC/BD) sin γ, or AC : BC = AD : DB. This theorem appears as Euclid VI.3, where it is proved as follows. In $\triangle$ABC (Fig. 4.4.14), $\angle$ACB is the bisected angle and BE∥CD. Call $\angle$ACD = $\angle$DCB γ, then (by parallels) $\angle$CBE = γ and (since $\angle$ACB is exterior to $\triangle$BCE) $\angle$CEB = γ, whence BC = CE. Now $\triangle$ACD ~ $\triangle$AEB, so that AB : AD = AE : AC, or (AB − AD) : AD = (AE − AC) : AC. But AB − AD = DB, AE − AC = CE = CB, so that DB : AD = BC : AC. Q.E.D.]

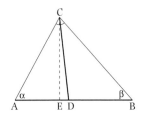

Fig. 4.4.12 Exercise 4.4.14.

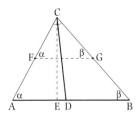

Fig. 4.4.13 Exercise 4.4.15.

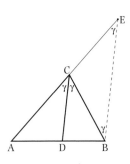

Fig. 4.4.14 Further to Exercise 4.4.15.

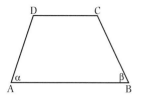

Fig. 4.4.15 Exercise 4.4.16.

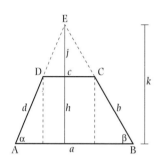

Fig. 4.4.16 Further to Exercise 4.4.16.

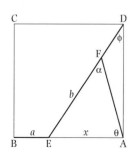

Fig. 4.4.17 Exercise 4.4.17.

4.4.16. A gentleman wishing to know how much it would cost him to fence a trapezoidal field at 6 pence a pole and unable to manage the calculation himself, applied to the *Ladies diary*. He gave the following information: length of base AB (Fig. 4.4.15) = a = 1432 links; $\angle DAB = \alpha = 34° 17'$; $\angle CBA = \beta = 54° 18'$; area of the field, Γ, = 2.75 acres. Can you solve his problem?

[Figure 4.4.16 completes the parent triangle by extending AD and BC to meet at E; since $\angle AEB = $ st.$\angle - \alpha - \beta$ is known, we have k by the law of sines: $k : a = \sin \alpha \sin \beta : \sin(\alpha + \beta)$. We wish to know h, since we can easily express the sides b, c, d, in terms of it: $d = h/\sin \alpha$, $b = h/\sin \beta$, $c = (2\Gamma/h) - a$. (This last comes from Ahmose's formula, $\Gamma = (a + c)h/2$.) To get h, first find $\Gamma' = $ area $\Delta DCE = ka/2 - \Gamma$; then, since the areas of similar triangles are as the squares of their altitudes, $j^2/k^2 = 2\Gamma'/ka$, $j = (2k\Gamma'/a)^{1/2}$, and, finally, $h = k - j$. To calculate, we must reduce all measurements to the same unit. You must know that 100 links = 4 poles (which we will abbreviate as 'p') and that 1 acre = $160p^2$; hence $a = 57.28p$ and $\Gamma = 440p^2$. Putting the numbers in your calculator, you find

$$k = \tan \alpha \tan \beta \tan \alpha/(\tan \beta + \tan \beta) = 26.39p,$$

$\Gamma' = 315.81p^2$, $j = 17.05p$, $h = 9.33p$; $b = 11.49p$, $c = 37.04p$, $d = 16.44p$; and the perimeter, ABCD, = 122.25p. At 6 pence a pole, this amounted to $3\ell.1s.3/2d.$, a not inconsiderable sum at a time when a professor at Oxford could live (though not well) on 100ℓ a year. (For English money, see problem 4.4.7.)

4.4.17. The following is offered as a demonstration, should any be required, that geometrical problems easy to state and grasp, and not difficult to set up, may nonetheless eventuate in very disagreeable equations; and also as a guide to action when that happens. The problem comes from the *Ladies diary* for 1797. 'Being employed to survey a field, which I was told was an exact geometrical square, but by reason of a river which runs through a part of it, I could only measure from the west corner of the south side nine chains, or BE [Fig. 4.4.17], and EF eighteen chains in a line from E towards D, and there took an angle EFA of 28°30', from hence the area of the field is required.'

Show that if x = EA and the known quantities are designated a = BE, b = EF, $\alpha = \angle EFA$, x satisfies the equation

$$x \sin \alpha [(2x^2 + 2ax + a^2)^{1/2} - b] = (x + a)(x^2 - b^2 \sin^2\alpha)^{1/2};$$

or, if $\phi = \angle EDA$, $x = a \tan \phi/(1 - \tan \phi)$, where ϕ satisfies the equation

$$\sin \phi (\text{ctn } \alpha \cos \phi + \sin \phi) = (b/a)(\cos \phi - \sin \phi).$$

[The law of sines applied to ΔEFA gives $\sin \alpha : x = \sin \theta : b$; and, applied to ΔAFD, $\sin \alpha : AD = \sin(\text{rt.}\angle - \theta) : FD$. But $AD = x + a$ and

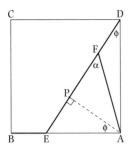

Fig. 4.4.18 Further to Exercise 4.4.17.

$FD = ED - b = [x^2 + (a + x)^2]^{1/2} - b$. Make the indicated substitutions for ϕ and θ; you will arrive at the first of the equations required.

The equation in ϕ may be derived via Fig. 4.4.18 where AP⊥ED; note that $\angle PAE = \angle FDA = \phi$ (both are complements of $\angle DEA$). Now $\sec \phi = EA/AP$, $\csc \phi = DA/AP$, so that $DA - EA = BA - EA = AP (\csc \phi - \sec \phi) = a$. Also, $\tan \phi = PE/PA$, $\text{ctn } \alpha = PF/PA$, so that $PE + PF = EF = b = PA(\tan \phi + \text{ctn } \alpha)$. Consequently, $(\tan \phi + \text{ctn } \alpha) : (\csc \phi - \sec \phi) = b/a$, which, with a little rearrangement, is the second of the equations required.]

4.4.18. Find a numerical solution to either of the equations demonstrated in the previous problem, using the values stated: $a = 9$, $b = 18$, $\alpha = 28°30'$. The best method is cut and try.

[The equation in ϕ requires that $\sin \phi(\sin \phi + 1.84 \cos \phi) = 2(\cos \phi - \sin \phi)$. We know that $\phi < 45°$, since the right side must be positive, and the cosine exceeds the sine for angles under $45°$. A good guess is $\phi = 30°$, since it gives easy numbers: $\sin 30° = 1/2$, $\cos 30° = 0.866$. The left side then equals 1.047; the right side, 0.732. Evidently, $\phi < 30°$. Try $\phi = 25°$: the left hand is 0.705, the right hand 0.966. We have overshot the mark: ϕ lies between $25°$ and $30°$, probably closer to $25°$. For $26°$, we have 0.916 as against 0.922; ϕ is just larger than $26°$, $26°2'15''$ to be over-exact. But we want to know not ϕ, but s, the side of the square ABCD. Since $s = a + x = a + s \tan \phi$, $s = a/(1 - \tan \phi) = 17.60$ chains, for an area just under 31 acres.]

4.4.19. Within a rectangular garden PQRS (Fig. 4.4.19), of length 4 chains and breadth 3 chains, lies a quadrilateral pond ABCD so arranged that the corners of the pond lie on the diagonals of the rectangle. We want to know the area of the pond, with the usual difficulty, that is, without measuring any side or angle of it. Instead, we measure the distances of the corners of the pond from those of the rectangle, which turn out to be 20, 25, 40, and 45 yards. Show that the area of the pond is, in general, $(xy \sin \alpha)/2$, where x and y are the diagonals of the quadrilateral and $\sin \alpha = \sin 2\beta$, where β is any of the

Fig. 4.4.19 Exercise 4.4.19.

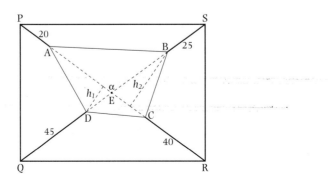

angles made by the diagonals of the rectangle with its sides. And show without using trig tables that in the special case under consideration, which comes from the *Ladies diary* of 1795, the area of the pond is 960 square yards. You will need to know that 1 chain = 22 yards.

[Let $\angle AEB = \alpha$, and drop the perpendiculars h_1, h_2 from D and B on the diagonal AC. Then Γ_1 = area $\triangle ADC = h_1 AC/2$ and Γ_2 = area $\triangle ABC = h_2 AC/2$; the area of the pond, $\Gamma = \Gamma_1 + \Gamma_2 = AC(h_1 + h_2)/2$. But $h_1 = DE \sin (\text{st.}\angle - \alpha)$ and $h_2 = EB \sin (\text{st.}\angle - \alpha)$, so $h_1 + h_2 = DB \sin \alpha$ and $\Gamma = (AC \times DB \sin \alpha)/2$. But $\alpha = \text{st.}\angle - 2\beta$, so $\Gamma = (AC \times DB \sin 2\beta)/2$. In our numerical example, $\triangle PSR$ is a (3, 4, 5) right triangle; hence PR = QS = 5 chains = 110 yd, and AC = 50 yd, DB = 40 yd. Also, $\sin \beta = 0.6$, $\cos \beta = 0.8$, and $\sin 2\beta = 2 \sin \beta \cos \beta = 0.96$. We have, finally, $\Gamma = (2000)(0.96)/2 = 960$ yd^2.]

4.4.20. A client asks you to determine the area of his quadrilateral field. He has made the task difficult by prohibiting you from taking measurements from anywhere except outside the field around two corners (A and D in Fig. 4.4.20), where there is a vacant lot; he and his neighbors grow tulips and don't want you and your gear crushing their flowers. He gives you an important piece of information: the corner C, opposite A, marked by a tall tree visible from A, is a right angle. The side CD is the longest of the four. Fortunately, you just (it is the year 1794) had read about a similar problem in the *Ladies diary* and you knew what to do: from A you determined the angular distance α of C from the side AB; you then walked a distance a along the line AB extended into the vacant lot until you hit the line CD (which you had had the foresight to extend) at E, and you measured the distance ED, b say. Show that the area, Γ, is

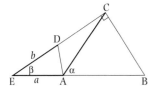

Fig. 4.4.20 Exercise 4.4.20.

$$\Gamma = \frac{a \sin \beta}{2}\left[\frac{a \sin^2 \alpha}{\cos \beta \sin^2(\alpha - \beta)} - b\right].$$

[$\Gamma = \triangle EBC - \triangle EAD = (\frac{1}{2})(EB \times EC \sin \beta - EA \times ED \sin \beta)$. The law of sines applied to $\triangle EAC$ gives $EC = a \sin \alpha/\sin(\alpha - \beta)$; applied to $\triangle EBC$, $EB = EC/\cos \beta$. The expression for Γ follows directly.]

4.4.21. The puzzler in the *Ladies diary* gave $\alpha = 42°28'$, $\beta = 30°16'$, $a = 520$ links, $b = 730$ links. (Recall that there are 100 links in the standard surveyor's chain four poles in length.) Find the acreage, remembering that 1 acre = 160 square poles.

[The tulip field occupied just over 7 acres.]

4.4.22. Do Exercise 3.5.24 in the Chinese manner, without using similar triangles.

[Complete the rectangles as indicated in Fig. 4.4.21; then rect. BLMT = rect. STVW, and rect. BLMT = rect. QRTU. But BLMT = $12y$, STVW = $O_2B \times C_2M = x(6 - y)$, and QRTU = RT $\times$ UT = $O_1B \times C_1M$

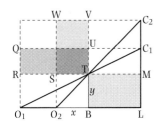

Fig. 4.4.21 Exercise 4.4.22.

Fig. 4.4.22 Exercise 4.4.23.
From Libbrecht, *Chinese
mathematics* (1973), 145–6.

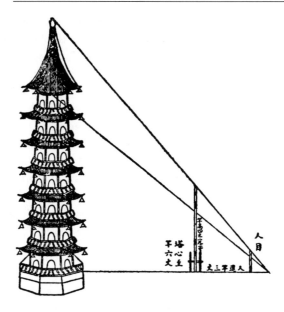

= $(x + 5)(3.5 - y)$. Therefore, from the equality of the rectangles STVW and QRTU, $x = 7 - 2y$; and this equation, substituted into rect. BLMT = rect. STVW, gives the quadratic $2y^2 - 31y + 42$, the solution of which is $y = 1.5$.]

4.4.23. The responsible authorities at last decided to replace the central pillar in a leaning pagoda (Fig. 4.4.22). To proceed, they needed to know its length. A geometer among them noticed a pole for hanging banners 60 feet from the base of the pagoda. A series of 14 rings, each a half foot in diameter, ran up the pole; the distance between the lowest points of nearest neighbors was 2.5 feet, and the lowest point of the lowest of them was 9.2 feet above the ground. An observer, whose eye was 4.8 feet above the ground, stood 30 feet behind the pole and sighted the roof of the pagoda in line with the upper end of the seventh hook. The central pillar extended above the roof and ended in a pinnacle. Find the height of the pagoda. [Libbrecht, *Chinese mathematics* (1973), 145–6.]

[The givens are displayed in Fig. 4.4.23: AD is the central pillar ending in the pinnacle A; B is the point on the roof sighted over the banner pole JE by the observer whose eye is at G. The height of the pole JE = $9.2 + 13(2.5) + 0.5 = 42.2$ feet; the middle of the seventh ring I is $6(2.5) + 0.5 = 15.5$ feet from the bottom of the lowest ring or 17.5 feet below the top of the pole; therefore HI = HJ – JI = 42.2 – 4.8 – 17.5 = 19.9 feet. Similar triangles give

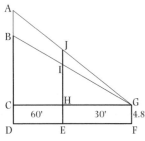

Fig. 4.4.23 Further to Exercise
4.4.23.

$$AC : HJ = CG : HG, \quad BC : CG = IH : HG,$$

or AC = $(90)(37.4)/30 = 112.2$ feet, BC = $(19.9)(90)/30 = 59.7$. We have for the height of the pagoda AD = 117 feet, for the roof, BD = 64.5 feet, and for the imposing pinnacle, 52.5 feet.]

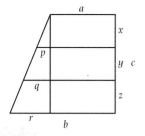

Fig. 4.4.24 Exercise 4.4.24.

4.4.24. Three brothers receive a legacy of a field in the form of a rudder (Fig. 4.4.24). They are to share equally. They decide that each should have a trapezoidal strip and run into trouble figuring out how to implement their agreement. Such problems were popular in Chinese and Japanese geometry. Can you determine how to divide the property given the bases of the rudder, a and b, and its height c? [Libbrecht, *Chinese mathematics* (1973), 103–4.]

[Let the unknown heights of the strips be x, y, z, the bases of the triangles, p, q, r. Then the first brother gets $F = ax + px/2$; the second, $S = ay + y(p + q)/2$; the third, $T = az + z(r + q)/2$. From the similar triangles, you have $p = rx/c$, $q = r(x + y)/c$; also, $r = b - a$, $z = c - x - y$. Putting $k = r/2c$, you have $F = ax + kx^2$, $S = ay + ky(2x + y)$, $T = (c - x - y)(k(x + y) + (a + b)/2)$, and, from the condition of the problem, $F = S = T$. That brings two equations and two unknowns. It is not worth the trouble of doing the algebra.]

4.4.25. Our penultimate problem is one of the best. Take any quadrilateral ABCD and draw squares on its sides (Fig. 4.4.25). Connect the centres of opposite squares by the lines PR, QS. Then PR is perpendicular to QS. Prove it.

[Nothing much can be done with the diagram as it stands. You must make triangles containing PR and QS to use Euclid's machinery. Follow your instincts and connect the centers of the squares, making the quadrilateral PQRS. Then, according to the problem, the diagonals of this quadrilateral are perpendicular.] What condition on the sides of a quadrilateral guarantee that its diagonals meet at right angles? The question can be answered with reference to Fig. 4.4.26. The Pythagorean theorem gives

$$f^2 + h^2 = x^2 + y^2 + w^2 + z^2 = e^2 + g^2,$$

that is, if the diagonals meet at right angles, the sums of the squares of opposite sides are equal. We need the converse of this proposition, which, fortunately, we have already demonstrated (Exercise 4.4.12).

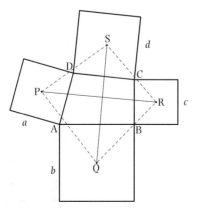

Fig. 4.4.25 Exercise 4.4.25

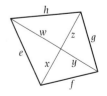

Fig. 4.4.26 Further to Exercise 4.4.25.

Does quad. PQRS have the property $PS^2 + QR^2 = PQ^2 + RS^2$? The law of cosines gives, in the representative case of PQ,

$$PQ^2 = AP^2 + AQ^2 - 2AP \times AQ \cos \angle PAQ$$
$$= a^2/2 + b^2/2 + 2ab \sin \angle A,$$

since AP, AQ, are half diagonals of squares of sides a, b, respectively, $\angle PAQ = \angle PAD + \angle A + \angle QAB = \text{rt.}\angle + \angle A$, and $\cos(\text{rt.}\angle + \angle A) = -\sin\angle A$. By a similar calculation,

$$QR^2 = (b^2 + c^2)/2 + bc \sin B, \quad RS^2 = (c^2 + d^2)/2 + cd \sin C,$$
$$SP^2 = (d^2 + a^2)/2 + ad \sin D.$$

Adding up you have

$$PQ^2 + RS^2 = (a^2 + b^2 + c^2 + d^2)/2 + ab \sin A + cd \sin C,$$

$$QR^2 + SP^2 = (a^2 + b^2 + c^2 + d^2)/2 + bc \sin B + ad \sin D. \qquad (4.4.2)$$

Since $ab \sin A + cd \sin C = bc \sin B + ad \sin D$ (APS 4.3.2), the sums of the squares of the opposite sides of the quadrilateral QPRS are equal, which, as we know, is the condition that the diagonals PR and QS meet at right angles.

4.4.26. We are not done with the last problem. Not only are PR and QS perpendicular, they are equal. The way is not easy. One useful move is to compute both diagonals twice to achieve some symmetry in the equations. With reference to Fig. 4.4.27,

$$PR^2 = (KA + a + BL)^2 + (RL - PK)^2,$$

with
$$DA = (d/2)\cos(135° - A) = (d/2)(\sin A - \cos A),$$
$$BL = (b/2)(\sin B - \cos B),$$
$$RL = (b/2)(\sin B + \cos B),$$
$$PK = (d/2)(\sin A + \cos A).$$

If you labor through the algebra, you will be able to confirm that

$$PR^2 = a^2 + b^2/2 + d^2/2 + ad (\sin A - \cos A) + ab (\sin B - \cos B)$$
$$- bd \sin (A + B).$$

Similarly, from

$$QS^2 = (IB + b + CH)^2 + (QI - SH)^2,$$

you can calculate

$$QS^2 = b^2 + a^2/2 + c^2/2 - ac \sin (B + C) + bc (\sin C - \cos C)$$
$$+ ab (\sin B - \cos B).$$

The calculations took the side AB extended to K and L, and perpendiculars dropped to it from P and R, to obtain PR, and a system based on the side BC to get QS. But we could also have calculated PR

Fig. 4.4.27 Exercise 4.4.26

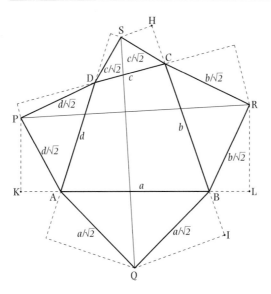

based on CD. The process follows the earlier exactly with c, C, and D in place of a, B, and A. The result:

$$PR^2 = c^2 + b^2/2 + d^2/2 + cd\,(\sin D - \cos D) + bc\,(\sin C - \cos C)$$
$$- bd \sin (C + D).$$

Add the two expressions for PR^2, noting that $\sin(A + B) = -\sin(C + D)$ because $\sin (2st.\angle - x) = \sin (st.\angle + st.\angle - x) = -\sin(st.\angle - x) = -\sin x$:

$$2PR^2 = a^2 + b^2 + c^2 + d^2$$
$$+ ad\,(\sin A - \cos A) + ab\,(\sin B - \cos B)$$
$$+ cd\,(\sin D - \cos D) + bc\,(\sin C - \cos C).$$

That begins to look encouragingly symmetric.

Now calculate QS based on AD; that amounts to replacing b by d, B by C, and A by D:

$$QS^2 = d^2 + a^2/2 + c^2/2 - ac \sin (A + D) + cd\,(\sin D - \cos D)$$
$$+ bc\,(\sin C - \cos C).$$

Add the two expressions for QS together, noting that $\sin (B + C) = -\sin (A + D)$:

$$2QS^2 = a^2 + b^2 + c^2 + d^2$$
$$+ ad\,(\sin A - \cos A) + ab\,(\sin B - \cos B)$$
$$+ cd\,(\sin D - \cos D) + bc\,(\sin C - \cos C).$$

You can see at a glance that $2PR^2 = 2QS^2$, PR = QS. Q.E.D.

4.4.27. The emperor Napoleon I liked to hobnob with mathematicians and sometimes thought he was one of them. The following theorem, if it is truly his, shows that his geometrical imagination

Fig. 4.4.28 Exercise 4.4.27
(Napoleon's theorem).

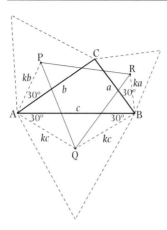

Fig. 4.4.28 Exercise 4.4.27
(Napoleon's theorem).

was not confined to military operations. Here is the theorem: if, on each side of a given triangle, an equilateral triangle is erected, their centers (the points of intersection of their altitudes) are the vertices of an equilateral triangle. In Fig. 4.4.28, ABC is the given triangle; P, Q, and R are the centers of the equilateral triangles constructed on its sides; and ΔPQR is to be proved equilateral.

[Since you are given two sides and the included angle in triangles QBA, QAP, and QBR, it makes sense to bring down the law of cosines on them. That gives

$$PQ^2 = k^2 \, (b^2 + c^2 - 2bc \, \cos(A + 60°)),$$
$$QR^2 = k^2 \, (a^2 + c^2 - 2ac \, \cos(B + 60°)),$$

where $k = 1/\sqrt{3}$, RB = ka, BQ = kc, QA = kc, AP = kb. You need only show that $PQ^2 - QR^2 = 0$. Now $PQ^2 - QR^2 = k^2[\text{I} + \text{II}]$, where I $= b^2 - a^2 = b^2(1 - a^2/b^2) = b^2(1 - \sin^2 A/\sin^2 B)$, the last step from the law of sines; and II $= -2c[b \cos(A + 60°) - a \cos(B + 60°)] = (-2b^2 \sin C/\sin^2 B)[\sin B \cos(A + 60°) - \sin A \cos(B + 60°)]$, where, again, the law of sines has been employed to eliminate the sides of the original triangle and reduce the problem to proving a trigonometrical identity. Since

$$\text{I} + \text{II} = (b/\sin B)^2[\sin^2 B - \sin^2 A - 2\sin C\{\sin B \cos(A + 60°)$$
$$- \sin A \cos(B + 60°)\}],$$

the expression in square brackets must be identically zero if Napoleon was right. You can confirm that he was by expanding $\sin C = \sin(A + B)$, $\cos(A + 60°)$, and $\cos(B + 60°)$, multiplying a bit, and recalling that $2 \cos 60° = 1$. If you go through the exercise, you will have PQ = QR. Similarly you can show that PQ = PR, and that, therefore, ΔPQR is equilateral, as the Emperor claimed.]

5 From polygons to pi

5.1. Some interesting circles

The world's oldest circle

Some four thousand years ago, on the Salisbury Plain in South West England, an ancient people built a structure now known as Stonehenge. The structure consisted of an internal horse shoe of huge stones surrounded by a circle of slightly smaller stones, 'small' being purely relative, since the smallest of them weigh 70 tons (Figs. 5.1.1 and 5.1.2). How the ancient builders managed to drag these huge boulders from great distances, erect them, and raise the horizontal lintels that connect them has puzzled historians for centuries. A few possibilities are indicated in Fig. 5.1.3.

Outside the 'sarsen circle'—so called because the stones come from a type of sandstone known as sarsen—are several rings marked

Fig. 5.1.1 Aerial view of Stonehenge. From Chippendale, *Stonehenge complete* (1983), 11.

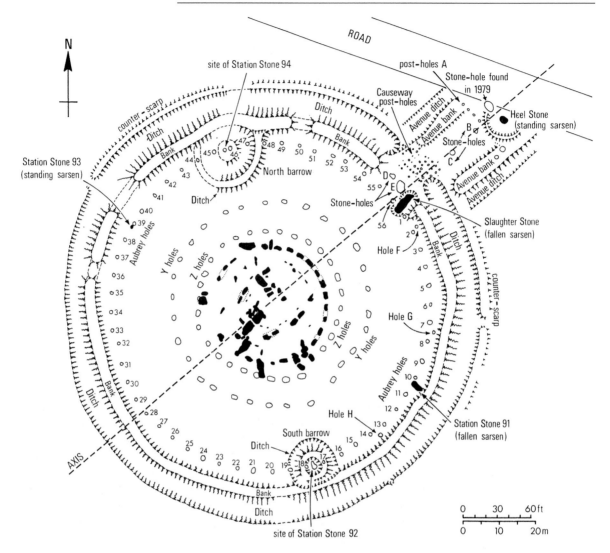

Fig. 5.1.2 Plan of Stonehenge. From Chippendale, *Stonehenge complete* (1983), 9.

out by holes now filled with debris. The outermost of these rings, named after John Aubrey, the British antiquary who first described them, has 56 holes. The number may have astronomical significance. Certainly much about the structure has to do with astronomy. For example, anyone standing at its center at sunrise on the longest day of the year can see the sun spring up from behind an outlying sarsen known as the heel stone. Among the many questions that Stonehenge raises in the mind of the practical geometer is how the builders set out the sarsen and Aubrey circles. A related question that might occur to a visitor who knows little geometry is how to find the unmarked center or centers of the circles as they now exist.

Fig. 5.1.3 A way to lay the stones. From Chippendale, *Stonehenge complete* (1983), 247.

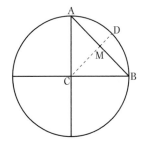

Fig. 5.1.4 To divide an arc with a rope.

The remaining lintels connecting the uprights in the sarsen circle stand 16 feet above ground; their vertical sides were curved so that, in the original structure, their inner faces marked out a circle 97 feet in diameter accurate to within an inch. The builders probably laid out the circle by tying a rope $97/2 = 48.5$ feet long to a post where they wished to place the center of their building; and then, keeping the rope taught, by making a complete circuit of the post. Sticks or stones placed along the walk would lie on the desired circle. The Aubrey circle presented a bigger challenge.

The same technique with a longer rope (the diameter of the circle is 285.feet) would lay out the circle. But how would the correct spacing of the 56 holes be achieved? The builders might have divided the circle into four equal parts or quadrants by two perpendicular diameters and divided each quadrant by connecting the extremes of the diameters (AB in Fig. 5.1.4), pacing out the length of the connecting line, and finding its midpoint M; the rope would then be stretched tight through M to touch the circle at D. The technique, repeated for each quadrant, would divide the circle into eight parts; since the task required placement of 56 holes, it remained to distribute $56/8 = 7$ holes along each of the eight arcs AD, DB, etc. Probably the builders placed these holes by trial and error. They did fairly well. The centers of the holes should be about 16 feet apart measured along the circle. The most serious error in placement is 19

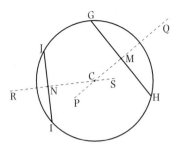

inches, on 1.6 feet, which amounts to 10% of the correct separation. You might try to divide a large circle into 56 equal parts using only a rope.

Since the post presumably set at the monument's center to hold the rope that defined the circles no longer exists, the location of the center must be deduced from the placement of the circles. Figure 5.1.4 suggests a possible method. Lay out any line connecting two points on the circle; such a line, like GH in Fig. 5.1.5, is called a chord. Draw the perpendicular bisector PQ of GH in the manner described earlier (Euclid, I.10); the center of the circle lies on PQ. (Recall that all points on the perpendicular bisector of a line are equidistant from the extremities of the line; the center of a circle must be the same distance from the ends of all chords, the special distance being the radius.) Now draw any other chord IJ and its perpendicular bisector RS; the center of the circle must lie on RS. Therefore the center of the circle, C, must be the intersection of PQ and RS.

Euclid's first proposition about circles is a method of finding their centers. He argues by reductio ad absurdum (Euclid, III.1). Draw any chord AB across the given circle and erect its perpendicular bisector CDE (Fig. 5.1.6). Now bisect the bisector, at F. This point, F, is the circle's center. For suppose that it is not, and that the center lies instead at G. Join G to A and B. Then AG = GB, since both lines are radii: and AD = DB, since D lies on the perpendicular bisector of AB. Therefore, since GD is common, $\triangle$ADG $\cong$ $\triangle$GDB, and the angles GDA and GDB must be equal. But $\angle$GDA+$\angle$GDB = straight angle; since they are equal, each must therefore be a right angle. But again, by construction $\angle$FDA < $\angle$GDA is a right angle. We have therefore two unequal angles each equal to a right angle. But that is impossible. Hence G must coincide with F. Q.E.D.

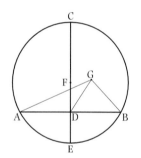

The converse of this proposition is also true: a diameter that bisects a chord cuts it at right angles, and a diameter perpendicular to a chord also bisects it (Euclid III.3). In Fig. 5.1.7, CD is the diameter, E the circle's center. Since $\triangle$EFA $\cong$ $\triangle$EFB by *s.a.s.* if CD bisects AB ($\angle$EAF = $\angle$EBF since $\triangle$AEB is isosceles) or if CD cuts AB at right angles ($\angle$EFA = $\angle$EFB = rt.$\angle$ and $\angle$AEF = $\angle$BEF as complements of

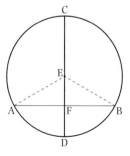

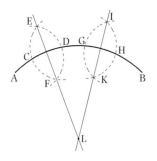

Fig. 5.1.8 The medieval way to the center of a circle.

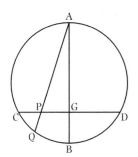

Fig. 5.1.9 An exercise on circles.

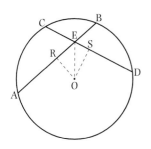

Fig. 5.1.10 Segments of intersecting chords are proportional (Euclid III.35).

the equal angles EAF, EBF), the converted proposition follows immediately from congruence. Of course III.3 holds for the radius ED as well as for the diameter CD.

The medieval masons had a way different from Euclid's for finding the lost center of a circle or circular arc. Let AB be the arc (Fig. 5.1.8). Choose any four points C, D, G, H, on it. With C as center, draw a circle through D; with D as center, draw a circle through C. Let the circles intersect at E, F; draw EF extended. Repeat the operation for G, H; let IK extended meet EF extended at L. L is the center sought.[1]

Let us play. Figure 5.1.9 has a chord CD bisected perpendicularly by diameter AB. Draw any line AQ cutting CD in P. Then (the properties of circles are marvelous, and inexhaustible) AP × AQ has the same value wherever P lies. The easy way to prove it is to use a pretty theorem of Euclid's that concerns pieces of intersecting chords (III.35). In Fig. 5.1.10, AB and CD are the chords; E, their point of intersection; and AE × EB = CE × ED, the pretty theorem. It follows from the Pythagorean theorem. Drop the perpendiculars OR, OS on the chords AB, CD respectively; then CS = SD, AR = RB (Euclid III.3). Call OE s, the radius OC, r. Then

$$SC^2 = r^2 - OS^2 = r^2 - (s^2 - SE^2),$$

$$SC^2 - SE^2 = r^2 - s^2,$$

$$(SC + SE)(SC - SE) = CE \times ED = r^2 - s^2.$$

Likewise,

$$(AR + RE)(AR - RE) = AE \times EB = r^2 - s^2.$$

Since things equal to the same thing are equal to each other, CE × ED = AE × EB, Q.E.D.

Now back to Fig. 5.1.9 and the proof that AP × AQ = constant independent of the position of P. We have

$$AP \times AQ = AP(AP + PQ) = AP^2 + AP \times PQ = AP^2 + CP \times PD$$
$$= AP^2 + CG^2 - PG^2 = AG^2 + CG^2 = AC^2.$$

But AC^2 is independent of the position of P along CD. Q.E.D.

A similar property is illustrated in Fig. 5.1.11, where AO = OB, PAOBQ being a diameter. Show that $AC^2 + BC^2$ is a constant independent of the position of C in the circle's circumference. The obvious move is to use the Pythagorean theorem:

$$AC^2 = h^2 + (2e + x)^2, \quad BC^2 = h^2 + x^2,$$

$$AC^2 + BC^2 = 4e^2 + 2(h^2 + x^2 + 2ex).$$

---

1. Shelby. *Gothic design* (1977), 121–2.

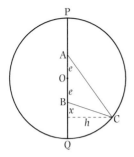

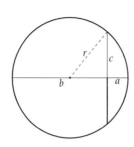

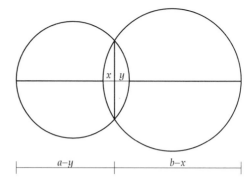

Fig. 5.1.11 A further exercise on circles.

Fig. 5.1.12 Indian bows and arrows (APS 5.1.1).

Fig. 5.1.13 Further to APS 5.1.1.

But $OC^2 = h^2 + (e + x)^2 = h^2 + x^2 + 2ex + e^2$, so $AC^2 + BC^2 = 4e^2 + 2(OC^2 - e^2) = 2(OC^2 + e^2)$, which is a constant independent of C, since OC is the radius of the circle and e is the fixed distance of A and B from the center of the circle.

APS 5.1.1. Remember the Indian mathematicians and their arrows? The product of the arrows a and b equals the square of the half-string of the bow (Fig. 5.1.12). The relationship follows immediately from the Pythagorean theorem, $r^2 = c^2 + (r - a)^2 = c^2 + a^2 - 2ar + r^2$, and $2r = a + b$, whence $c^2 + a^2 = a(a + b)$ or $c^2 = ab$. The Indian mathematicians also defined the 'arrows of the interaction' of two circles. Let the circles have diameters a, b and their intersection have arrows x, y (Fig. 5.1.13). By using the equation for the arrows of a single circle, show that

$$x = \frac{t(a - t)}{a + b - 2t}, \quad y = \frac{t(b - t)}{a + b - 2t},$$

where $t = x + y$ is the overlap of the circles.[2]

One of nature's smallest circles

The rainbow makes a double use of the circle. The bow itself seems to be made up of a series of differently colored arcs with a common center. The common center usually lies beneath the ground, on an extension of the line joining the center of the sun to the head of the observer (Fig. 5.1.14). The concentric colored arcs are the first expression of the circle in the construction of the rainbow. The second is the shape of the raindrops that create the bow by reflecting the sun's rays to the observer's eye. The geometry of the

2. Jha, *Aryabhata* (1988), 201, 204–5.

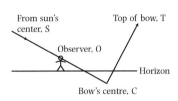

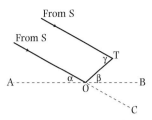

Fig. 5.1.14 Geometry of the rainbow.

Fig. 5.1.15 Further to the geometry of the rainbow.

phenomenon for the raindrop located at the top of the bow T is indicated in Fig. 5.1.15. Here AB is a horizontal line through the observer's eye; $\alpha = \angle SOA$, the sun's altitude (angular height above the horizon); $\beta = \angle TOB$, the angular height of the top of the bow above the horizon; and $\gamma = \angle STO = \angle TOC$, the angle through which the droplet at T deviates the sun's rays to send them to the eye. You can see that

$$\gamma = \alpha + \beta. \qquad (5.1.1)$$

By vertical angles, $\gamma = \angle TOC = \beta + \angle BOC = \alpha + \beta$.

The angle γ always has the same value, about 42°. Hence a rainbow cannot appear if the sun is more than 42° above the horizon. The value $\gamma = 42°$ is a consequence of geometry that can be derived easily with the help of a pocket calculator that has trigonometrical functions. It requires only Snel's law, which, as we know, states that the sine of the angle of incidence of the ray on the droplet is proportional to the sine of the angle of refraction, that $\sin i = n \sin r$.

The itinerary of a sun ray through the droplet at T is indicated in Fig. 5.1.16. C is the droplet's center, P the point at which the ray strikes the droplet. (Of course every point on the surface of the droplet facing the sun is illuminated; we are following one arbitrary ray in order to find out how the angle of deviation, γ, changes with the angle of incidence, i.) We have $\angle CQP = r$, since ΔCQP is isosceles

Fig. 5.1.16 Action of water droplets in the formation of the rainbow.

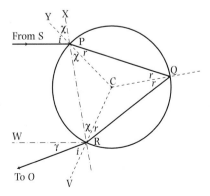

(CP = CQ are radii); also $\angle RQC = \angle CQP = r$, since the ray is reflected at Q and, by a fundamental law of optics, the angle of reflection ($\angle RQC$) equals the angle of incidence ($\angle CQP$); and, finally, $\angle CRQ = \angle RQC = r$, since $\triangle RQC$ is isosceles. Outside the droplet, $\angle ORV = i$ by Snel's law. Draw WR∥SP. Then $\angle WRO = \gamma$, the angle of deviation defined in Fig. 5.1.15.

To obtain the relation between γ and i, draw RP and continue it to X. Call $\angle CPR$ and $\angle CRP$ χ. Since CRV is a straight line, $\gamma = \text{st.}\angle - i - \angle PRW - \chi$. Now, $\angle XPS = i + \chi$ ($\angle XPY = \angle CPR$ by vertical angles); also, $\angle XPS = \angle PRW$ (PS∥RW); therefore

$$\gamma = \text{st.}\angle -2i - 2\chi.$$

But from $\triangle PQR$, $2\chi = \text{st.}\angle - 4r$. We have, at last,

$$\gamma = \text{st.}\angle -2i - 2\chi = 4r - 2i. \tag{5.1.2}$$

Tangency. A critical geometrician might object that this ray tracing rests on an identification so far unproven, namely that the angles formed by the rays and the radii are the angles of incidence and refraction. For that to hold, the radius would have to be perpendicular to the surface of the drop. The direction of a curve at any point is defined as the direction of the line that just touches it there: in the case of a circle, 'touching' means meeting in such a way that however far prolonged in either direction, the line does not cut the circle. Euclid proves that the radius to the touching line (or as it is usually called, the *tangent*) at the touching point makes a right angle with it (Euclid III.18). The argument appears from Fig. 5.1.17, where AB is the tangent, CP the radius. Suppose that $\angle CPB \neq \text{rt.}\angle$. Then draw the correct perpendicular from C to AB; let it fall at D. Since the side opposite the right angle in a right triangle must be longer than either of the other two sides (Euclid I.47), PC > CD. But PC = CE < CD. Our assumption that $\angle CPD$ is not a right angle has led to the absurdity that PC must be both greater and less than CD. Therefore the assumption must be wrong and $\angle CPD = \text{rt.}\angle$. Q.E.D.

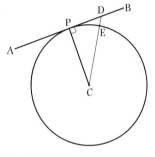

Fig. 5.1.17 Tangency.

With this business secure, let us calculate γ as we vary i. That amounts to examining the pathways of rays that strike the droplet at various points P on its surface. The remarkable outcome will be that the droplet focuses at the observer only rays that it deviates through an angle $\gamma \sim$ ('approximately equal to') 42°, which corresponds to $i \sim 60°$. For other angles i, the droplet reflects rays in various directions; but around $i = 60°$, it deviates all rays by the same amount, the observed angle γ. It is this focusing that produces the rainbow. The calculation requires only equation (5.1.2), Snel's law, and the information, derived from experiment, that n, the index of refraction of water, is 1.333.

Table 5.1.1 Focusing by droplets

$i°$	$\sin i/n$	$r°$	$4r°$	$2i°$	$\gamma°$
10	0.130	7.48	29.94	20	9.94
15	0.194	11.20	44.78	30	14.78
20	0.257	14.87	59.47	40	19.47
.	.	.	.	.	.
.	.	.	.	.	.
.	.	.	.	.	.
45	0.532	32.12	128.47	90	38.47
50	0.575	35.08	140.31	100	40.31
55	0.615	37.92	151.67	110	41.67
60	0.650	40.52	162.07	120	42.07
65	0.680	42.84	171.34	130	41.34

Table 5.1.2

$i°$	$\sin i/n$	$r°$	$4r°$	$2i°$	$\gamma°$
57	0.629	38.55	155.95	114	41.95
58	0.636	39.51	158.03	116	42.03
59	0.643	40.02	160.07	118	42.07
60	0.650	40.52	162.07	120	42.07
61	0.656	41.00	164.02	122	42.02
62	0.662	41.48	165.92	124	41.92

Proceed as in Table 5.1.1. Evidently γ passes through a maximum around 60°. The effect is very strong (Table 5.1.2) The value $n = 1.333$ for water is an average. Each color of the sun's rays is refracted by a different amount and so is strongly reflected, or focused, at a slightly different angle. That splits the bow into its several colors. Were all the colors refracted at the same angle, the bow would be an intense arch of white light. The secondary bow sometimes seen above the primary one, and with its colors reversed, derives from the strong focusing of rays that have undergone two reflections within the water droplets.

The purist might object, and rightly, that the geometry of the rays and the droplets considered so far applies only to the droplet at T, which is in the plane of the paper. How do the other drops act? As appears from Fig. 5.1.18, they act exactly as the droplet at T. The droplet T′ is supposed to be located in the bow outside the plane of the paper, in which the observer O and the center of the bow stand. The intense reflection now takes place in the plane ST′O but with the same deviation $\gamma = \angle ST'O$ and the same angle $\beta = \angle T'OC$ between the intensely reflected ray and the line OC connecting the observer

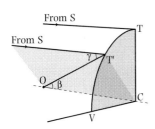

Fig. 5.1.18 Circular form of the rainbow.

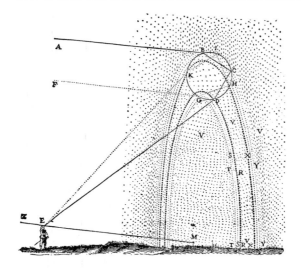

and the bow's center. This beautiful theory of the formation of the rainbow is very old. The French philosopher and mathematician René Descartes invented it over 350 years ago, shortly after he had learned about Snel's law. You will immediately understand the diagram with which he explained it to his contemporaries (Fig. 5.1.19), and know that it is wrong: the bow cannot stand at right angles to the horizon.

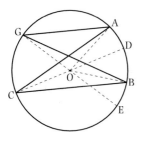

Fig. 5.1.20 Circumferential
angles are half the
corresponding central angles
(Euclid III.20).

Circumferential angles. Let us return to the geometry of the ray inside the droplet, redrawn in Fig. 5.1.20. Angles at the circumference, like AGB and ACB, have a very useful property: they are equal if they 'subtend' or 'intercept' equal arcs, regardless of where they lie on the circumference; and, further, they are equal to half of the angle at the center of the circle that subtends the same arc. This is the content of Euclid III.20. Its proof is easy. Draw the diameters COD and GOE. Since $\angle$AOD is external to $\triangle$AOC, $\angle$AOD = $\angle$ACO + $\angle$CAO = 2$\angle$ACO, since $\triangle$COA is isosceles. Also, $\angle$DOB = 2$\angle$OCB. Adding the pieces, we have $\angle$AOD + $\angle$DOB = $\angle$AOB = 2($\angle$OCB + $\angle$OCA) = 2$\angle$ACB. The same argument works for $\angle$BGA = $\angle$OGA − $\angle$OGB = ($\angle$AOE − $\angle$EOB)/2 = $\angle$AOB/2. Hence the circumferential angles AGB and ACB, being each equal to half of $\angle$AOB, are equal to one another.

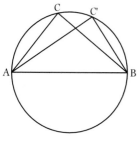

Fig. 5.1.21 Angles in
semicircles are right angles.

Now take the special case $\angle$AOB = st.$\angle$ (Fig. 5.1.21). Then all angles such as ACB, AC'B, inscribed in a semi-circle are right angles (Euclid III.31). This fact allows an alternative derivation of the relationship between the arrows and the half-string. The semi-circle ABC (Fig. 5.1.22) contains a right angle at B. Drop the perpendicular from B to AC at D. Now $\angle$BAC = $\angle$DBC, since both are complements of $\angle$ABD; similarly, $\angle$ABD = $\angle$BCD. From the similar triangles ADB, BDC, BD : AD = DC : BD, or BD^2 = AD × DC. The half chord BD is

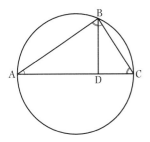

Fig. 5.1.22 The arrows again.

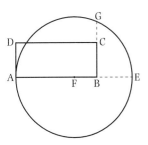

Fig. 5.1.23 To square a rectangle.

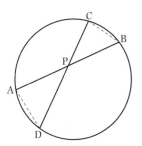

Fig. 5.1.24 Euclid III.35 via Euclid III.20 (APS 5.1.2).

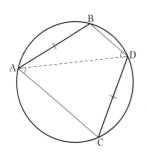

Fig. 5.1.25 Exercise on parallel chords

said to be a mean proportional between the longer and shorter pieces into which it divides the diameter, irrespective of the position of D.

The relationships implicit in Fig. 5.1.22 suggest an easy construction for squaring a rectangle, should you have any interest in doing so. 'Squaring a rectangle' means finding a square equal in area to a given rectangle. Let ABCD (Fig. 5.1.23) be the given rectangle. Extend AB to E, making BE = BC. Find the midpoint of AE, at F. With F as center and AF = FE as radius draw circle AGE. Extend side BC of the rectangle to intersect the circle at G. Then, according to the previous proposition, $BG^2 = AB \times BE = AB \times BC$, whence BG is the side of the square equal in area to the given rectangle. There is no similar construction for a square equal in area to a circle.

That an angle inscribed in a semi-circle is always a right angle is not only a useful fact. The poet Dante took Euclid III.31 as the exemplar of geometrical truth, on a par with logical necessity, and spoke of it in Paradise. There an angel implies that no reasonable person could entertain the possibility that a triangle without a right angle could be drawn in a semicircle, 'se del mezzo cerchio far si puote/Triangol sì, che un retto non avesse.'[3]

APS 5.1.2. Euclid III.35 follows immediately from III.20 without the bother of algebra and the Pythagorean theorem. In Fig. 5.1.24, AB, CD are the intersecting chords, whose parts, according to III.35, satisfy the relationship AP × PB = CP × PD. Join A and D and B and C: ∆APD ~ ∆BPC. We have ∠DPA = ∠BPC (vertical angles); ∠ADP = ∠CBP (they are angles at the circumference subtending the same arc AC); and ∠DAB = ∠DCB (by I.32, and also because they both subtend arc DB). From the similarity of ∆DAP and ∆BCP, AP : CP = PD : PB. Q.E.D.

APS 5.1.3. Arc AB = arc CD in Fig. 5.1.25. Can you see that the equality implies that AC‖BD? The proof follows immediately by joining A and D (or B and C) and using Euclid I.29 and III.20.

APS 5.1.4. The equal circles around A and B have their centers on one another's circumference (Fig. 5.1.26). Draw any line CD‖AB making a chord in both circles, as shown in the figure. It follows that ABEC and ABFD are parallelograms, where E, F are the intersections of CD with the circles. Prove it by dropping the perpendiculars AP = EQ (they are equal since CE‖AB). Also AC = EB (radii of equal circles). Hence ∆CPA ≅ ∆EQB (*h.s.*) and ∠EBQ = ∠PCA. But ∠PCA = ∠CAG, where G is the intersection of AB extended and the circle centered at A; hence ∠CAG = ∠EBA and,

3. Dante, *Paradiso, xiii*: 101–2, ed. Singleton (1975), *1*, 146–7.

Fig. 5.1.26 Exercise on
intersecting circles (APS 5.1.4)

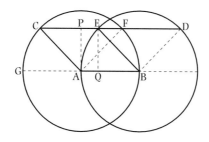

consequently, CA∥EB (I.29). Since CA is equal and parallel to EB, ABEC is a parallelogram. Q.E.D. The proof that AFDB is also a parallelogram follows precisely the same route.

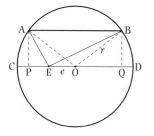

Fig. 5.1.27 Another exercise on chords (APS 5.1.5).

APS 5.1.5. In Fig. 5.1.27, E is any point on the diameter CD and AB is any chord parallel to CD. Can you show that $AE^2 + BE^2 = CE^2 + ED^2$? Since you must make use of the fact that the theorem relates to a circle, put in features peculiar to circles, like the center O and the radii OA = OB = r. Also, drop the equal perpendiculars AP = BQ = h. Let OE = e. Then

$$CE^2 + ED^2 = (r - e)^2 + (r + e)^2 = 2(r^2 + e^2),$$

$$AE^2 + BE^2 = h^2 + (OP - e)^2 + h^2 + (OQ + e)^2$$
$$= 2h^2 + 2OP^2 + 2e^2 = 2(r^2 + e^2),$$

Q.E.D. The penultimate step required PO = OQ, which follows from the congruency (h.s.) of triangles OAP and OBQ.

The simple form of interlaced equal circles whose centers stand on one another's circumferences, as in Fig. 5.1.26, leads to an architectural application of great historical and artistic importance. Let CD be the common chord (Fig. 5.1.28). If you describe arcs from F to G and from E to H with C and D as centers and radii twice that of the original circles, you will have the oval of Fig. 5.1.29; the joins are smooth at E, F, G and H because there the circles of radius r are tangent internally to those of radius $2r$. A great many ovals of this

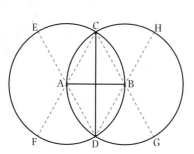

Fig. 5.1.28 Construction of ovals.

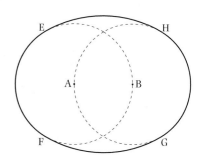

Fig. 5.1.29 An oval from interlaced circles.

Fig. 5.1.30 An oval from touching circles.

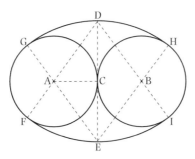

kind may be made, for example that of Fig. 5.1.30, where the generating circles centered at A and B touch at C and the centers D and E of the arcs FI and GH lie a distance $4r/3$ above and below C along the common tangent at C. (You can obtain DE = R = $8r/3$ from ΔACE, in which AE = $R - r$, AC = r, and CE = $R/2$.)

These ovals became popular designs for *piazze* in Italy during the sixteenth century. The most famous occurs in the great 'square' of St Peter's in Rome, the work of Giovanni Lorenzo Bernini, who thus completed Michelangelo's layout for the cathedral. Figure 5.1.31 shows the thing itself; Fig. 5.1.32 indicates its relationship to the oval depicted in Fig. 5.1.28. Note that the arc GH = 120° around B; arc EH = 60° around D; arc EF = 120° around A; and arc FG = 60° around C. The points I, J, K, L were defined architecturally, not geometrically.

Fig. 5.1.31 Saint Peter's Square and Bernini's colonnades, Rome. From Kitao, *Circle and oval* (1974), no. 95.

Fig. 5.1.32 The construction of Bernini's colonnades.

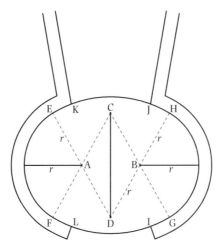

In the original design, a building occupied part of the arc LI; Mussolini tore it down in the 1930s to give an unobstructed view of St Peter's from the river Tiber.[4]

2. Touching, again

Generalities

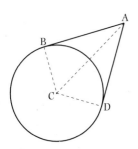

Fig. 5.2.1 The line from the center of a circle to an external point bisects the angle between the tangents dropped on the circle from the point.

One of the most important affairs in geometry, as in other pleasure-able activities of life, is touching. As we know, touching in geometry refers to the meeting of a line and a circle, or between circles, without intersection. Touching lines or curves are said to be 'tangent' in the sense of 'going off on a tangent', moving away from a point without engaging it, like AB, AD, in Fig. 5.2.1. There are fewer possibilities for flying off tangentially in geometry than in everyday life, where they are legion, and too often seized. For example, from a point inside a circle, you can drop no tangent upon it; from a point outside you can manage two, AB = AD. To see how to manage, draw CA, CB, CD. Since ∠ABC = ∠ADC = rt.∠ (radii are perpendicular to tangents, as proved earlier at Fig. 5.1.17), ΔABC ≅ ΔADC by *h.s.* (BC = CD = radius). Hence ∠BAC = ∠CAD: the line drawn from the center of a circle to any external point bisects the angle made by the tangents dropped on the circle from that point.

The converse of this theorem also holds: the bisector of the angle made by the two tangents dropped from a point on a circle passes through the center of the circle. This proposition may be demon-

4. Muller, *Oval* (1967), iv-v; Kitao, *Circle and oval* (1974), 31–4, 39–42, 107, 109, Figs. 41–2, 51, 56.

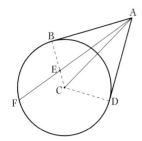

Fig. 5.2.2 The bisector of the angle between the tangents passes through the center of the circle.

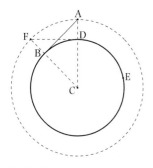

Fig. 5.2.3 To drop a tangent (Euclid III.17).

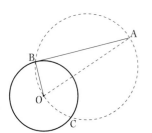

Fig. 5.2.4 Another method for dropping tangents.

strated by *reductio ad absurdum*. In Fig. 5.2.2, let AF be the angle bisector, supposed not to pass through the circle's center C. Join C to B, D, and A as before: ΔABC ≅ ΔADC, which implies ∠BAC = ∠DAC. But, by hypothesis, ∠BAE = ∠DAE. Now, ∠BAC = ∠BAE + ∠EAC, and ∠DAC = ∠DAE − ∠EAC, which equations can be satisfied only if ∠EAC = − ∠EAC, that is, if ∠EAC = 0. But in this case, AF passes through C. Q.E.D.

The logical mind will have objected long before this that we have been drawing tangents to circles without showing how to do so. Euclid takes up the problem in III.17, after he too had made use of tangents without demonstrating how to drop one. In Fig. 5.2.3, the given circle is BDE, the point from which the tangent falls, A. Join A with the circle's center C; D marks the point where the line from A to C cuts the circle. With radius CA, draw a circle enclosing the original circle; raise a perpendicular at D and continue it to the external circle at F. Join C and F; the point B where CF cuts the given circle is the point of tangency. To confirm this construction, it is only necessary to show that ∠ABC = rt.∠. We have CB = CD (radii of the smaller circle) and CA = CF (radii of the larger circle). Therefore, since ∠ACB is common, ΔFDC ≅ ΔABC by s.a.s. Consequently, ∠ABC = ∠FDC = rt.∠. Q.E.D.

There are many ways besides Euclid's to drop tangents on circles. Legendre preferred, and may have introduced, the method depicted in Fig. 5.2.4: Join A with the center of the circle O and draw an auxiliary circle through A and O centered at the midpoint of AO. The intersections of the auxiliary circle with the given one identify the tangent points. The business is as plain as a pikestaff: ∠ABO, being an angle in a semicircle, is a right angle.[5]

A most important property of externally touching circles is that the the line connecting their centers passes through the point of tangency (Euclid III.12). Assume that AB (Fig. 5.2.5) does not go through P. Draw QR ⊥ BP at the point of tangency, continue BP to C, and connect A and P; since A is by hypothesis the center of the larger circle, AP also should be perpendicular to QR. But that is impossible unless BP passes through A; for otherwise we would have two different lines, PC and PA, both perpendicular to the same line at the same point. A similar proposition holds for internal tangency (Euclid III.11, Fig. 5.2.6). The circle centered at B touches circle center A at P. Suppose the line from B to P extended to F does not go through A. Join A and F, and draw the straight line QR tangent to the circles at P. Now AP⊥QR, because QR touches circle center A; and BP (and therefore FP) ⊥QR, because QR touches circle center B.

5. Cf. Jones, *MT, 37* (1994), 10.

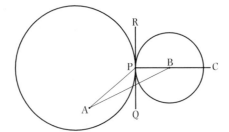

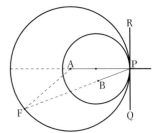

Fig. 5.2.5 The line of centers of two circles touching externally passes through the point of tangency (Euclid III.12).

Fig. 5.2.6 The line of centers for circles touching internally.

But then FP must coincide with AP extended; otherwise we would be in the same predicament about perpendiculars as before. Q.E.D.

APS 5.2.1. Figure 5.2.7 presents circles touching externally at P. Draw any two lines AB, CD through the point of tangency P, and prove that AC is parallel to BD. What defining characteristic of parallelism will you use and how will you introduce the fact that you are dealing with circles? A suggestion: if AC‖DB, the alternate interior angles, ∠1 and ∠2, must be equal; as for the circles, you might use that excellent dodge about the arcs subtended by circumferential angles. Draw the radii OC, O′D. Then (III.20) ∠1 = ∠COP/2 and ∠2 = ∠DO′P/2. But ΔCOP ~ ΔDO′P: both are isosceles with the same base angles, since ∠OPC = ∠O′PD. Hence the equality of the central angles, ∠COP, ∠DO′P, and of the corresponding circumferential angles, ∠1, ∠2, Q.E.D. Incidentally, you have also shown that OC‖O′D, since ∠COP and ∠DO′P are alternate interior angles.

APS 5.2.2. A problem similar to the preceding one was set on the Mathematical Tripos at Cambridge University in 1802. The examinee faced two touching circles with parallel diameters (Fig. 5.2.8) and had to show that the straight lines joining the opposite extremities of the diameters pass through the point of tangency. It will be enough to show that one of them, BC, does.

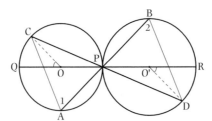

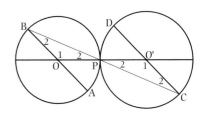

Fig. 5.2.7 Exercise on touching circles (APS 5.2.1).

Fig. 5.2.8 Another exercise on touching (APS 5.2.2).

Successful examinees joined C and B to P without assuming that BPC is a straight line: that is what was to be proved. Since AB∥CD, ∠BOP = ∠PO'C (I.29); and since triangles BOP and CO'P are isosceles, the angles marked '2' are equal. Now ∠BPO' = ∠1 + ∠2, so ∠BPC = ∠1 + 2 ∠2 = st. ∠, as appears from ΔBOP. Q.E.D.

APS 5.2.3. Now for some remote touching. Let the centers of two circles, of radii a and b, lie a distance s apart, where $s > a + b$. Draw tangents to each circle from the center of the other, as in Fig. 5.2.9. Show that the chords x, y are equal. [Evans, *MT, 31* (1938), 114.] Nothing to it. Since ∠XOP is the complement of both ∠OXP and ∠OO'B, ΔOPX ~ ΔOBO', whence $(x/2) : b = a : s$. Similarly, $(y/2) : a = b : s$, and, by division $x : y = 1$. Q.E.D.

APS 5.2.4. Three circles touch one another and a line externally, as shown in Fig. 5.2.10. Let the radii a, b of the first and third circles be given. A straightforward exercise in symmetry reveals that, if $b = n^2 a$, where n is any number, the radius x of the middle circle is na.

First you need to make some symmetrical triangles by dropping perpendiculars OB, PC, QD to the common tangent AD. (A is the intersection of QO extended and DB extended.) To save writing, let AB = y. From the similar triangles,

$$\frac{a}{y} = \frac{x}{y + \text{BC}} = \frac{b}{y + \text{BC} + \text{CD}}.$$

You have four unknowns and two equations. Too little. So draw OR⊥PC. From the right triangle ORP, $\text{OP}^2 = \text{PR}^2 + \text{BC}^2$. But

Fig. 5.2.9 Exercise on the common tangents to two circles (APS 5.2.3).

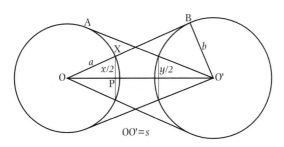

Fig. 5.2.10 Exercise on touching circles (APS 5.2.4).

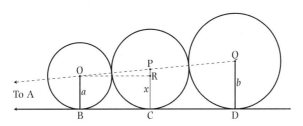

PR = $x - a$ and, because the line of centers of two touching circles goes through the point of tangency, OP = $a + x$. You have therefore

$$BC^2 = (a + x)^2 - (x - a)^2 = 4ax,$$

and, similarly, $CD^2 = 4bx$. With these values for BC and BD, the equations derived from the similar triangles can be solved for x, with the result that

$$x = \frac{a\sqrt{b} + b\sqrt{a}}{\sqrt{a} + \sqrt{b}}.$$

But we are given $b = n^2a$; and, therefore, $x = (n + n^2)a/(n + 1) = na$. Q.E.D.

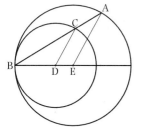

Fig. 5.2.11 Another exercise on touching circles (APS 5.2.5).

APS 5.2.5. If circle center D touches circle center E internally at B, and AB is any chord cutting the smaller circle at C, then AE∥CD. Figure 5.2.11 shows the situation. You will not find it difficult to show that △CDB ~ △AEB and, therefore, that ∠AEB = ∠CDB, whence AE∥CD. Hint: the triangles in question both are isosceles.

APS 5.2.6. Two unequal circles touch externally (Fig. 5.2.12); AB is their common tangent. Show that the circle with AB as diameter passes through the point of contact P, where it is tangent to the line of centers of the first two circles. All that needs to be done is to raise the common tangent PC and show that AC = CP = CB. We have △CAE ≅ △CPE by *h.s.* (EP = a), whence AC = CP; likewise △CPF ≅ △CBF and CP = CB. Q.E.D.

To vary matters, allow two equal circles to intersect so that the radii to a point of intersection (P in Fig. 5.2.13) are the tangents there. Then draw any secant through P that meets both the circles, as at A and B. No matter how you draw AB, you have $AP^2 + PB^2$ = constant. That is because, owing to the right angle at OPQ, △LOP ≅ △MPQ; hence $LO^2 = PM^2$. Then $LP^2 + PM^2 = LP^2 +$

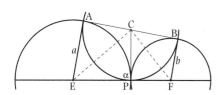

Fig. 5.2.12 The tangent to two touching circles at their point of tangency bisects their common tangent (APS 5.2.6).

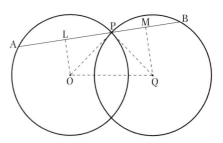

Fig. 5.2.13 A variation on **Fig. 5.2.12** (APS 5.2.6).

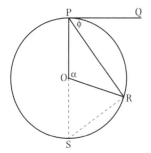

Fig. 5.2.14 Angles between secants and chords (APS 5.2.7).

$LO^2 = r^2$. But $LP = AP/2$, $PM = PB/2$; $AP^2 + PB^2 = 4r^2$, a constant as required.

APS 5.2.7. Many miscellaneous relationships among angles made by tangents, secants and chords fall out quickly using Euclid III.20. Figure 5.2.14 gives the case of a tangent (PQ): if PR is any chord, you can see that the angle ϕ, between the chord and the tangent, is half the angle α subtended by the chord at the center. Since $\angle OPR$ is rt.$\angle - \phi$, and $\angle PSR = $ rt.$\angle - \angle OPR$ ($\angle PRS = $ rt.$\angle$), $\angle PSR = \phi$. Then, by Euclid III.20, $\phi = \alpha/2$. Q.E.D.

APS 5.2.8. Figure 5.2.15 gives the case of two secants, which define chords ST, PR subtending angles β and α respectively, at the center. You are to show that ϕ, the angle between the secants, is half the difference between α and β. Let $\gamma = \angle TOR$. Then $\phi = $ st.$\angle - \angle P - \angle R = $ st.$\angle - (\angle SPT + \angle TPR) - (\angle TRS + \angle SRP) = $ st.$\angle - (\beta/2 + \gamma/2) - (\beta/2 + $ st.$\angle - (\alpha + \beta + \gamma)/2) = (\alpha - \beta)/2$. Q.E.D.

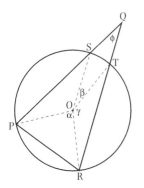

Fig. 5.2.15 Angles between secants (APS 5.2.8).

APS 5.2.9. Finally, and neatest of all, the angle of intersection ϕ between two chords is half the sum of the angles subtended by the ends of the chords at the center. In Fig. 5.2.16, AB, CD are the chords. Join B and C. Then $\phi = \angle BCP + \angle PBC$ (ϕ is an external angle of $\triangle BPC$). But $\angle BCP = \angle BCD$ subtends the arc BD as does α; therefore $\angle BCP = \alpha/2$. Similarly $\angle PBC = \beta/2$, and $\phi = (\alpha + \beta)/2$. Q.E.D.

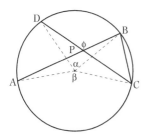

Fig. 5.2.16 Angles between intersecting chords (APS 5.2.9).

The diagrams 5.2.1 and 5.2.2 suggest other situations easy to investigate and pleasant to know. Here are a few. If, from a point outside a circle, any line CE (Fig. 5.2.17) is drawn through the circle's center to the other side, cutting it at D; and another line, CA is drawn through cutting the circle at B, then CE : CA = CB : CD. Draw DB and EA to produce two triangles, which turn out to be similar. Their similarity follows from the facts that $\angle EAB$ subtends arc BDE and that $\angle BDE$ subtends arc EAB; together, therefore, they subtend the entire circumference, which is to say that they are supplementary. (Remember, an angle at the circumference subtends half the arc that it would at the center.) So $\angle EAB = $ st.$\angle - \angle BDE$. But $\angle BDE = $ st.$\angle - \angle BDC$, hence $\angle EAD = \angle BDC$, and, since $\angle C$ is common, $\triangle BDC \sim \triangle EAC$. The corresponding parts give CE : CA = CB : CD, as required.

From the proposition just demonstrated, you can find immediately a proportion between the longest and shortest parts of two secants from the same point to the same circle. In Fig. 5.2.18,

$$CE \times CD = CA \times CB,$$
$$CE \times CD = CF \times CG,$$

from which, for two secants CBA and CGF, CA : CG = CF : CB.

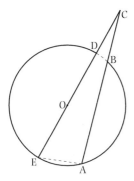

Fig. 5.2.17 A circle makes similar triangles from a secant from a given point and the line from that point to the center of the circle.

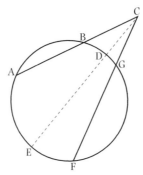

Fig. 5.2.18 A circle cuts the secants from the same point into proportional segments.

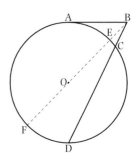

Fig. 5.2.19 The situation when one secant becomes a tangent (Euclid III.36).

Table 5.2.1 Terms about circles

Radius	A straight line from the center of a circle to its circumference
Diameter	A straight line through the center equal to twice the radius
Secant	A straight line that cuts a circle twice
Tangent	A straight line that touches a circle
Circumference	The perimeter of a circle
Arc	A portion of a circumference
Chord	A line from one point of the circumference of a circle to another that does not pass through the center

If instead of two secants we have a secant and a tangent, the situation of Fig. 5.2.19 holds (Euclid III.36). Let AB touch the circle at A; then $AB^2 = BC \times BD$. But this is evident, no? Just draw BEOF. Then $BC \times BD = BE \times BF = (BO - r)(BO + r) = BO^2 - r^2 = AB^2$.

Descartes and the princess

A famous problem in geometry is to draw a circle tangent to three others given in position, as in Fig. 5.2.20. Since we know the radii, a, b, c, say, and the distances AB, BC, CA, we can easily set up an equation that contains the unknown radius x of the required circle centered at O. For, from Euclid III.12, $AO = a + x$, $BO = b + x$, $CO = c + x$; with these formulas and the givens, the areas of triangles AOB, BOC, and COA can be written in terms a, b, c, and x. And, since the sum of the areas of these traingles equals the known area of $\triangle ABC$, the problem is solved. In principle. But, as the philospher Descartes wrote to a young woman who had sent him the solution, 'this way seems to me to involve so many superfluous multiplications that I would not undertake to get through them in three

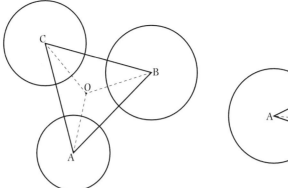

Fig. 5.2.20 To draw a circle tangent to three others.

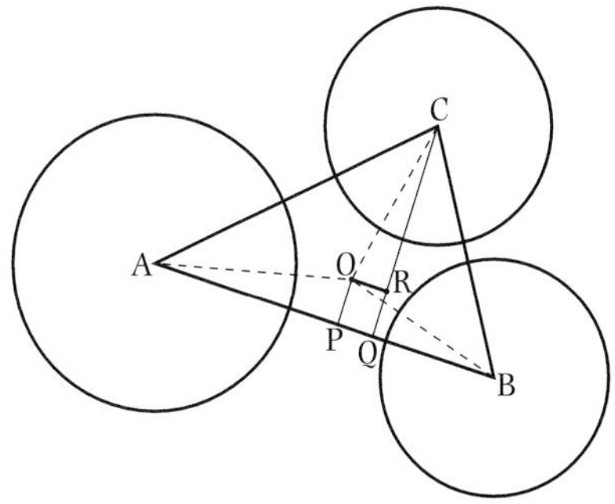

Fig. 5.2.21 Descartes' analysis of the three-circle problem.

months'.[6] The formula for the area of a triangle in terms of the lengths of its sides is given in §5.2 after APS 5.2.11 and in §6.1

Descartes' correspondent, Princess Elisabeth of Bohemia, had a very sharp mind. She had used it earlier to pose profound questions about his philosophy, which he could not answer satisfactorily. She used mathematics as a sedative, to keep her mind off family troubles. Her father had lost his estates during the Thirty Years War, which still raged when Elisabeth first questioned Descartes in 1643; and her uncle, Charles I of England, would soon lose his head as well as his kingdom. Doing the problem of the three circles by her method does help to dull the mind.

Descartes had a better way. 'I make sure when analyzing a question in geometry that the lines I use to resolve it are either parallel or intersect at right angles, as far as possible.' That way he needed only two tools: the Pythagorean theorem and the proposition that the sides of similar triangles are proportional. To proceed in this way—which is the characteristic approach of analytical or Cartesian geometry—one must not scruple to introduce many unknown lines into the problem, for example, the perpendiculars OP, CQ, OR in Fig. 5.2.21. 'I do not fear at all to suppose many unknown quantities in order to reduce the question so that it depends only on these two theorems; on the contrary, I prefer to suppose more than fewer. For then I see everything I do more clearly ...' By applying the Pythagorean theorem to triangles APO, BPO, and CRO Descartes obtained OP and OR in terms of x, but to extract x from them would have been too much work. No doubt you will agree with him, if you set up the problem his way, that to go further 'does not serve to

6. Descartes, *Correspondance avec Elisabeth* (1989), 81.

Fig. 5.2.22 The three-circle problem when the circles touch.

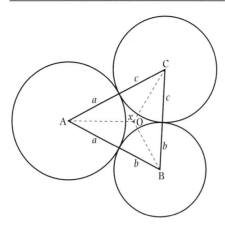

cultivate or amuse the mind, but only to exercise the patience of a laborious calculator'.[7]

He reckoned without Elisabeth, who sent him a complete solution. 'I was not only astonished to see it [he wrote her], but I can't refrain from adding that I was also filled with joy'—no doubt because he knew how to improve it. Elisabeth's solution apparently lacked elegance (it has not survived) and did not contain only the given radii and distances. That at least is the natural interpretation of Descartes' remark that 'to obtain a theorem to serve as a general rule to resolve similar problems, it is necessary to keep to the end the notation used as the beginning, or, if changes are made to ease the calculation, the original notation must be reinstated.'[8] This excellent advice cannot be followed without great prolixity in the general case. However, the special case that the three circles touch, as shown in Fig. 5.2.22, can be worked out neatly with a little effort.

Figure 5.2.23 redraws 5.2.22 with the unknown radius x and Descartes' perpendiculars from Fig. 5.2.21. By eliminating OP between the triangles APO, BPO, you will have

$$(a + x)^2 - \text{AP}^2 = (b + x)^2 - (a + b - \text{AP})^2,$$

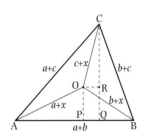

Fig. 5.2.23 Detailed geometry of the case in Fig. 5.2.22.

from which, after a little manipulation, you will find

$$\text{AP} = \frac{a^2 + ab + ax - bx}{a + b}.$$

Since, as Descartes pointed out to Elisabeth, c plays the same part in the algebra of ΔABC as x does in that of ΔAOB, he wrote down AQ by replacing x by c in the expression for AP:

$$\text{AQ} = \frac{a^2 + ab + ac - bc}{a + b}.$$

7. Descartes to Elisabeth, November 1643, ibid., 79, 83.
8. Ibid., 85, 86.

He than gave his equation for x in terms of a, b, c, omitting all the calculations, no doubt so as to give Elisabeth the pleasure of discovering the steps for herself.

He may have proceeded as follows. In $\triangle ORC$,

$$(c + x)^2 = CR^2 + OR^2 = (AQ - AP)^2 + (CQ - OP)^2. \quad (5.2.1)$$

From the preceding values of AQ, AP,

$$OR = (a - b)(c - x)/(a + b). \quad (5.2.2)$$

To get CQ and OP you will find it convenient to use the following trick, one of those that occur to you if, as Descartes directed Elisabeth to do, you keep your eyes open when you calculate. From $\triangle APO$,

$$OP^2 = (a + x)^2 - AP^2 = \frac{(a + x)^2(a + b)^2 - (a^2 + ab + ax - bx)^2}{(a + b)^2}.$$

This looks like a lot of work. But notice that

$$(a + x)^2 (a + b)^2 = [(a + x)(a + b)]^2 = (a^2 + ab + ax + bx)^2.$$

You then have for OP an expression of the form

$$OP^2 = \frac{(M + N)^2 - (M - N)^2}{(a + b)^2} = \frac{4MN}{(a + b)^2},$$

when $M = a^2 + ab + ax$, $N = bx$. So,

$$OP^2 = \frac{4bx(a^2 + ab + ax)}{(a + b)^2}. \quad (5.2.3)$$

CQ has the same form as OP with c in place of x.

You now need to insert relations (5.2.2) and (5.2.3) and the equivalent to (5.2.3) for CQ into equation (5.2.1) and simplify. The algebra will cost you some pains. You will have occasion to use the trick illustrated in the preceding paragraph, which will help get you to

$$ab(a + b)(c + x) - cx(a^2 + b^2) = 2ab[cx(a + b)^2 + (c + x)(a + b) + cx]^{1/2}.$$

You must now square this equation to remove the root. 'Tis a job for an Elisabeth. In simplifying the result, you will have the pleasure of dividing out a factor of $(a + b)^2$; recall in this connection that $a^2 - b^2 = (a + b)(a - b)$. If all goes well, you will get, as Descartes did,

$$x^2[a^2b^2 + a^2c^2 + b^2c^2 - 2abc\,(a + b + c)] - 2x(abc)(ab + ac + bc) + (abc)^2 = 0.$$

Note that a, b, c enter this equation on an equal footing, as they should: the problem accords all of the given circles the same status. Note also that the equation is only quadratic in x, which should be a pleasant surprise after all the squaring, and so can be solved for x.

Descartes did not bother, but settled for a numerical example. Putting $a = 2$, $b = 3$, $c = 4$ (it of course does not matter which letter you set equal to which number), he told Elisabeth that

$$x = -\frac{156}{47} + \sqrt{\frac{31104}{2209}},$$

'if I've not made a mistake in the calculation I've just done.'[9] Was he right?

Scriptions

Inscription. To inscribe a circle within a plane figure means to draw a circle that touches each and every side of the figure. The preceding propositions indicate how to draw such a circle within a triangle. For let the given triangle be ABC and the circle sought centered at D (Fig. 5.2.24). Then, since from each vertex A, B, C, two tangents fall on the circle, the circle's center lies on the bisector of each angle of the triangle. Hence D, the so-called 'incenter' (for 'center of the inscribed circle') sits at the intersection of the angle bisectors.

Euclid's proof (IV.4) brings together arguments earlier made informally, and so deserves repeating here. Let △ABC be the given triangle (Fig. 5.2.24). Bisect the angles at B and C; the bisectors intersect at D. From D drop the perpendiculars DE, DF, DG onto the sides of the triangle. The proof of the proposition that EGF is the desired circle follows from the demonstration that ED = DG = DF = r, the circle's radius. The right triangles BFD and BED are congruent (*a.s.a.*), since all their corresponding angles are equal and they have a common side BD; hence DE = DF. Also the right triangles DFC and DGC (*a.s.a.*); hence DG = DF = DE. Therefore EFG is a circle. Furthermore, since DE, DF, DG are perpendicular to the sides, the circle touches the triangle internally as required. Q.E.D.

What is the radius of a circle that is inscribed in a right triangle? To make the question sporting, let us answer it in the ancient Chinese manner. Assume the inscription accomplished, as in Fig. 5.2.25, which shows the angle bisectors OA, OB, OC and the radii perpendicular to the sides OD, OE, OF. The coloring of the parts (indicated here in shading and shown on Plate III) follows the prescription in Liu Hui's commentary on the *Jiu zhang suanshu*. Now double the parts and rearrange them as in Fig. 5.2.26, in which, for convenience, the horizontal dimension is made half that of Fig. 5.2.25.

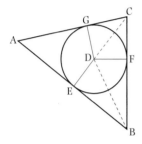

Fig. 5.2.24 To inscribe a circle in a triangle.

9. Ibid., 88. The explicit solution is

$$x = \frac{ab + ac + bc - 2[abc(a + b + c)]^{1/2}}{ab/c + ac/b + bc/a - 2(a + b + c)}.$$

Fig. 5.2.25 To find the radius of the incircle in the Chinese manner.

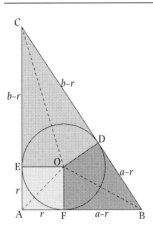

Fig. 5.2.26 Further to Fig. 5.2.25.

The area of the entire long rectangle is EO[(CE + EA) + (CD + DB) + (BF + FA)] = $r(a + b + c)$, where r is the radius of the incircle and a, b, c designate the sides of the triangle as usual. But the long rectangle has an area twice that of $\triangle ABC$, since each part of the triangle occurs twice in the rectangle. We have therefore $2(ab/2) = r(a + b + c)$, or

$$r = \Gamma/[a + b + c)/2] = \Gamma/s, \qquad (5.2.4)$$

where Γ signifies the triangle's area and $s = (a + b + c)/2$ is its semi-perimeter.

Another expression for Γ can be read immediately from Fig. 5.2.25: BC = c = CD + DB = $b - r + a - r = b + a - 2r$, from which you collect $r = (a + b - c)/2 = (s - c)$. This expression and equation (5.2.4) must be equal; equating them, you have $\Gamma = s(s - c)$. The equality holds only for a right triangle, as you can see by writing out both sides in terms of a, b, and c and applying the Pythagorean theorem. The generalized form for Γ in terms of s, valid for all triangles and known as *Hero's formula*, will be presented momentarily. Equation (5.2.4), however, is valid for all triangles, as appears from Fig. 5.2.27. You have by inspection

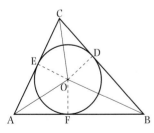

Fig. 5.2.27 Another method for finding the radius of the incircle.

$$2\Gamma = OD \times BC + OE \times AC + OF \times AB = r(a + b + c),$$

and, again. $$\Gamma = rs.$$

APS 5.2.10. Can you show that the area of a right triangle equals the product of the segments of the hypotenuse made by the perpendicular on it from the incenter? This amounts to demonstrating that, in Fig. 5.2.25, $\Gamma = (b - r)(a - r)$. Hint: expand the right side and use equation (5.2.4).

APS 5.2.11. A little more algebra will show that, for any triangle, $1/h_a + 1/h_b + 1/h_c = 1/r$, where the h's represent the altitudes dropped on the corresponding sides (h_a is the altitude from vertex A on side a, etc.) The relation follows directly from equation (5.2.4) and the usual formulas for the area of a triangle, $\Gamma = ah_a/2 = bh_b/2 = ch_c/2$. You might also want to confirm that

$$(h_a h_b h_c)\, abc = (a + b + c)^3\, r^3.$$

But then maybe you do not want to. [After Halsted, *Mensuration* (1881), 170.]

Circumscription. To circumscribe a circle about a plane figure means to find the circle (if any) that passes through all the vertices of the figure. In the case of a triangle, the sides will be chords of the circumscribing circle. We know that the perpendicular bisector of a chord passes through the center of the circle containing it; hence the center of the circumscribing circle (the 'circumcenter') lies at the intersection of the bisectors of two chords (Euclid III.1). Euclid (IV.5) distinguishes three cases, according to whether the intersection falls within the triangle, on one of its sides, or outside the triangle (Fig. 5.2.28). In each case, we must show that FA = FB = FC = R. In the first case, $\triangle ADF \cong \triangle BDF$ (s.a.s.), since DF is the perpendicular bisector of AB), whence FA = FB. Similarly, FA = FC. Q.E.D. You should do the other cases as exercise. Note that the three cases correspond to an acute, right (Euclid III.31), and obtuse angle at A, respectively.

It remains to find an expression for the radius of a circle circumscribed around a triangle to match equation (5.2.4). Let the center of the circle be O (Figure 5.2.29) and the altitude onto c, h_c. Then $\Gamma = ch_c/2$. From the similar triangles ADC and OEC, $h_c : b = a : 2R$.

Fig. 5.2.28 To circumscribe a circle about a triangle (Euclid IV.5).

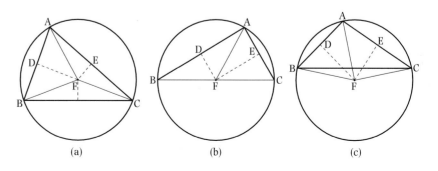

(a)   (b)   (c)

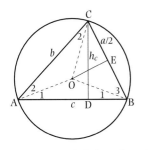

Fig. 5.2.29 To find the radius of the circumcircle.

(The triangles are similar because $\angle 1 + \angle 2 + \angle 3 = $ rt.$\angle$ by Euclid I.32 and $\angle COE = $ rt.$\angle - \angle 3$. Hence $\Gamma = cba/4R$, or

$$R = abc/4\Gamma, \qquad (5.2.5)$$

which is the analogue desired.

We might have proceeded less elegantly and calculated h_c in terms of the sides of the triangle. The Pythagorean theorem applied to $\triangle ADC$ and $\triangle BDC$ gives

$$h_c^2 = \frac{2a^2b^2 + 2b^2c^2 + 2a^2c^2 - a^4 - b^4 - c^4}{4c^2}. \qquad (5.2.6)$$

Therefore, $4c^2h_c^2 = 16\Gamma^2 = N$, the numerator of the fraction on the right of equation (5.2.6). As an anticipation of things to come, let us rework N:

$$\begin{aligned} N &= -[(b+c)^2 - a^2][(b-c)^2 - a^2] \\ &= (b+c+a)(b+c-a)(b+a-c)(a+c-b) \\ &= 16s(s-a)(s-b)(s-c), \end{aligned}$$

where $s = (a+b+c)/2$ is the semi-perimeter introduced earlier. Since $N = 16\Gamma^2$,

$$\Gamma = \sqrt{s(s-a)(s-b)(s-b)},$$

which is Hero's formula for the area of a triangle. This is the expression from which Elisabeth set out to solve the problem given by Descartes. Its original geometrical proof is offered in §6.1.

APS 5.2.12. Show that the product of the radius of the incircle of a triangle, r, and that of the circumcircle, R, is twice the product of the triangle's sides divided by its perimeter: $rR = 2abc/(a+b+c)$.

Escription. An escribed circle touches one side of a triangle and the other two sides extended. The concept will be clear from Fig. 5.2.30, where the circle is escribed to side a and to sides b and c extended. Naturally, we want to know its radius in terms of the area and perimeter of the triangle. You can anticipate an easy time, since the quantity of interest, the radius r_a, is the altitude associated with

Fig. 5.2.30 To escribe a circle to a triangle.

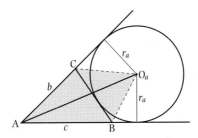

triangles built on the extended sides, so we will not have to pass through the trigonometric functions or their equivalents to express r_a in terms of areas.

The shaded area in Fig. 5.2.30 can be expressed in two different ways, as $\triangle ABO_a + \triangle ACO_a$ and as $\triangle ABC + \triangle BO_aC$. We read directly from the figure that $\triangle ABO_a = r_ac/2$, $\triangle ACO_a = r_ab/2$, $\triangle BO_aC = r_aa/2$. Let the area of $\triangle ABC$ be Γ, as usual. Then

$$(r_a/2)(b + c - a) = \Gamma,$$

or

$$r_a = \Gamma/(s - a), \tag{5.2.7}$$

where s once again is the semi-perimeter. The radii of the two other escribed circles have the same form as equation (5.2.7), with b or c in place of a.

The center of an escribed circle is easily found at the intersection of an internal angle bisector with two external ones. (That these bisectors intersect was shown in §3.3.) In order that O_a be the center sought, it must be equidistant from BC, AB extended, and AC extended. This follows by definition, which requires O_a to be equidistant from the sides of the angles bisected by the lines that meet there. O_b and O_c are located similarly. The sides of the triangle with vertices at the centers of the three escribed circles go through the vertices of the original triangle (Fig. 5.2.31). You may enjoy proving it.

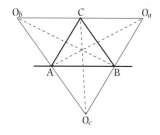

Fig. 5.2.31 The sides of the triangle formed from the centers of the escribed circles pass through the vertices of the original triangle.

APS 5.2.13. Prove the useless formula, $rs^2 = r_ar_br_c$, where r is the radius of the incircle, s the semi-perimeter, and the other r's are the radii of the escribed circles. Hint: use Hero's formula.

APS 5.2.14. The length of the tangent from A to the circle escribed to a is s; from B, $s - c$; from C, $s - b$. The situation is pictured in Fig. 5.2.32. Let the length of the tangent from A, AE, be x; then AF = AE = x. Also, BG = BE, CG = CF, since B and C, like A, lie at the intersection of tangents. So a = BG + CG = CF + BE = $x - b + x - c$, from which $2x = a + b + c = 2s$. Q.E.D.

APS 5.2.15. To exhaust your interest in and patience with escription, try to show that the area of the triangle whose vertices

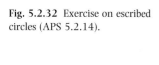

Fig. 5.2.32 Exercise on escribed circles (APS 5.2.14).

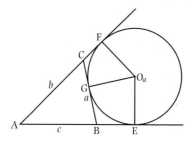

are the centers of the escribed circles is $abc/2r$, where r is the radius of the incircle. The problem is set up in Fig. 5.2.31. The area Λ ('lambda') of $\Delta O_a O_b O_c$ is the sum of the areas of ΔABC, $\Delta BO_a C$, $\Delta AO_b C$, and $\Delta AO_c B$. Hence

$$\Lambda = \Gamma + \frac{1}{2}\ (r_a a + r_b b + r_c c),$$

since the radii of the escribed circles are the altitudes of the external triangles on the bases a, b, c. Equation (5.2.7) allows substitution for the radii r_a, r_b, r_c:

$$\Lambda = \frac{\Gamma}{2}\left(2 + \frac{a}{s-a} + \frac{b}{s-b} + \frac{c}{s-c}\right).$$

The rest is algebra. From equation (5.2.4), $\Gamma = rs$; from Hero's formula, $(s-a)(s-b)(s-c) = r^2/s$. Therefore

$$\Lambda = \frac{1}{2r}[2(s-a)(s-b)(s-c) + a(s-b)(s-c) + b(s-a)(s-c)$$

$$+ c(s-a)(s-b)] = abc\,/\,2r,$$

Q.E.D.

Cyclic quadrilaterals

Every triangle has an incircle and a circumcircle. That is not true of every quadrilateral. Even a good symmetric one, like a parallelogram, can not be inscribed in a circle unless it is a rectangle. Inscriptable quadrilaterals are called 'cyclic.' The condition they must meet is easily determined. Within a given circle draw the arbitrary quadrilateral ABCD containing the circle's center (Fig. 5.2.33): the opposite angles sum to a straight angle, that is $\angle A + \angle C = \angle B + \angle D$ (Euclid III.22). The demonstration leaps from the figure: the sum of the arcs subtended by each pair of opposite angles is the whole circumference; therefore (Euclid III.20), each pair is supplementary. That not every quadrilateral has this property may be demonstrated easily. Consider Fig. 5.2.34, with right angles at A and B, an obtuse angle at C and an acute angle at D. Then $\angle ACD + \angle ABD >$ st.$\angle$, $\angle CAB + \angle CDB <$ st.$\angle$.

Cyclical quadrilaterals are a useful tool in the geometer's kit because they allow a quick hunt for equal angles in the virtual circumscribing circle. For example, the demonstration of the coincidence of the altitudes of a triangle follows immediately from the recognition of two inscriptable quadrilaterals. In Fig. 5.2.35, AE and BF are altitudes meeting at O; you are to show that CO when continued to D makes a right angle with AB. You note that quad. FOEC is

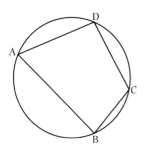

Fig. 5.2.33 A cyclic quadrilateral.

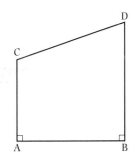

Fig. 5.2.34 A non-cyclic quadrilateral.

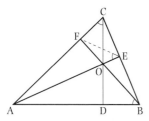

Fig. 5.2.35 The coincidence of altitudes via cyclic quadrilaterals.

inscriptable because it contains opposite right angles at E and F; imagining F, O, E, and C to lie on the circumference of a circle, draw chord FE and deduce that the circumferential angles FEO and FCO, which subtend the same chord FO, are equal. The figure contains another inscriptable quadrilateral, AFEB: its right angles at E and F, standing on the same hypotenuse, can be inscribed in the same semicircle with AB as diameter. Supposing now that A, B, E, and F lie on a circle, ∠ABF = ∠FEA since both subtend the chord AF. The upshot is ∠ACD = ∠FCO = ∠FEO = ∠FBA. From ∠ACD = ∠FBA and the common angle at A, ΔADC ~ ΔAFB, and, consequently, ∠ADC = ∠AFB = rt.∠. Q.E.D. Very neat, though perhaps not very obvious.

Simson's line. Here is another striking analysis by inscriptable quadrilaterals. Robert Simson, the influential Scottish editor of Euclid, noticed one day that the feet of perpendiculars dropped from any point P on a circumcircle onto the sides of the circumscribed triangle lie on a straight line. In Fig. 5.2.36, L, N, and M are the feet of the perpendiculars. Join P with A and B. The argument by cyclical quadrilaterals shows that ∠PNM + ∠PNL = st.∠, which implies that LNM is a straight line.

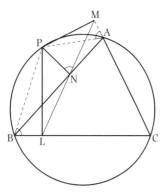

Fig. 5.2.36 Simson's line.

Quadrilateral PMAN is cyclic, since the angles at M on N are right angles by construction. Imagine it inscribed in a circle (Fig. 5.2.37); ∠PNM = ∠PAM, since both subtend arc PM. Also quad. PBCA is patently cyclic, since it is inscribed in a circle (fig. 5.2.36); hence ∠PBL = st.∠ − ∠PAC. Now ∠PBL = ∠B is equal to one of the two angles of interest to us, ∠PNM. The equality follows because (Fig. 5.2.36) ∠PAM is the supplement of ∠A, which is the supplement of ∠B; hence ∠PBL = ∠B = ∠PAM = ∠PNM.

It remains to show that ∠PBL = st.∠ − ∠PNL. That would be true if quad. PBLN is cyclic. Is it? Well, you have two right angles, BLP and BNP, standing on the same diameter BP. Q.E.D.

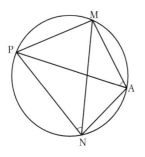

Fig. 5.2.37 Application of cyclic quadrilaterals to Simpson's line.

Since LNM is a straight line, PLM is a triangle, sometimes called the *pedal triangle* because it is defined with the help of the feet (Latin *pedes*) of perpendiculars. Whenever you have a lot of perpendiculars you might expect that something interesting can be proved via cyclic quadrilaterals. A good example is a property of the so-called *orthic triangle* (orthogonal, from the Greek for right angle) made by connecting the feet of the altitudes of a triangle. In Fig. 5.2.38, DEF is the orthic triangle; the curious property is that the altitudes AE, BF, and CD bisect its angles. We have all sorts of virtual cyclic quadrilaterals. Take FABE and FODA. From the first, ∠EAB = ∠EFB; from the second, ∠DFB = ∠OFD = ∠OAD = ∠EAB = ∠EFB: BF bisects DFE. In the same way, invoking other pairs of inscriptable quadrilaterals, you can show that CD bisects ∠FDE and AE bisects ∠DEF. Q.E.D.

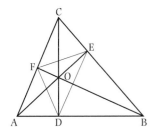

Fig. 5.2.38 The orthic triangle.

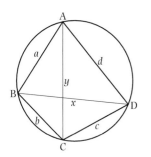

Fig. 5.2.39 Ptolemy's theorem by the cyclic quadrilateral.

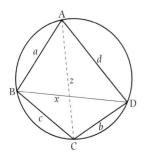

Fig. 5.2.40 Paramesvara's generalization of Ptolemy's theorem.

Ptolemy's theorem, again. You may recall that by a clever artifice (Fig. 4.3.4), Ptolemy showed that the product of the diagonals of a cyclic quadrilateral equals the sum of the products of its opposite sides; as expressed in Fig. 5.2.39, $xy = ac + bd$. The same expression can be obtained without artifice from the defining property of a cyclic quadrilateral. The law of cosines gives

$$x^2 = a^2 + d^2 - 2ad \cos A = b^2 + c^2 - bc \cos C.$$

But the cyclic character of quad. ABCD mades $\angle C = \text{st.}\angle - \angle A$. The previous equation may therefore be solved for $\cos A$:

$$2 \cos A \, (bc + ad) = (a^2 - c^2) + (d^2 - b^2), \tag{5.2.8}$$

which, after a little algebra, yields

$$x^2 = \frac{(ab + cd)(ac + bd)}{bc + ad}.$$

A similar calculation for the other diagonal gives

$$y^2 = \frac{(bc + ad)(bd + ac)}{cd + ab}.$$

Multiply the last two equations together:

$$x^2 y^2 = (ac + bd)^2,$$

which is Ptolemy's theorem, squared.

An Indian mathematician of the early fifteenth century, Parameśvara, had a brilliant idea for obtaining the preceding expressions for the squares on the diagonals of cyclic quadrilaterals without much work.[10] He observed that a third diagonal is defined by exchanging two adjacent sides of the quadrilateral, as in Figs 5.2.40 and 5.2.41. The diagonal x remains in the first figure and diagonal y in the second, since they subtend the same sides as in the original; but a third diagonal, z, is also required in both cases. Parameśvara applied Ptolemy's theorem to Figs 5.2.39, 5.2.40, and 5.2.41:

$$xy = ac + bd, \qquad xz = ab + cd, \qquad yz = ad + bc.$$

These equations can be solved readily for x, y, z, giving the previous equations for x^2 and y^2, and, for z^2,

$$z^2 = \frac{(ad + bc)(ab + cd)}{ac + bd}.$$

10. Gupta, *Hist. math.*, 4 (1977), 67–74.

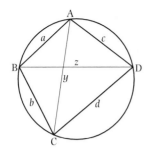

Fig. 5.2.41 Parameśvara's expression for the area of a cyclic quadrilateral.

Parameśvara also found a beautiful rule relating the three diagonals to the area, Q, of the quadrilateral and the radius R of its circumcircle. He wrote the equivalent of

$$Q = x(h + j)/2,$$

where h, j are the altitudes of $\triangle ABD$ and $\triangle CBD$, respectively (Fig. 5.2.42). But we know from the analysis leading to equation (5.2.5) that $h = ad/2R$, $j = bc/2R$. Therefore,

$$Q = (ad + bc)x/4R.$$

Now switch c and d, as in Fig. 5.2.41, and apply Ptolemy's theorem: $yz = ad + bc$. We have Parameśvara's rule,

$$Q = xyz/4R.$$

While playing with the cyclic quadrilateral, the Indian mathematicians found a generalization of Hero's formula that fit it. The generalization is exactly what you would expect:

$$Q = \sqrt{(s - a)(s - b)(s - c)(s - d)}, \tag{5.2.9}$$

although you might not have foreseen its restriction to cyclic quadrilaterals. The problem is easy to set up. From Fig. 5.2.42,

$$Q = \triangle BAD + \triangle BCD = x(h + j)/2.$$

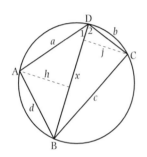

Fig. 5.2.42 Parameśvara's generalization of Hero's formula to cyclic quadrilaterals.

But $h = a \sin \angle 1 = ad \sin A/x$ (by the law of sines applied to $\triangle BAD$), and $j = b \sin \angle 2 = bc \sin C/x$; also, since ABCD is cyclic, $\angle A = \text{st.}\angle - \angle C$, and $\sin C = \sin A$. Altogether,

$$Q = \frac{ad + bc}{2} \sin A.$$

What to do for $\sin A$? Take it from the expression for $\cos A$ in equation (5.2.8). That gives

$$Q^2 = \frac{(ad - bc)^2}{16(bc + ad)^2} \times$$
$$[4(bc + ad)^2 - (a^2 - c^2)^2 - 2(a^2 - c^2)(d^2 - b^2) - (d^2 - b^2)^2].$$

Some algebra is needed to reduce this last expression to equation (5.2.9).

5.3. Regular polygons

Special cases

Circumscription and inscription always work for a regular polygon, that is, for a polygon all of whose sides and all of whose angles are equal. The square is easy. To inscribe a circle in a given square,

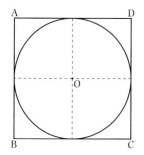

Fig. 5.3.1 To inscribe a circle in a square (Euclid IV.8).

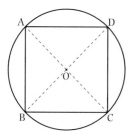

Fig. 5.3.2 To circumscribe a circle about a square (Euclid IV.9).

Fig. 5.3.3 The Pentagon, seat of the U.S. Defense Department. From Borklund, *Department* (1968), after p. 86.

erect the perpendicular bisectors of adjacent sides; their intersection locates the center of the inscribed circle, the radius of which is half the side of the square (Fig. 5.3.1). To circumscribe a circle about a square (Euclid IV.9), draw the square's diagonals; their intersection locates the center of the circumscribed circle, the radius of which is half the diagonal of the square (Fig. 5.3.2). You may wish to show that the area of a square inscribed in a semicircle is two-fifths the area of the square inscribed in the whole circle. But these are small things. Let us rise to something exotic.

The pentagon. The regular pentagon (a regular polygon with five sides) is perhaps best known from the shape of the headquarters of the U.S. defense establishment. This gigantic building (it occupies 29 acres and has a perimeter of a mile) was completed in 1943. Figure 5.3.3 gives an idea of its size and geometry. Coincidentally, the pentagon intrigued the military engineers of the Renaissance, who recognized in it an efficient plan for a fortification; cannon placed at the vertices could command the walls at convenient angles.[11] Figure 5.3.4 reproduces a drawing by Leonardo da Vinci, who designed forts as well as frescos, of a method of laying out pentagonal walls. Figure 5.3.5 shows an idealized realization of the same fort 150 years later, in a book on applied mathematics by Johann Christoph Sturm, the first professor in all the Germanies to teach experimental physics.

11. Kline, *Math. thought* (1990), *1*, 234; Gille, *Engineers* (1966), 149.

Fig. 5.3.4 Leonardo's method of
laying out pentagonal walls.
From Gille, *Engineers* (1966),
149.

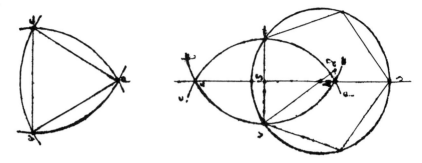

Fig. 5.3.5 An idealized
pentagonal fort, in Johann
Christoph Sturm, *Mathesis
compendaria* (1714). From
Bennett and Johnstone,
Geometry of war (1996), 78.

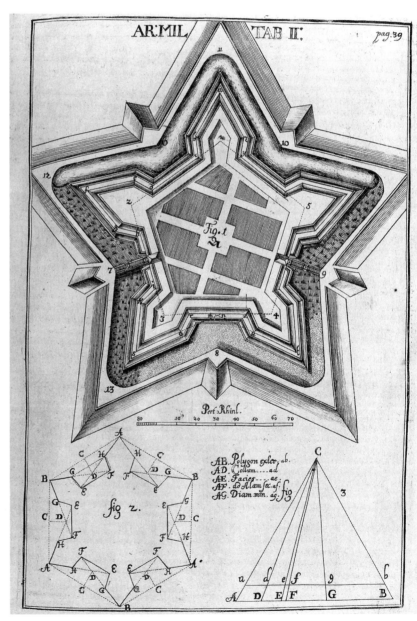

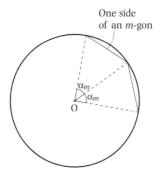

Fig. 5.3.6 Definition of an *n*-gon.

Fig. 5.3.7 The angle subtended by one side of a pentagon.

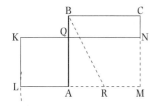

Fig. 5.3.8 Geometrical determination of the divine proportion (Euclid II.11).

The construction of a regular *n*-gon comes down to finding or making an angle α equal to $(2 \text{ st.}\angle)/n$ as appears from Fig. 5.3.6. For the square, $\alpha = (2 \text{ st.}\angle)/4 = \text{rt.}\angle$, which presents no problem. For the pentagon, $\alpha = (2 \text{ st.}\angle)/5 = 72°$. How to proceed? The construction, which appears to be a discovery of the Pythagoreans, excited great interest among ancient geometers. It comes to this: you must draw an isosceles triangle the base angles of which are each twice the remaining angle. That is, the Pythagoreans found how to draw $\triangle$ABC in Fig. 5.3.7. Since $5\phi = \text{st.}\angle$, $\phi = 36°$ and 2ϕ is the angle wanted.

The Pythagorean discovery related to the *divine proportion*, that is, to the division of a line into two parts so that the square on the larger part equals the rectangle made from the whole line and the smaller part—or, to speak algebraically, so that the square of one part equals the product of the line and the other part. Euclid shows how to divide a line that way in II.11. You might recognize the figure (5.3.8): it is one of those sand drawings that convinced the shipwrecked philosopher Aristippus that the Rhodians were civilized (frontispiece). AB is the given line, ALKQ the square on part of it, QNCB the rectangle made by the whole of it (BC = AB = *a*) and the remaining part QB. Euclid's geometrical solution completes the square ABCM, bisects AM at R, and locates the corner of the desired square at L on MA extended by making RL = BR. The area of the square is, accordingly, $(\text{BR} - \text{AR})^2 = a^2 (\sqrt{5} - 1)^2/4$, that of the rectangle $\text{BC} \times \text{BQ} = a^2 (3 - \sqrt{5})/2$; in a word, they are equal.

Euclid's solution may be paraphrased algebraically as the extraction of the unknown side of the square *x* from the equation $x^2 = a(a - x)$, which expresses the geometrical condition sought, $\text{AC}^2 = \text{AB} \times \text{BC}$. We would do the algebra by completing the square:

$$x^2 + ax = a^2, \quad (x + a/2)^2 = 5a^2/4, \quad x = (a/2)(\sqrt{5} - 1). \tag{5.3.1}$$

Fig. 5.3.9 The divine proportion as expressed by the Parthenon in the ratios DE : EA and AC : DA. From Ghyka, *Geometry of art and life* (1977), pl. LVII.

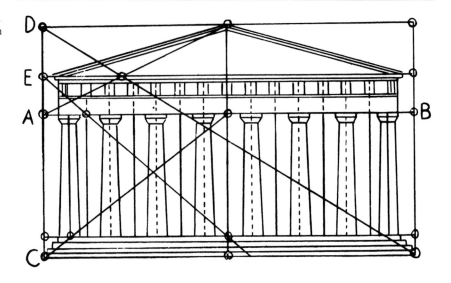

Fig. 5.3.10 Leonardo's drawing of the human body as specified by Vitruvius. The circle is centered on the navel, the square on the genitals; the circle's radius is in divine proportion to the square's side. From MacCurdy, *Note books* (1938), 1, 225. The legs are spread to 60°; the knees, penis, and nipples divide the height into equal quarters. Ibid., 225–6, and Millon and Lampugni, *Renaissance* (1994), 306.

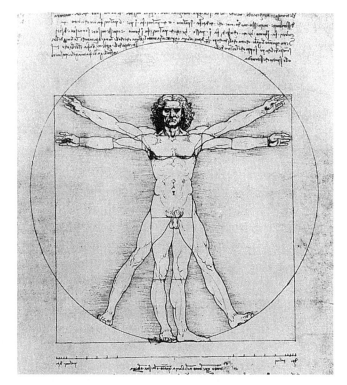

The divine proportion, so called because the ideal of classical beauty seemed to satisfy it (Figs 5.3.9, 5.3.10), has the numerical value $x/(a - x) = a/x = 2/(\sqrt{5} - 1) = (\sqrt{5} + 1)/2 = 1.618$. You can obtain $\sqrt{5}$ from a table of square roots or approximate it by the method introduced in §4.1. The obvious guess, that $5 = (2.2 + e)^2$, gives 2.236, which is accurate to four figures.

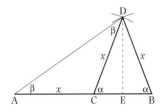

Fig. 5.3.11 Geometric method for obtaining the central angles of a pentagon.

Now take a line AB divided at C in divine proportion (Fig. 5.3.11). Erect on BC as base the isosceles triangle BDC whose sides equal AC = x. We are to prove that the isosceles triangle wanted is ΔADB, that is, that AD = AB and that ∠ABD = ∠ADB = 2∠DAB. (The following proof paraphrases Euclid IV. 10.) First, to prove AB = AD = a, invoke Euclid I.47:

$$AD^2 = DE^2 + AE^2 = (BD^2 - BE^2) + (AC + BE)^2$$
$$= x^2 + x^2 + x(a - x) = x^2 + ax = a^2,$$

Q.E.D. Here x has been written for AC = BD and $(a - x)$ for 2BE = BC; the last step uses the defining equation (5.3.1).

Let ∠DCB = ∠DBC = α. Then since, as just proved, ΔABD is isosceles, ∠ADB = ∠ABD = ∠BCD = α. And, since ΔACD also is isosceles, ∠DAC = ∠ADC = β (say). At last we have $\alpha = 2\beta$ (α is external to ΔACD) and, consequently, 5β = st.∠, whence $2\beta = \alpha = 72°$ is the central angle, earlier called 2ϕ (Fig. 5.3.7), of the regular inscribed pentagon. Hence, to inscribe a pentagon in a circle of any desired size, make an angle ABD as in the foregoing construction, taking B as the circle's center (Fig. 5.3.12). Let the sides of the angle cut the circle in E, F. Join E and F; EF is one side of the pentagon. Then, with compass at E and radius EF, cut the circle at G; EG is another side. Continue to H, I, and back to F to complete the figure.

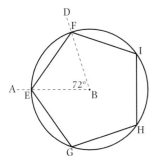

Fig. 5.3.12 To inscribe a pentagon in a circle.

Euclid's construction of the pentagon (IV.11) is more elegant and, perhaps, more suggestive of the manner in which it was discovered. He circumscribes an isosceles triangle ABC with base angles twice the remaining angle in a circle and then bisects the angles (Fig. 5.3.13), continuing the bisectors to the circumference at D and E. The points A, D, B, C, and E are the vertices of the pentagon. All the chords AD, DB, etc., are intercepted at the circumference by the equal angles ACD, DCB, etc., all of which equal ∠BAC, which is 36°. But the same chords intercept twice these angles, or 72°, as seen from the center (Euclid III.20).

To obtain a pentagon circumscribed about a circle, Euclid (IV.12) uses the divisions obtained in Fig. 5.3.13 (or in IV.11) to produce Fig. 5.3.14. The external pentagon is formed from the intersections of the tangents drawn to the circle where it meets the vertices of the internal polygon. We must show that β, the angle subtended by one (and therefore by all) the sides of the external pentagon, equals α, the angle subtended by the sides of the internal pentagon. Let FB be the radius of the circle and IJ the tangent at B. We know that FJ lies equidistant from the sides of BFC and therefore bisects it; hence $\alpha/2$ = ∠BFJ = $\beta/2$. This last step requires that BF bisect ∠IFJ, which it does: because the tangents dropped from an external point to a circle equal one another, ΔIFB ≅ ΔBFJ by *h.s.*

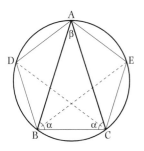

Fig. 5.3.13 The same another way (Euclid IV.11).

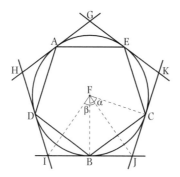

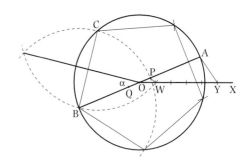

Fig. 5.3.14 To circumscribe a pentagon about a circle (Euclid IV.12).

Fig. 5.3.15 A medieval approximation to a pentagon.

A practical approximate method for drawing a pentagon may be devised from the observation that sin 36° = 0.588, which differs from 0.6 by only 2 per cent. If you ignore the difference, you can find the vertices by dividing a radius, OA say, into five equal parts. Use the method of parallels: draw a line OX (Fig. 5.3.15) at a convenient angle to OA and mark off five equal intervals on OX with your compass; join the end Y of the last division to A and raise a parallel to AY at the first point of division W to cut OA at P; OP = OA/5, as you can see immediately from the similar triangles OWP, OYA. With your compass at B, the end of the diameter AOB, draw an arc of radius BP intersecting the circle at C; BC is the side of the inscribed pentagon, approximately. For an exact solution, α = 36°; in this approximation, BC = $2a \sin \alpha$ = 6a/5, or sin α = 0.6, α = 36°52. As you can see by drawing through B an arc of a circle centered at C and by completing the pentagon by stepping BC around the circle centered on O, this approximation must have been the one Leonardo used to construct Fig. 5.3.4.

The medieval masons, who frequently needed fast approximate constructions for regular polygons, used the method indicated in Fig. 5.3.16. They drew the equal circles with centers A and B a common radius r apart; set out the common chord CD, ⊥AB; described an arc of radius r centered at D cutting one of the circles at J and CD at E; and extended JE until it intersected the other circle at K. From K they drew an arc of radius r cutting CD extended at L; and from L an arc of the same radius cutting the first circle at M. Then ABKLM was the desired pentagon.[12] By construction, all five sides of ABKLM equal r. How close do its angles come to equality?

Consider first the angles at A and B (= β). From $\triangle$NBK, sin γ = NB(sin∠KNB)/r. But by construction, ∠BAD = ∠ADJ = 60°, and

12. *Shelby, Gothic design* (1977), 115–18.

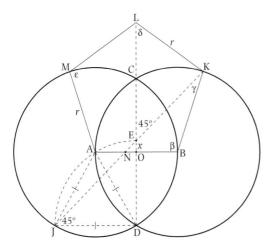

Fig. 5.3.16 Geometrical justification of **Fig.** 5.3.15.

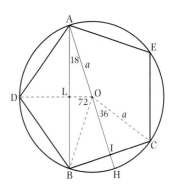

Fig. 5.3.17 To calculate the side of a regular pentagon.

$\angle ADO = 30°$ (sin $\angle ADO = 1/2$); hence $\angle EDJ = $ rt.$\angle$ and ΔJDE is an equilateral rt.Δ. Consequently $\angle KNB = \angle NJD = 45°$. Also, $x = EO = NO = ED - OD = r \,(1 - \sqrt{3}/2)$ and $NB = r/2 + x = (3 - \sqrt{3})/r/2$. Putting the values for NB and $\angle KNB$ into the equation for sin γ, you get $\gamma = 26.63°$ and $\beta = 180° - \gamma - 45° = 108.37°$. Next, $\angle MLK = 2\delta$, where sin $\delta = EK$ (sin $45°$)/r. But $EK = NK - x \sqrt{2}$, and $NK = $ (sin β) r/sin $45°$, making $2\delta = 109.17°$. Finally $\angle LMA + \angle LKB = 2\epsilon = 540° - 109.17° - 2(108.37°)$, or $\epsilon = 107.04°$. The mason's pentagon therefore had all its sides equal, two angles of 107°, two of a little over 108°, and one of a little over 109°, rather than 108° for all.

It remains to calculate the length of the side s_5 of a properly inscribed pentagon in terms of the radius of its circle. The easy way is to suppose $\angle EBF$ in Fig. 5.3.12 to be bisected, to see that $EF = s_5 = 2a \sin 36°$, and look up sin 36° in the tables. But that is no challenge. A more instructive way is to use the golden ratio, $\mu = 1.618$, and, for the purpose, to redraw Fig. 5.3.13 so as to introduce the circle's diameter AOH (Fig. 5.3.17). Since $\angle AOB = 144°$, $\angle BOH = 36°$, which makes $\angle BAH = 18°$ and $\angle ALO$ a rt.$\angle$. Consequently, triangles LAO and LBO being congruent, $AL = LB = AB/2$. Now by construction, $AB = \mu s_5$. From the right triangles ADL, ALO,

$$s_5^2 = AL^2 + (a - LO)^2, \qquad LO^2 = a^2 - AL^2.$$

Eliminating LO and then setting $AL = \mu s_5/2$, you should find $AL = s_5^2 - s_5^4/4a^2$ and

$$s_5 = a(4 - \mu^2)^{1/2} = a\left(\frac{5 - \sqrt{5}}{2}\right)^{1/2}. \qquad (5.3.2)$$

An Arab mathematician named Abū Kāmil looked at this problem over a thousand years ago. He found by the grace of God ('He the mighty and glorious') things inaccessible to everyone 'advanced in the science of calculation and geometry' of whom he had ever heard.[13] He had discovered a way to the side of the regular pentagon and decagon, inscribed and circumscribed, without passing through the golden ratio or the construction of Euclid's isosceles triangle with base angles twice the third angle. Instead, Abū Kāmil invoked Ptolemy's theorem about cyclic quadrilaterals, which, applied to Fig. 5.3.17, gives

$$AE \times BC + AB \times EC = AC \times BE, \qquad (5.3.3)$$

where AC and BE are the implied diameters of the quadrilateral ABCE. Evidently (or by symmetry) AC = BE = AB, so equation (5.3.3) is equivalent to

$$s_5^2 + ABs_5 = AB^2.$$

But we have already calculated $AB^2 = 4AL^2 = 4(s_5^2 - s_5^4/4a^2)$ without reliance on μ or Euclid's triangle; with this expression for AB, we recover equation (5.3.2)

To obtain s_{10}, the side of the inscribed decagon, Abū Kāmil chose as his diameter AOM parallel to the decagon side EH (Fig. 5.3.18). (That two sides of a regular pentagon and one of the corresponding decagon fill a semi-circle is indicated in Fig. 5.3.17, where s_{10} = BH subtends a central angle of 36°, as it must.) Ptolemy's theorem gives

$$2as_{10} + s_5^2 = AH \times EM = AH^2 = 4a^2 - s_5^2,$$

since $\angle AHM$ is a rt.$\angle$ (Euclid III.31). With the value of s_5 from equation (5.3.2),

$$s_{10} = a(\sqrt{5} - 1)/2.$$

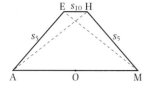

Fig. 5.3.18 To calculate the side of a regular decagon.

The hexagon. We rise to the six-sided figure (the hexagon). Here the central angle is one-sixth of two straight angles or one-third of a straight angle, or any of the angles of an equilateral triangle. One possible construction for a regular hexagon begins with a circle inscribed in an equilateral triangle and bisects its angles, as indicated in Fig. 5.3.19. Another method derives from the observation that, since the triangles GOF, FOI, etc., formed by radii and the sides of the hexagon, are equilateral, each side equals a radius. Therefore take any point on the circle's circumference (for example, A in Fig.

13. Lorch, in Folkerts and Hogendijk, *Vestigia* (1993), 216 (quote), 217, 234.

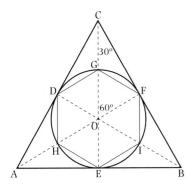

Fig. 5.3.19 To draw a hexagon.

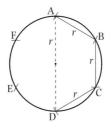

Fig. 5.3.20 To draw a hexagon, another way.

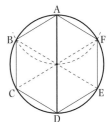

Fig. 5.3.21 To draw a hexagon, a third way.

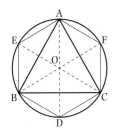

Fig. 5.3.22 To draw a hexagon, a fourth way.

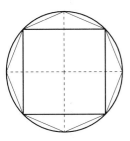

Fig. 5.3.23 To draw an octagon.

5.3.20), open the compass to the radius of the circle, and cut the circumference at B; move the compass to B, and repeat, this time marking the circle at C, and so on. A regular hexagon appears by connecting the marks. Since one half the hexagon occupies 1 st.∠, the point D lies opposite point A. Euclid's construction (Euclid IV.15) begins with a diameter AD. From A a circle with radius equal to that of the given circle is drawn, cutting the original circle in B and F (Fig. 5.3.21). The diameters from B and F meet the circle in C and E. The points A, B, C, D, E, F are the vertices of the regular hexagon.

A final method points to a generalization essential for finding the area and circumference of a circle. Begin with the inscribed equilateral triangle ABC (Fig. 5.3.22). Bisect the equal central angles AOB, BOC, and AOC and produce the bisectors to meet the circumference at E, D, and F. The inscribed hexagon has vertices at A, E, B, D, C, and F. Similarly (Fig. 5.3.23) an octagon can be produced from a square. The corresponding circumscribed figures can be constructed in the manner shown in detail for the pentagon.

A star hexagon is made by joining every other vertex in the manner indicated in Fig. 5.3.24 to give the form PQRSTD. If you draw an axis PS (a diameter of the circumcircle) and two parallel satellites (AC and BE extended) any equal distance x from it, then AB = CD. This peculiar theorem may well be the subject of discussion by Euclid and his students in Raphael's famous painting, *The School of Athens*; at least, all the lines except CD appear in a detail of the painting (Plate IV).[14] The theorem, which might be original with Raphael, is easy to prove. Let RT = $2s$. Then, from the symmetry of the figure, the sides of all the six small equilateral triangles with vertices at P, Q, R, S, T, and D equal $2s/3$. We have $q : (q + p) = 1 : 3$, where q and $(q + p)$ are the altitudes of the small and large triangles, respectively;

14. After Fichtner, *Verborgene Geometrie* (1984), 13–15, 20–2.

Fig. 5.3.24 A star hexagon, with indication of Raphael's theorem.

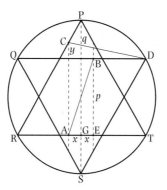

and, since $(q + p) = s\sqrt{3}$, $p = 2s/\sqrt{3}$, $q = s/\sqrt{3}$. Now $AB^2 = p^2 + 4x^2$, $CD^2 = (s + x)^2 + y^2$. From the similar triangles CAR and PGR, $(y + p) : (q + p) = (s - x) : s$, whence $y = \sqrt{3}(s/3 - x)$. You have after substitution,

$$AB^2 = p^2 + 4x^2 = 4s^2/3 + 4x^2,$$

$$CD^2 = (s + x)^2 + (s/\sqrt{3} - x\sqrt{3})^2 = AB^2.$$

Q.E.D

It would not do to leave the hexagon even provisionally without remarking that it shares with the equilateral triangle and the sphere uniquely among regular polygons the capacity to fill the plane, leaving no spaces. It therefore appears commonly in tiled walls and floors. Plates V–VI show some elegant examples.

The octagon. Architects of the classical tradition made frequent and eloquent use of the octagon. A powerful example is the base, or drum, of the cupola that Filippo Brunelleschi raised over the cathedral, aptly called the Duomo, in Florence. The gigantic structure has stood firm for over 500 years (Plate VII).

The third President of the United States, Thomas Jefferson, who prided himself on his architectural taste (he designed the main buildings of the University of Virginia as well as his own home and the State House in Richmond, Virginia) liked to put in octagons wherever he could (Fig. 5.3.25). Sometimes he had room only for six sides, or even fewer; so he devised (or discovered in his reading) a way to draw three sides of an octagon given the length of the chord subtending them. Fig. 5.3.26 shows his prescription, in his own hand.[15] The text reads:

To draw 3 sides of an octagon on the subtense a.b. geometrically.
[1] Bisect it by the line d.e.
[2] Take c.a. and lay it off towards d. at f.

15. Adams, *Jefferson's Monticello* (1983), pl. 28.

Fig. 5.3.25 A three-bay octagonal room designed by Thomas Jefferson for his house in Monticello, Virginia. The busts represent Benjamin Franklin, John Paul Jones, the Marquis de Lafayette, and George Washington. McLaughlin, *Jefferson* (1988), 280; photograph by R. Lautman/Monticello.

[3] On the center f. with radius f.a. describe the quadrant a.g.b.
[4] On the center g. with the radius g.a. describe the arc a.h.i.b.
[5] This arc cuts a.f. and b.f. at the angles of the octagon required.

Is it so? Does Jefferson's recipe produce an octagon?

To check, we need to show that the sides AH, HI, IB are equal and that each subtends an angle of $360°/8 = 45°$ at G, the center of the circumcircle. Figure 5.3.27 shows the piece of Jefferson's octagon inscribed in its circle. Let's follow his steps.

[1] Bisect AB at C by the line DE, which, since AB is a chord, must pass through the as yet unknown center G.
[2] Let AC = CB be a; then CF also is a.
[3] with F as center, draw an arc with radius AF cutting EF at what is the now-located center G; $FG = FA = FB = a\sqrt{2}$.
[4] with AG = r as radius, draw the circle AHIB intersecting AF at H and FB at I.
[5] H and I are the corners sought.

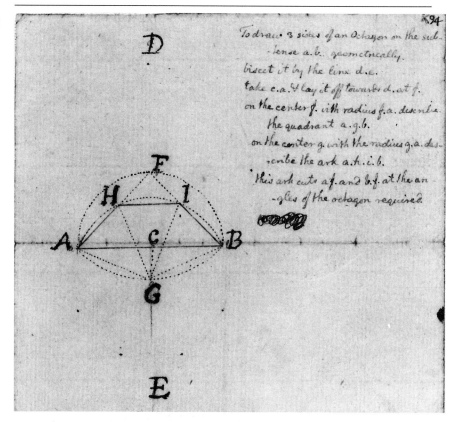

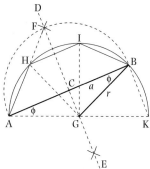

Fig. 5.3.27 Geometrical justification of Jefferson's method.

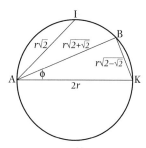

Fig. 5.3.28 Numerology of the octagon.

If the construction works, $\phi = \angle BAK$ must be one quarter of a right angle, i.e. 22.5°, since it is half $\angle BGK$ (Euclid III.20), which subtends at the center the putative side BK. Furthermore, if $\phi = 22.5°$, so does $\angle ABG$ (ΔAGB is isosceles), which makes $\angle GBI = 67.5°$ ($\angle ABF = 45°$ by construction); and thus, since $\angle GIB = \angle GBI$ (GB and GI are radii), the central angle IGB is 45°. Similarly you can show that $\angle AGH = 45°$ and, by subtracting $\angle AGH + \angle IGB + \angle BGK$ from the straight angle AGK, that $\angle HGI = 45°$. To prove the missing essential piece, that $\phi = 22.5°$, observe that $\angle FAG = \angle FGA$ (because FA = FG by construction) = rt.$\angle - \phi$ (because $\angle ACG$ = rt.$\angle$ by construction). You have then in ΔFAG, st.$\angle$ = 2rt.$\angle + \angle AFG - 2\phi$. But DE by construction bisects the right angle (it is inscribed in a semicircle) at F, so $\phi = 45°/2$.

The octagon lends itself to numerological indulgences. Figure 5.3.28 redraws figure 5.3.27 to show some of the relationship among the lines. We have AK = 2r. The side BK = $2r \sin \phi = 2r(1 - \cos^2\phi)^{1/2} = r\sqrt{2 - \sqrt{2}}$. (Remember $\cos^2\phi = (1 + \cos 2\phi)/2$ and $\cos 45° = \sqrt{2}/2$.) The subtense AB = $r\sqrt{2 + \sqrt{2}}$, as appears from applying the Pythagorean theorem to ΔABK. The subtense AI = $r\sqrt{2}$. You can discover many relations among these quantities, for example, $AI^2/BK \times AB = \sqrt{2}$.

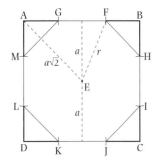

Fig. 5.3.29 A medieval method of drawing an octagon.

A good, quick, medieval way to draw an octagon is to make a square ABCD (Fig. 5.3.29), find its center E, and open your compass to half the square's diagonal, AE.[16] Then from A cut AB at F so that AF = AE; from B, cut BA at G, so that GB = AE; GF is a side of the octagon, which can be completed by treating the remaining sides of ABCD as you did AB. That FG is a side of an octagon inscribed in a circle of radius r appears by writing

$$FG = 2(AF - a) = 2a(\sqrt{2} - 1), FG^2/4 = r^2 - a^2,$$

whence $FG^2 = (2 - \sqrt{2})r^2 = s_8^2$.

APS 5.3.1. It is easy to show that the area of an inscribed hexagon is three-quarters that of a circumscribed one on the same circle. Look at Fig. 5.3.30. AB is a side of the inscribed, CD that of the circumscribed hexagon, both subtending 60° at the center O. Let Γ_i, Γ_c, be the areas. Then

$$\frac{\Gamma_i}{\Gamma_c} = \frac{OF \times FB}{OE \times ED} = \frac{r\cos 30° \times r\sin 30°}{r \times r\tan 30°} = \cos^2 30° = \frac{3}{4},$$

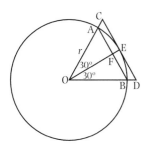

Fig. 5.3.30 Exercise on hexagons (APS 5.3.1).

Q.E.D.

APS 5.3.2. The side of an inscribed square is a mean proportional between the side of an inscribed equilateral triangle and that of a circumscribed hexagon. Can you prove it? The side of the square, AB, is $2r\sin 45°$ (Fig. 5.3.31); of the triangle, CDE, $2r\sin 60°$; of the hexagon, $2r\tan 30°$ (Fig. 5.3.30). You have

$$\frac{2r\sin 60°}{r\sqrt{2}} = \frac{r\sqrt{2}}{2r\tan 30°},$$

as required, since $4\cos 30° \tan 30° = 2$.

APS 5.3.3. The area of an inscribed octagon equals the area of the rectangle made by the sides of the inscribed and circumscribed squares. The area of the octagon is $8\triangle BGK$ (Fig. 5.3.27) $= 8r^2 \sin 22.5° \cos 22.5°$ (see Fig. 5.3.32) $= 4r^2 \sin 45° = 2r^2\sqrt{2}$. The side of the inscribed square is $r\sqrt{2}$, of the circumscribed square, $2r$. We have $2r^2\sqrt{2} = 2r \times r\sqrt{2}$, Q.E.D.

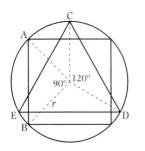

Fig. 5.3.31 Relationship among regular polygons (APS 5.3.2).

Euclid concludes his discussion of inscribed and circumscribed figures by a curiosity, a 15-gon (Euclid IV. 16). In Fig. 5.3.33, AC is the side of an inscribed equilateral triangle, AB that of an inscribed pentagon. If the entire circle were divided into 15 parts, the arc AB would cover three of them (5 sides times 3) and the arc AC five of them (3 sides times 5). Therefore, the arc BC would cover two.

16. Shelby, *Gothic design* (1977), 120–1.

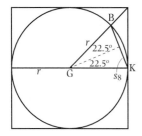

Fig. 5.3.32 Exercise on an octagon (APS 5.3.3).

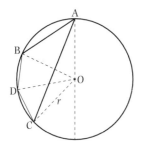

Fig. 5.3.33 To inscribe a 15-gon (Euclid IV.16).

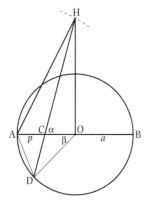

Fig. 5.3.34 Renaldini's approximation to the side of an n-gon.

('Cover' means that the arc subtends an angle at the center of the circle that would embrace 3, 5, or 2 sides of the 15-gon.) Now bisect $\angle BOC$ and continue the bisector to cut the circumference at D. Then $\angle BOD = (2 \text{ st.}\angle)/15$ and BD is the side of the regular 15-gon.

Renaldini's approximation. A mathematician of the seventeenth century, Carlo Renaldini, who, you may well think, deserves his obscurity, suggested a way to draw the side of an inscribed n-gon swiftly though, alas, only approximately in most cases. The method (Fig. 5.3.34): divide the diameter $AB = 2a$ into n parts and let $AC = p = 4a/n$ be two of these parts; describe an arc centered at A of radius $2a$ until it intersects at H with the perpendicular raised at O; draw a line from H through C to strike the circle at D; AD is the side of an n-gon, sometimes. To check the value of the method, you might calculate β, the central angle subtended by the putative side AD.[17]

It's a job for the law of sines. From $\triangle OCD$,

$$\sin \alpha = a \frac{\sin(\alpha - \beta)}{a - p}.$$

Putting x for p/a into the preceding equation, we have

$$1 - x = \cos \beta - \text{ctn } \alpha \sin \beta = \cos \beta - \sin \beta \times (1 - x)/\sqrt{3},$$

where $\text{ctn } \alpha = OC/OH = (a - p)/\sqrt{3}a = (1 - x)/\sqrt{3}$. The indicated algebra produces a very disagreeable expression for β in terms of x:

$$\sin \beta = \frac{\sqrt{3}[(1 - 2x^2 + 4x)^{1/2} - (1 - x)^2]}{x^2 - 2x + 4}. \tag{5.3.4}$$

Try a few values of $n = 4a/p = 4/x$ in equation (5.3.4): You may arrive at the following results (β_R signifies Renaldini's β, β_C the correct value):

Table 5.3.1 Renaldini's approximations

n	x	$\sin \beta$	β_R°	β_C°
3	4/3	$\sqrt{3}/2$	120	120
4	1	1	90	90
6	2/3	$\sqrt{3}/2$	60	60
8	1/2	0.7034	44.7	45
12	1/3	0.5071	30.5	30

17. Cantor, *Vorl.* (1907), 3, 21–2, who ascribes the method to Antoine de Ville and to 1628; Kästner, *Geometrische Abhandlungen*, 1 (1790), 266–81.

The construction works perfectly for the equilateral triangle, square, and hexagon, well for the octagon and 12-gon, and then becomes progressively worse as the number of sides increases. At the limit, when n is very large, the ratio of the circumference of a circle to its diameter implied by the construction reaches $2\sqrt{3} = 3.4644$, which is worse than the biblical value 3. Renaldini's construction was resurrected as a school exercise in 1911, in the style of the *Ladies diary*, by a Mary Sabin of Denver, who did not let on that someone had thought of it before.[18]

The polygon and the cathedral

You have met an example of polygons in cathedrals in the octagonal drum on which the vast dome of the Florentine cathedral rests (plate VIII). Another place to seek polygons is in Gothic windows, where, however, you must look hard, since they are usually present only spiritually.

The polygon in the window. Figure 5.3.35 is a typical presentation of virtual polygons. Perhaps the best way to see the polygon symmetry is to imagine the centers of the touching circles making up one of the four-lobed figures, or quadrifoils. Now imagine the centers of neighboring circles connected by straight lines. The result is a square

Fig. 5.3.35 The choir of the cathedral at Sées (ca. 1300). From Binding, *Masswerk* (1989), 83.

18. Sabin, *School sci. math.*, 11 (1911), 164 (February, the problem proposed), 364–5 (April, the problem solved).

Fig. 5.3.36 To draw a quadrifoil.

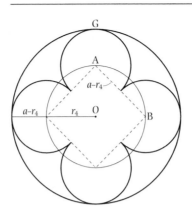

Fig. 5.3.37 Tracery in the cathedral of Amiens (early fourteenth century). From Binding,

inscribed in a virtual circle on which the centers lie. If you make this circumscribing circle manifest, you will be able to calculate its radius and thereby find the rule for drawing the tracery of Fig. 5.3.35, which depicts a window in the choir of the cathedral of Sées dating from around 1300.

Figure 5.3.36 is the sort of drawing you will make. Since the construction has the symmetry of a square, you know that the centers of the touching circles must be 90° apart as seen from the center O; since you are trying to find the radius $r_4 = OA = OB$ of the loci of the centers, you will have to fiddle to draw the circles centered on A and B so as to be tangent to one another and to the circle center O, and also to have their centers on mutually perpendicular lines. The calculation that will relieve you from further fiddling, at least in the case of the quadrifoil, may be accomplished easily by the obvious step of joining A and B. Then, setting the radius $OG = a$, you have from right ΔAOB,

$$2r_4^2 = AB^2 = [2(OG - r_4)]^2 = 4(a - r_4)^2,$$

from which $r_4 = a(2 - \sqrt{2}) \sim 0.585a$. If you check your cut-and-try drawing, you should find $OA \sim 0.6a$. If you are perplexed about the expression for AB in the previous equation, remember that the circles centered at A and B are tangent and that, therefore, AB is twice the circles' radius, or $2(a - r_4)$.

Now that you have the method, you can obtain the general case and make windows with any number of lobes. Figure 5.3.37, from the cathedral of Amiens, has eight. The general case gives rise to Fig. 5.3.38, where α_n is the angle subtended by one side of the virtual polygon. Drop the perpendicular OP, which bisects both α_n and AB (ΔAOB is isosceles). You have $\sin \alpha_n/2 = PB/OB = (a - r_n)/r_n$, and, solving for r_n,

$$r_n = a/(1 + \sin \alpha_n/2).$$

Table 5.3.2 Radii defining n-foils

n	α_n	$\alpha_n/2$	$\sin \alpha_n/2$	r_n
3	120°	60°	$\sqrt{3}/2$	$0.536a$
4	90°	45°	$\sqrt{2}/2$	0.585
5	72°	36°	0.588	0.630
6	60°	30°	1/2	0.667
8	45°	22.5°	0.383	0.723
12	30°	15°	0.383	0.794

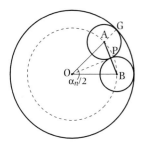

Fig. 5.3.38 To draw an n-foil.

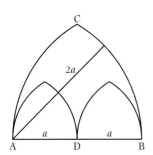

Fig. 5.3.39 An equilateral arch with two subsidiary lights.

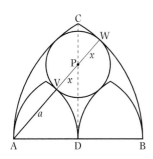

Fig. 5.3.40 An equilateral arch with a large circular window.

For the quadrifoil, $\alpha_n = 90°$ and $r_n = a/(1 + \sqrt{2}/2)$, which, after multiplying the top and bottom of the right side by $(1 - \sqrt{2}/2)$ returns the previous result, $r_4 = a(2 - \sqrt{2})$.

It might be useful to construct a table of r_n for the cases frequently met in practice (Table 5.3.2). You see that the radius gradually enlarges; it is impractical to go beyond a 12-foil except for very large windows.

The window in the cathedral. Circular windows usually stand in or betwen Gothic arches, fitted, of course, so as to touch three circular arcs. You have seen examples (Figs. 2.3.14–2.3.15, 5.3.35, 5.3.37). The geometry of the simplest case is illustrated in Figs. 5.3.39 and 5.3.40. The first reproduces the standard arch you encountered in connection with Euclid I.1 and fits it with two smaller arches each with half the span of the larger and tangent to it and to one another. The second figure adds a complete circle tangent to all three Gothic arcs. Its center obviously lies on the line of symmetry CD. Can you find its radius x?

From Euclid III.11 and III.12, you know that A, V, and P, and also A, P, and W are collinear. Hence AVPW is a straight line. But AP = $a + x = 2a - x$, so $x = a/2$. To find the center P, note that symmetry requires that it lie on CD and Pythagoras says that it stands a distance PD = $[(a + x)^2 - a]^{1/2} = a\sqrt{3}/2$ above AB. You have all the information necessary to draw the circle centered at P. Try it. The form of Fig. 5.3.40 occurs in the windows of the bishops' palace at Sens (Fig. 5.3.41) and, more simply, in the abbey of Saint-Jean-des Vignes in Soissons (Fig. 5.3.42).

Now consider the form in Fig. 5.3.43. The small arcs have radius $a/3$; hence $2x + a/3 = a$, $x = a/3$. The big circle sitting on the point of the middle small arch (which occurs in any odd number of small arches) does not make an overly pleasing design although it goes well among the flying buttresses of Bourges Cathedral (Fig. 5.3.44). Usually Gothic architects subdivided their large arches into an even number of small ones, as at Amiens (Fig. 5.3.37), which continues

Fig. 5.3.41 The bishop's palace
at Sens (1235–40). From
Binding, *Masswerk* (1989), 59.

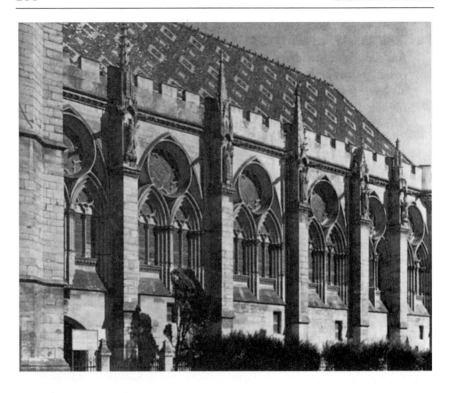

Fig. 5.3.42 A window in the
abbey of Saint-Jean-des-Vignons
in Soissons (ca. 1240). From
Binding, *Masswerk* (1989), 61.

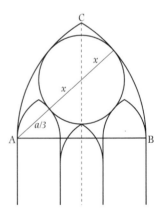

Fig. 5.3.43 An equilateral arch
with three subsidiary lights and
a circular window.

Fig. 5.3.44 The plan of
Fig. 5.32.43 realized, at the
cathedral of Bourges. From
Bony, *French Gothic cathedrals*
(1983), 206.

the pattern of Fig. 5.3.40 by dividing each smaller arch into two
others. You can easily draw it, since you know that the radius of a
circular window in this design is one-quarter that of the enclosing
arcs. The scheme was used to great effect in the cathedral at
Strasbourg (Fig. 5.3.45). The division was sometimes repeated, to
give eight long slender windows as in the east end of Lincoln
cathedral (Fig. 5.3.46).

Another way—apart from replication in subdivision—to obtain an
even number of sub-arches is by direct division of the major arch.
That gives an odd number of circular windows. Figure 5.3.47 shows
the result for division by four, which is perhaps the largest practicable, since beyond that the subarches become unaesthetically small; it
yields the pleasant curvilinear figure GHIJK. Can you show that the
radius of the equal circles centered at P and Q is $7a/20$, and that of
the circle centered at Q is $8a/20$, where $AB = 2a$? [$PW^2 = (2a - x)^2 - (3a/2)^2 = (x + a/2)^2 - (a/2)^2$, and $2a - y = a/2 + 2x + y$, where x, y,
are the radii of the circles centered at P, Q, respectively.]

Fig. 5.3.45 Four lights with
three roundels, as realized in the
cathedral of Strasbourg. From
Bony, *French Gothic cathedrals*
(1983), 393.

Fig. 5.3.46 Eight lights with seven roundels, as realized at Lincoln Cathedral. From Bony, *French Gothic cathedrals* (1983), 418.

5.4. Not so easy as pie

The circumferences of circles stand in a constant ratio to their diameters, just as the perimeters of squares bear a constant ratio to their sides. It is hard to conceive the relationship otherwise. The ratio is easy to calculate for squares (it is $2\sqrt{2}$), but surprisingly difficult for circles. It cannot be expressed as a number, even an irrational one, and consequently has been given its own special symbol, the Greek letter π ('pi'). The invention of a procedure for estimating its value to any desired degree of accuracy was one of the major achievements of Greek geometry. The achiever was Archimedes.

Archimedes' method

Archimedes' method for estimating the circumference and area of a circle has two steps. The first establishes that the area is that of a right triangle one leg of which equals the circumference, and the

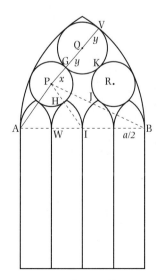

Fig. 5.3.47 Scheme of four lights and three circles.

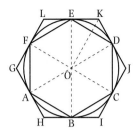

Fig. 5.4.1 Archimedes' method applied to hexagons.

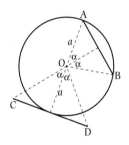

Fig. 5.4.2 Side and central angle of inscribed and circumscribed *n*-gons.

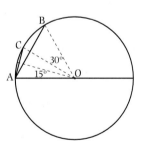

Fig. 5.4.3 Calculation of the side of a 12-gon.

other the radius, of the circle. In symbols, $A = rC/2$, area = (1/2) radius × circumference. The second step calculates C in terms of r by an approximation involving the perimeter p_i of an inscribed, and the perimeter p_e of a circumscribed polygon of 96 sides. Since $p_i < p_e$, the exact value of the circumference of a given circle must lie between the perimeters of its circumscribing and inscribing polygons. Figure 5.4.1 illustrates this proposition in the case of a hexagon. ABCDEF is the inscribed hexagon. Since each of its sides equals the radius a of the circle, its perimeter, p_i, is $6a$. The side of the circumscribed hexagon, GHIJKL, is $2a \tan 30° = 2a/\sqrt{3}$. Therefore, $p_e = (12/\sqrt{3})a$ $= 4a\sqrt{3}$. If you estimate the root in the way introduced earlier, you have $\sqrt{3} = 1.7323$, so that C lies between $6a$ and $6.9292a$.

To know that a number lies between 6 and 7 may not be thought sufficient. To do better, you need only to increase the number of sides in the polygons. The relevant geometry for an n-gon is indicated in Fig. 5.4.2. AB is a side of the circumscribed figure, CD one of the circumscribed figure. The angles AOB and COD each equal $360°/n$, making $\alpha = 180°/n$. The length of the side s_i of the inscribed n-gon is $s_i = AB = 2a \sin \alpha$; that of the side s_e of the circumscribed n-gon is $s_e = CD = 2a \tan \alpha$. The circumference therefore lies between $p_i = ns_i = 2na \sin \alpha$ and $p_e = ns_e = 2na \tan \alpha$. For the case Archimedes treated, $n = 96$, $\alpha = 180°/96 = 1.875°$. At this juncture, a modern geometer would punch a calculator to find $\sin 1.875° = 0.03272$ and $\tan 1.875° = 0.03274$, and thus that $p_i = 6.2821\,a$ and $p_e = 6.2854\,a$. A sturdier type, in the ancient tradition, would have had to calculate the trigonometrical values by slogging, as follows.

A 96-gon can be created from a hexagon by successive bisection of the central angles. In Fig. 5.4.3, AB is the side s_i of the internal hexagon; its bisector, extended to the circumference at C, marks a vertex of a side AC of a dodecagon (a 12-gon). We have for side AC, $s_i = 2a \sin 15°$. Since we know the values of the trigonometric functions for 30° ($\sin 30° = 1/2$, $\cos 30° = \sqrt{3}/2$, $\tan 30° = 1/\sqrt{3}$), we can obtain those for 15° by invoking the formulas for the functions of the sums of angles. It is easiest to use the formula for the cosine:

$$\cos 2x = \cos(x + x) = \cos^2 x - \sin^2 x = 1 - 2 \sin^2 x,$$
$$\sin^2 x = (1 - \cos 2x)/2.$$

Therefore $\sin^2 15° = (1 - \cos 30°)/2 = (1 - \sqrt{3}/2)/2$. With $\sqrt{3} = 1.7323$, $\sin^2 15° = 0.06692$. The root can be extracted by our usual method of approximation or by the calculator to give $\sin 15° = 0.2588$. For the 12-gon, therefore, $p_i = 12 \times 2a(.2588) = 6.1392a$.

Another division in the manner indicated in Fig. 5.4.3 gives a 24-gon, another, a 48-gon; and a third, Archimedes' goal, the 96-gon.

Table 5.4.1 Archimedes' calculation of π

Figure	$\alpha = 180°/n$	$1 - \cos 2\alpha$	$\sin \alpha$	$p_i = 2 na \sin \alpha$
Hexagon	30°	0.5	0.5	$6a$
12-gon	15°	0.1339	0.2587	$6.2099a$
24-gon	7.5°	0.0341	0.1305	$6.2652a$
48-gon	3.75°	0.008555	0.06540	$6.2784a$
96-gon	1.875°	0.002141	0.03272	$6.2821a$

Table 5.4.2 Archimedes' calculation of π, continued

Figure	$\alpha = 180°/n$	$\sin \alpha$	$\tan \alpha$	$p_e = 2 na \tan \alpha$
Hexagon	30°	0.5	0.5773	$6.9276a$
12-gon	15°	0.2558	0.2679	$6.4296a$
24-gon	7.5°	0.1305	0.1316	$6.3168a$
48-gon	3.75°	0.06540	0.0655	$6.2880a$
96-gon	1.875°	0.03272	0.03274	$6.2854a$

The calculation can be arranged as in Table 5.4.1. As we see from the increase in the values of p_i, the circumference of a circle of radius a is larger than $6.2821a$.

Archimedes fixed an upper bound to the circumference by calculating p_e for a 96-gon. Since $p_e = 2na \tan(180°/n)$, the calculation can be arranged in a manner similar to the foregoing (Table 5.4.2). As appears from the tendency of the values of p_e, the circumference of a circle of radius a is less than $6.2854a$.

The ratio of the circumference of a circle of radius a to its diameter $2a$ has been designated by π since the early eighteenth century. Presumably William Jones, who introduced it,[19] chose the Greek letter equivalent to Roman p, the first letter in periphery or perimeter. Archimedes' calculation bracketed π between $p_i/2a = 3.1411$ and $p_e/2a = 3.1427$. Now $1/7 = 0.1429$ and $10/71 = 0.1408$. Archimedes delivered his result in the form

$$3\ 10/71 < \pi < 3\ 1/7,$$

which is one of the most famous relationships in mathematics. The approximate value, $\pi = 3\ 1/7$, became standard among the Roman agrimensores and their lineal descendents the medieval masons.[20]

---

19. Beckmann, *History of π* (1971), 145.
20. Shelby, *Gothic design* (1977), 65, 121.

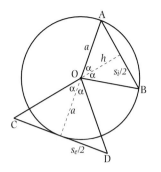

Fig. 5.4.4 Area of an *n*-gon.

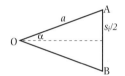

Fig. 5.4.5 The side of a circumscribed *n*-gon in terms of its central angle.

As we know, Archimedes equated the area of a circle to half the product of its radius by its circumference, or

$$\text{area} = a(\pi 2a)/2 = \pi a^2.$$

This expression, which he obtained by *reductio ad absurdum*, can be derived easily from the area of the inscribed and circumscribed polygons. Figure 5.4.4 indicates the quantities needed to calculate the areas of the polygons. The area A_e of the external *n*-gon is $n \times \text{area} (\Delta OCD) = n(s_e a/2)$. Hence

$$A_e = ap_e/2.$$

Similarly, $A_i = n \times \text{area}(\Delta OAB) = n(s_i h/2)$, or, since $h = a \cos \alpha$,

$$A_i = (ap_i \cos \alpha)/2.$$

The area of the circle thus stands bracketed between $ap_e/2$ and $ap_i \cos \alpha/2$.

It is easy to see that as *n* becomes very large, and, consequently, the length of the side s_i becomes very small, $\cos \alpha$ approaches the value of 1 as closely as you please. For consider ΔAOB of Fig. 5.4.4, redrawn in Fig. 5.4.5. You have

$$\cos^2 a = 1 - \sin^2 \alpha = 1 - s_i^2/4a^2.$$

As s_i becomes very very small in comparison with *a* (that is, as *n* becomes very very large), $s_i^2/4a^2$ becomes negligible in comparison with 1; $\cos^2\alpha = 1$, whence $\cos \alpha = 1$. Thus for very large *n*, $A_i \sim ap_i/2$. The approximation can be made as good as you please merely by increasing the value of *n*.

We now have the areas of the polygons inscribed and circumscribed about a circle equal to one-half the product of their perimeters with the radius of the circle. The relation holds exactly for the circumscribed polygon and as exactly as one wishes for the inscribed polygon in the case of indefinitely large *n*. But the regime of large *n*—of polygons with a very large number of sides—is where their perimeters approach as closely as one pleases to the circumference of the circle. It is the region where $p_i \sim p_e \sim 2\pi a$.

Archimedes' indirect demonstration that the area of a circle is half the product of its circumference and its radius works with a double reduction to absurdity.[21] Let A_c be the area of the circle and assume that it exceeds πa^2. Let the difference be *E*. But this is impossible, as you can see by imagining the number of sides of the inscribed polygon to increase until its area A_i falls short of A_c by an amount $G < E$. (You can always make $G < E$, no matter how small *E* is sup-

---

21. Archimedes, 'Measurement of a circle', Prop. 1, in *Works*, ed. Health (1952), 90–1.

posed to be, by increasing the number of sides.) You must satisfy simultaneously

$$A_c = \pi a^2 + E, \qquad A_c = A_i + G, \qquad E > G,$$

or

$$A_i = \pi a^2 + (E - G) > \pi a^2.$$

Now $A_i = (ap_i\cos\alpha)/2$ and $\pi a^2 = aC/2$; since $p_i\cos\alpha < p_i$, and $p_i < C$, $A_i < \pi a^2$. But A_i cannot be both $>$ and $< \pi a^2$; hence the hypothesis that A_c exceeds πa^2 must be rejected.

To complete the demonstration that $A_c = \pi a^2$, you must show that the hypothesis $A_c = \pi a^2 - F$ also leads to disaster. You now multiply the sides of the circumscribing polygon until the difference between A_e and A_c is $H < F$. You must satisfy simultaneously

$$A_c = \pi a^2 - F, \qquad A_e = A_c + H, \qquad F > H,$$

or

$$A_e = \pi a^2 - (F - H) < \pi a^2.$$

But $A_e = ap_e/2 > aC/2$ since $p_e > C$, so that $A_e > \pi a^2$. Since A_e cannot be both $>$ and $< \pi a^2$, we must also give up the hypotheses that A_c falls short of πa^2.

Other methods

Our Egyptian scribe Ahmose had a way of estimating the areas of circles that correspond to a value of π of $4(8/9)^2 = 3.1605$. At least that follows from his instruction, 'subtract from the diameter its ninth part, and square the remainder'.[22] A clue to the derivation of this remarkable formula, which has been known for 4000 years, occurs in Ahmose's illustration of the problem (Fig. 5.4.6). If the illustration is interpreted as an irregular octagon inscribed in a square, it can be redrawn more accurately as in Fig. 5.4.7.

Ahmose says that if the side of the square is 3 l.u., the area of the octagon derived from it is $3 \times 3 - 4(1 \times 1)/2 = 7$ a.u. Since he uses numbers as examples of a general formula, it may not misrepresent his thought to write the area of the octagon as $d^2 - (2/9)d^2$, that is, to write d, the length of the side of the square, for its exemplary value 3. Now as appears from Fig. 5.4.8, the circle inscribed in a square has an area very close to that of Ahmose's octagon. Since the circle's diameter is d, we have

$$\text{area of} \bigcirc = \pi d^2/4 \sim \text{area of octagon} = d^2(1 - 2/9).$$

---

22. Gillings, *Mathematics* (1982), 139–46.

Fig. 5.4.6 Egyptian diagram for the area of a circle. From Chace *et al.*, *The Rhind papyrus* (1927), problem 48, vol. 2, plate 70.

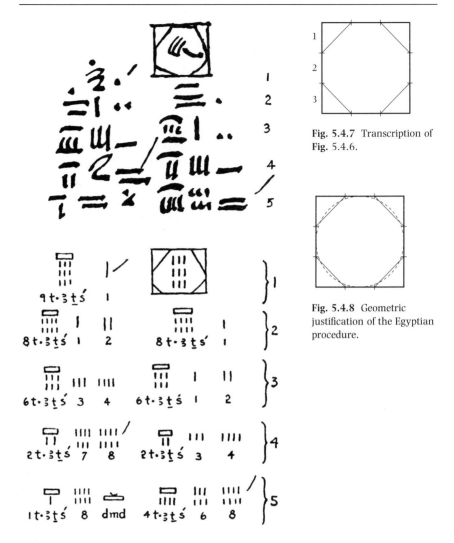

Fig. 5.4.7 Transcription of Fig. 5.4.6.

Fig. 5.4.8 Geometric justification of the Egyptian procedure.

This relation yields $\pi = 3.1111$. Ahmose did better than that. Wishing to express the area of a circle as the area of a square, he replaced $d^2(1 - 2/9)$ with $d^2(1 - 1/9)^2 = d^2(1 - 2/9 + 4/81)$. With this new rule, $\pi = 3.1605$, closer than 3.1111 to the correct value to five figures, 3.1416. Whether Ahmose just guessed luckily, whether he knew the equivalent of $(1 - x)^2 = 1 - 2x + x^2$, or whether he knew by measurement that $(8d/9)^2$ came closer than $d^2(1 - 2/9)$, is not known.

There is a possibility, worth recording for its geometrical interest, that Ahmose (or his source) arrived at his value of π by adapting a game with beads.[23] Start with a bead of radius r at the center C (Fig. 5.4.9). Fit a ring of beads around the center one; these beads will

23. Joseph, *Crest*(1991), 84.

Table 5.4.3 The bead game and π

	No. beads	Total	Greatest integer	R
Center	1	1	1	
Ring 1	2π	$2\pi + 1$	7	$3r$
Ring 2	4π	$6\pi + 1$	19	$5r$
Ring 3	6π	$12\pi + 1$	38	$7r$
Ring 4	8π	$20\pi + 1$	64	$9r$

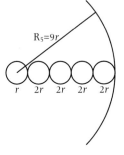

Fig. 5.4.9 A possible route to the Egyptian value of π.

have their centers on a circle of radius $2r$, which has a circumference of $2\pi(2r)$. The ring therefore can accommodate a maximum of 2π beads, which amounts in practice to 6. A second ring, outside the first and touching it, can receive the centers of at most 4π beads, that is, 12. Proceeding in this way, you can fill up four rings, with radius R, as in Table 5.4.3. Hence a circle of radius $9r$ can contain 64 beads of radius r. (The final arrangement will not have concentric rings, as in the calculation, since the bits left over in each ring combine together to make up whole beads.) But so will a square of side $16r$. Therefore, roughly, $256r^2 = \pi(81r^2)$, or $\pi = 3.1605$.

Other ancient values of π did not match Ahmose's in accuracy. Despite their long sojourn in Egypt, the Jews preferred a value of $\pi = 3$, or so people deduce from the Old Testament, which, however, for all its virtues, is not a book on geometry. The passage that implicitly defines π comes in the first Book of Kings (vii.23), which runs as follows in the *New English Bible*: 'He then made the Sea of cast metal; it was round in shape, the diameter from rim to rim being ten cubits; it stood five cubits high, and it took a line thirty cubits long to go round it.' We have circumference/diameter = π = 3. The 'he' in the passage was Hiram of Tyre, King Solomon's sculptor in bronze; the 'Sea', a raised bath or swimming pool, allowed for comfortable bathing, since a cubit equalled about a foot and a half.

The Vedic scriptures describe a circular altar with a radius $r = a/2 + (1/3)(a/\sqrt{2} - a/2)$, where a^2 is the altar's area. The text therefore implies a value of π deducible from

$$\pi r^2 = \pi (2 + \sqrt{2})^2 a^2/36 = a^2,$$

or

$$\pi = 36/(4 + 4\sqrt{2} + 2).$$

Using the standard Indian value $\sqrt{2} = 1 + \frac{1}{3} + \frac{1}{3} \times 4$, one finds a Vedic $\pi = 108/35 = 3.0857$.[24]

---

24. Kulkarni, *Layout* (1987), 43–4.

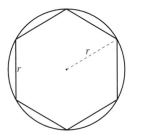

Fig. 5.4.10 A value of π derived from the hexagon.

The approximation $\pi = 3$ amounts to setting the circumference of a circle equal to the perimeter of its inscribed hexagon. (Since the side of the hexagon equals the radius, the approximation makes $2\pi r = 6r$, or $\pi = 3$.) The ancient Babylonians looked at the picture (Fig. 5.4.10) and knew they could do better. Their improvement slumbered for millennia in a royal sepulchre filled with clay tablets, which French archaeologists dug up in 1936 at Susa, some 200 miles east of Babylon. One tablet uses the value $\pi = 3$; but another lays down, without explanation, that the perimeter p_6 of a hexagon is 0.96 the circumference of the circumscribing circle.[25] In this approximation, $2\pi r = p_6/0.96 = r/0.16$, or $\pi = 3.125$. In explanation, we might suppose that the Babylonian geometers, noting that π exceeds 3, estimated the excess at an eighth, and computed backwards to get $p_6 = 0.96$ of the circumference.

The ancient Chinese practiced Archimedes' method. Liu Hui, who, as we know, worked in the third century AD, sweated over a polygon of 3072 sides to obtain $\pi = 3.14159$, a number still used as the best short approximation. A calculation of the fifth century, with a polygon of 24 576 sides (!), yielded 3.141 592 920 3. But what distinguished the Chinese pi-men (and also, after them, the Indians), was the strange value $\pi = \sqrt{10}$. Some say that it derived from the guess that the square of the circumference equals about five-eights of the square of the perimeter of the circumscribing square, that is, $(2\pi r)^2/(8r)^2 = 5/8$, which does indeed produce $\pi = \sqrt{10}$. It is perhaps more likely that someone (see Exercise 5.5.27 for a possible identification) noticed that the values of π obtained by successive doublings of the sides of the inscribed polygon increased from less than 3 in the case of the square to over 3.14 for a 3072-gon and, being decimally inclined, supposed that, with enough doubling, it might reach $3.16 \sim \sqrt{10}$.[26]

When European mathematicians of the Renaissance became acquainted with Archimedes, they too thought they could improve upon their original. Not that Archimedes' method was in any way formally deficient: by taking polygons with enough sides, the calculator can obtain π to as many decimal places as he or she desires. But Archimedes' method produces better decimals only slowly. Viète resorted to a polygon of 393 216 sides and got only to nine decimals (3.141 592 653), or one decimal for over 40 000 sides,[27] which many did not consider a good bargain. To double the decimals, to 20, a German computor, Ludolf van Ceulen, had to invoke a figure

25. Neugebauer, *Exact sciences* (1957), 46–7; Bruins, Acad. sci., Amst., *Proceedings* (1950).

26. Needham, *Science, 3* (1970), 100–1; Joseph, *Crest* (1991), 188–96; Libbrecht, *Chinese math.* (1973), 37; Hobson, *Squaring the circle* (1913), 23. Compare Exercise 5.5.27.

27. Beckmann, *History of π* (1971), 95.

with 60×2^{29} (that is, 60 times 2 times itself 29 times) sides. His grateful colleagues consequently sometimes called π 'Ludolf's number'.[28] Fortunately, there are better ways than Archimedes'. Those who sought them included Christiaan Huygens, a student of Descartes and rival of Newton, who, among much else, discovered the rings of Saturn, invented an accurate clock, and wrote about the inhabitants of the moon. He published his improvement over Archimedes in 1654, at the age of 24, because, he said, he had obtained some new insights into 'the ancient problem of squaring the circle, the most famous of all [geometrical problems] to those who do not understand mathematics'.[29] Huygens' ingenious and intricate calculation will be set out in Chapter 6.

Lunatics

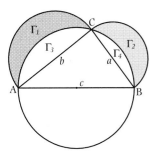

Fig. 5.4.11 Hippocrates' lunes.

Despite repeated failures, adventuresome geometers over the ages have tried their hands at squaring the circle. One of the earliest and most promising attempts was the work of Hippocrates of Chios, not the great physician but a man distinguished enough in his own line to be named in the old texts as the first to compile an *Elements of geometry*. Hippocrates' attempt at circle-squaring involved 'lunes', moon-shaped areas bounded by arcs of circles.[30]

Figure 5.4.11 presents lunes made between circles and semicircles described on the sides of a triangle ABC inscribed in half of the primary circle. Hippocrates showed that the sum of the areas $\Gamma_1 + \Gamma_2$ of the lunes is just equal to the area of $\triangle$ABC. It was a moral victory: by demonstrating that the area of a curvilinear figure could be equal to that of a rectilinear figure, he made it appear plausible that the circle itself could be squared. To prove that the sum of the shaded areas equals $\triangle$ABC, you must show that

$$\Gamma_1 + \Gamma_2 = \pi(b/2)^2/2 - \Gamma_3 + \pi(a/2)^2/2 - \Gamma_4 = \triangle ABC.$$

But $\Gamma_3 + \Gamma_4 = \pi(c/2)^2/2 - \triangle$ABC. Hence

$$\Gamma_1 + \Gamma_2 = (\pi/8)(a^2 + b^2 - c^2) + \triangle ABC = \triangle ABC$$

via the Pythagorean theorem, since $\angle$ACB is inscribed in a semicircle. Q.E.D.

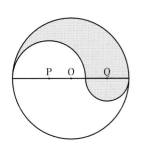

Fig. 5.4.12 Exercise on areas of circles (APS 5.4.1).

APS 5.4.1. Let P be any point on the diameter of a circle centered at O (Fig. 5.4.12). Describe semicircles on the pieces of the diameter as shown. Then the sum of the circumferences of the circles centered on P and Q equals the circumference of the parent

28. Hobson, *Squaring the circle* (1913), 27.
29. Huygens, *Oeuvres* (1888), *12*, 114.
30. Heath, *Hist. Greek math.* (1921), *l*, 170, 183–200.

circle and the shaded area is to the unshaded as the diameter of the smaller to that of the larger internal semi-circle. Really. The circumferences of the circles centered at O, P, and Q are $C_o = 2\pi(p + q)$, $C_p = 2\pi q$, $C_Q = 2\pi q$, whence the demonstration of the first assertion. For the second,

$$\Gamma_s = \pi[(p + q)^2 - p^2 + q^2]/2,$$

$$\Gamma_u = \pi[(p + q)^2 - p^2 + q^2]/2,$$

and we have for the ratio sought,

$$\Gamma_s + \Gamma_u = q(p + q)/p(p + q) = q/p,$$

Q.E.D.

A few centuries after Hippocrates, someone found out how to square the circle, or so said Boethius, who wrote early in the sixth century AD. Unfortunately, he said further, the demonstration was too long to reproduce.[31] He thus transmitted along with fragments of Euclid the encouragement to find the old quadrature or invent a simpler one. Europeans kept at the hopeless task until well into the eighteenth century. Finally, in 1775, the mathematicians at the Paris Academy of Sciences, fed up with requests to judge solutions submitted to them, refused to have anything further to do with circle-squarers. The subject has not been respectable for over 200 years.

Some interesting geometry came out of attempts at circle-squaring. One example, an approximate rectification of the circumference, can stand for them all. It is the handiwork of a Jesuit priest, a contemporary of Huygens and Newton, who served the King of Poland as a mathematician and librarian. This redoubtable rectifier bore the resounding name of Adam Amandus Kochansky. He knew that his construction was only an approximation. It is given in Fig. 5.4.13, where $\angle BOE = 30°$, BG is tangent to the circle at B, DL = $3r$,

Fig. 5.4.13 Kochansky's approximate rectification of a circle.

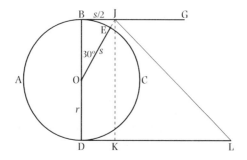

31. De Morgan, *Budget* (1954), *I*, 42.

Table 5.4.4 Euclidean propositions in Chapter 5

II.1	To cut a given line in mean and extreme ratio
III.1	To find the center of a given circle
III.3	A radius perpendicular to a chord bisects it
III.11	The line through the centers of two circles tangent internally passes through the point of contact
III.12	The analogue of III.11 for circles tangent externally
III.17	To draw a tangent to a given circle from a point outside it
III.18	The radius to a tangent at the point of contact is perpendicular to it
III.20	The angle subtended by an arc at the center is double that subtended by the same arc at the circumference
III.22	The opposite angles of an inscriptable quadrilateral are supplementary
III.31	An angle inscribed in a semicircle is right; in less than a semicircle, acute; in more than a semi-circle, obtuse
III.35	If two chords cut one another, the product of the segments of the one equals the product of the segments of the other
III.36	If a secant and a tangent fall from the same point to a circle, the square of the tangent equals the rectangle formed by the whole secant and the portion between the point and the circle
IV.1	To fit a given straight line less than a diameter into a circle
IV.4	To inscribe a circle in a triangle
IV.5	To circumscribe a circle about a triangle
IV.8	To inscribe a circle in a square
IV.10	To construct an isosceles triangle each of whose base angles is double the remaining one
IV.11	To inscribe a regular pentagon in a circle
IV.12	To circumscribe a regular pentagon about a circle
IV.15	To inscribe a regular hexagon in a circle
IV.16	To inscribe a regular 15-gon in a circle

r being the radius, and JL very nearly rectifies (that is, makes a straight line equal to) the semicircle BCD. It is not hard to show that $JL^2 = (40/3 - 2\sqrt{3})r^2$ and that, therefore, $JL/r = 3.14153$. JL obviously rectifies any rational fraction or multiple of a semicircle; thus the arc BE, corresponding to 30°, is close to JL/6, whereas the entire circumference ABCD is about equal to 2JL. The not-hard demonstration: Let OJ = s; then, because $\angle BOJ = 30°$, BJ = $s/2$, and, by Euclid I.47, $s/2 = r/\sqrt{3}$. Now draw JK∥BD; then $JL^2 = KJ^2 + (DL - DK)^2 = 4r^2 + r^2(3 - 1/\sqrt{3})^2 = (40/3 - 2\sqrt{3})r^2$. Q.E.D.

Kochansky liked to calculate. He ran his approximation to π out to 15 decimal places, which differed from the best value of π then established by Archimedes' procedure by 0.000 593 148, or by one part in a little over 16859 parts. 'I say, therefore,' said Kochansky, 'that the periphery thus found differs from the closest approximation to the true Archimedian value by less than the ratio of one to ten times the current year 1685 ... and by more than the ratio of one to ten times the year 1686 next to follow.' From these calendrical considerations, he adds, anyone can easily write down the value of π correctly to 10 decimals, 'since what is grasped by the reason is easily retained in the memory'. You have a mnemonic for π good to 11 figures: $\pi = 40/3 - 2\sqrt{3} + 1/16850$. You need only remember Kochansky's construction and the fact that he wrote in 1685.[32]

5.5. Exercises

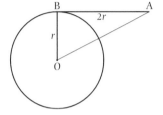

Fig. 5.5.1 Exercise 5.5.1.

5.5.1. In an extended diameter of a circle, find a point from which the tangent equals the diameter.

[As usual in construction problems, assume the solution (Fig. 5.5.1) and look for a property that can be constructed from the givens. In this case the property is the length of OA = $\sqrt{5}r$. To construct this length, draw a rectangle with sides r, $2r$; its diagonal is $\sqrt{5}r$.]

5.5.2. Describe a circle with center in one leg of a right triangle that touches the hypotenuse and passes through the right angle.

[Again assume the task done (Fig. 5.5.2) and look for a constructible property. An obvious move is to join C and O and to put in the radius OD⊥AC. (The center O must lie on AB because BC is a tangent.) You see that ΔDOC ≅ ΔBOC (h.s.), whence ∠DCO = ∠BCO. Hence the construction: the bisector of ∠C meets the leg AB at the center of the circle required.]

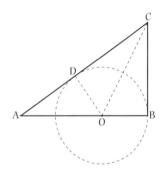

Fig. 5.5.2 Exercise 5.5.2.

5.5.3. A is any point on a diameter extended, BO⊥AO, PC the tangent at P where AB intersects the circle (Fig. 5.5.3). Can you show that PC = AC?

[You should always pick a strategy before starting to demonstrate. A plausible strategy here is to try to prove ∠PAC = ∠CPA; for then ΔPCA would be isosceles and PC = AC. Now ∠PAC is the complement of ∠OBP; you need only show that ∠CPA is so also. Draw OP⊥DC. You have ∠CPA = ∠BPD = rt.∠ − OPB = rt.∠ − ∠OBP, as required.]

5.5.4. Circle center P is internally tangent to circle center O and passes through O. Take D as any point on the smaller circle and draw a line from the point of tangency (Fig. 5.5.4) through D to meet the larger circle at C. Then AD = DC.

[You might reason thus: if AD = DC, AC = 2AD; by construction, AB = 2AO; hence AC : AB = AD : AO. To prove it, you need similar

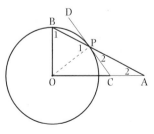

Fig. 5.5.3 Exercise 5.5.3.

32. Kochansky, *Acta eruditorum*, 1685, 397–8; a typographical error in the original has been corrected. Cf. Cantor, *Vorlesungen*, 3 (1898), 20–1.

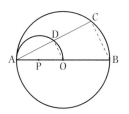

Fig. 5.5.4 Exercise 5.5.4.

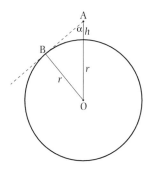

Fig. 5.5.5 Exercise 5.5.5.

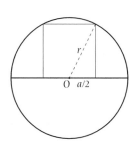

Fig. 5.5.6 Exercise 5.5.7.

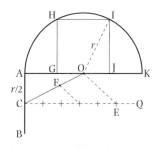

Fig. 5.5.7 Further to Exercise 5.5.7.

triangles containing AC, AB, AD, and AO. Nothing easier: draw OD, BC; ΔABC ~ ΔAOD, since ∠ADO = ∠ACB = rt.∠ (they are inscribed in semicircles) and ∠CAB is common.]

5.5.5. A square of side $2a$ is divided into n^2 equal smaller squares. Suppose circles inscribed in each cell. The total area of all these n^2 circles equals the area of a circle inscribed in the original square.

[The big circle has area πa^2; the side of a cell is $2a/n$, the area of its inscribed circle $\pi a^2/n^2$; the total area of the small circles is $n^2(\pi a^2/n^2) = \pi a^2$.]

5.5.6. When traveling in Northern India, the Arab philosopher Al-Biruni invoked a new way to measure the circumference of the earth. 'It does not require walking in deserts,' he said, in recommending it. Instead, you must climb a high mountain. [Berggren, *Episodes* (1986), 142–3.] Can you guess his method?

[A (Fig. 5.5.5) is the mountain top, from which you measure α, the angle between the vertical and the farthest point on the horizon; your gaze then falls along the tangent AB. You determine the height of the mountain h by the standard surveying technique described in §3.3. Then the only unknown in the equation $\sin \alpha = r/(r + h)$ is r. Al-Biruni supposed or measured his mountain to be 625.05 cubits high and found (he said) $\alpha = 89°26'$; together these numbers gave a value for r in perfect agreement with a suspicious earlier 'measurement' involving a mountain 5.5 Arab miles high. Knowing that an Arab mile is 4000 cubits, calculate the value of a used in the romance of the 5.5 mile mountain and the circumference of the earth according to Al-Biruni. Columbus knew some of the Arabs' results, which he preferred to Christian ones, since, by assuming incorrectly that they referred to Italian miles, he derived from them additional evidence that the world was smaller than anyone else imagined. Cf. Jordan, *Zeits. Verm.*, *18* (1889), 100–9, and Hammer, ibid., *38* (1909), 721–3.]

5.5.7. Inscribe a square in a semicircle. As indicated in Fig. 5.5.6, the construction needed is $a/2 = r/\sqrt{5}$.

[Draw AB = r perpendicular to the diameter AK (Fig. 5.5.7). Bisect the radius AB = r at C. Join O and C. Then OC = $(\sqrt{5}/2)r$. But you want $a/2 = r/\sqrt{5}$, that is, 2OC/5. To divide OC in fifths, draw CQ at a right angle to AB and, opening your compass to any convenient distance, mark off five equal divisions on CQ. Join the fifth or last division, at E, to O, and draw the parallel to EO through the third division, intersecting CO at F. Since, by similar triangles, OF = 2OC/5 = $r/\sqrt{5}$ = $a/2$, mark off OG on OA = OF; draw a semicircle of radius r around O; raise the perpendicular to AO at G until it intersects the semicircle at H; H is a corner of the desired square.]

Fig. 5.5.8 Exercise 5.5.8.

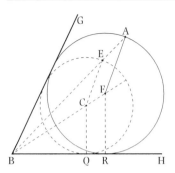

5.5.8. Draw a circle that passes through a given point and touches two given straight lines. A possible construction is shown in Fig. 5.5.8: A is the given point, BG and BH the given lines extended until they intersect. Draw any circle touching BG and GH; let the circle be centered at C. Draw BA cutting the arbitrary circle at E. If a line is made to pass through A parallel to CE, it cuts BC or BC extended at the center F of the circle sought. Prove that this construction does indeed solve the task.

[Because FA∥CE, ΔBCE ~ ΔBFA, whence CE : FA = BC : BF. Draw the radii CQ, FR to the points Q, R where the circles touch BH; since ΔBCQ ~ ΔBFR, BC : CQ = BF : FR. Hence FA : FR = CE : CQ = 1, or FA = FR, and point A lies on the circle centered on F. Q.E.D.]

5.5.9. James Glenie, of the Royal Engineers, specialized in fortification, on which he exercized his strong talent for geometry. He overestimated his powers, however; after fighting with his commander over the best way to fortify English harbors, he resigned his commission and set up as a tutor in mathematics. Too few of his fellow citizens were both willing and wealthy enough to engage his services, and he died in poverty. During his military career he had concocted to a few pretty theorems, one of which, as follows, he set as a problem in the *Ladies diary* for 1778. [From Leybourn, *3*, 35.]

In Fig. 5.5.9, AB is any chord, and HJ any diameter, in the circle centered at O. The points E and F are the intersection of prependiculars raised at A and B with HJ; at the point G, lines from E and F make equal angles with the line AB. Glenie asked for a proof that EG + GF = HJ. You will find the proof easy if you complete the rectangle ABCD and draw its diagonal AC = HJ = the circle's diameter and show that GF∥AC. (Since, similarly, diagonal DB∥EG, Glenie's theorem has some semipractical consequences: a billiard ball resting against one cushion AD can be sent to the diametrically opposite point F off cushion AB by aiming along EG∥DB; or, if AB is a mirror, the point E will be seen at G by an eye at F.

[From the similar triangles EAG, FBG, EG : GF = EA : FB, whence (EG + GF) : GF = (EA + FB) : FB; from the congruent triangles (*a.s.a.*)

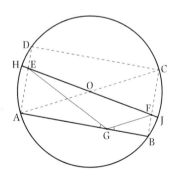

Fig. 5.5.9 Exercise 5.5.9.

EOA and COF, EA = CF; therefore EG + GF = (GF/FB)(CF + FB) = CB(GF/FB). This last expression equals AC. This follows from the similarity of triangles ABC and GBF, which can be demonstrated easily: since CB = CF + FB = EA + GB × EA/AG = EA + (EA/AG)(AB − AG) = AB(EA/AG), CB : AB = EA : AG and ΔABC ~ ΔGAE ~ ΔGBF. The similarity of ΔABC and ΔGBF gives CB : AC = FB : GF, or EG + GF = CB(GF/FB) = diameter of the circle, Q.E.D.]

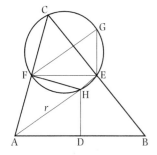

Fig. 5.5.10 Exercise 5.5.10.

5.5.10. While the British colonies in America declared their independence, the *Ladies diary* made its contribution to history by offering its readers the following theorems. Let H be the center of the circle circumscribing the triangle ABC (Fig. 5.5.10); HD and HF, perpendiculars (and hence perpendicular bisectors) of the sides AB, AC. Draw a circle through H, F, and C. The theorems: (1) HE, where E is the intersection of the circle HFC with side CB, is perpendicular to BC (2) the chord FE is parallel to AB (3) the diameter of the circle FG equals AH, the radius of the circle circumscribing ΔABC (4) GE is parallel and equal to HD. Prove these theorems of 1776. No construction lines are required beyond the diagram as given. [*Ladies diary* (1776), in Leybourn, *3*, 2.]

[(1) HE⊥BC because FCEH is a quadrilateral inscribed in a circle and ∠CFH = rt.∠ by construction; hence ∠CEH = rt.∠ (2) FE∥AB because FE joins the midpoints of the sides AC, BC (3) because ΔFCE ~ ΔACB, FE = AB/2, ΔFCE has half the perimeter of ΔABC; therefore the radius of its circumscribing circle is half that of ΔABC's (4) ∠FEG = rt.∠ (it is inscribed in a semicircle); hence GE⊥FE∥AB, so GE∥HD; that GE = HD follows from the congruence of the right triangles FEG and ADH.]

5.5.11. The bisectors of the angles formed by the extended opposite sides of an inscribable quadrilateral meet at right angles. Referring to Fig. 5.5.11, show that QS⊥RT, where QS bisects ∠DQC and RT bisects ∠DRA.)

Fig. 5.5.11 Exercise 5.5.11.

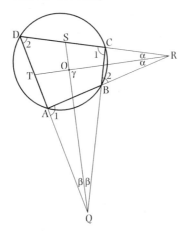

[From the inscriptability, $\angle QDR = \angle CBR$, $\angle QAR = \angle QCD$. Hence, from $\triangle BCR$, $\angle 1 = \angle 2 + 2\alpha$, and, from $\triangle AQB$, $\angle 1 + \angle 2 + 2\beta =$ st.$\angle$; by eliminating $\angle 2$, you have $\angle 1 = $ rt.$\angle + (\alpha - \beta)$. Now γ, being external to $\triangle OTQ$, is $\beta + \angle RTA = \beta + \angle 1 - \alpha = (\beta - \alpha) + $ rt.$\angle + (\alpha - \beta)$; whence $\gamma = $ rt.$\angle$, Q.E.D.

5.5.12. Prove Euclid III.32, namely, that if AB is the diameter of a circle meeting the tangent EF at B (Fig. 5.5.12) and D is any point on the circle, $\angle BAD = \angle FBD$; and, further, if C is any point on the arc DB, $\angle DCB = \angle EBD$.

[Since $\angle ABF = \angle ADB = $ rt.$\angle$, $\angle DBA$ is the complement both of $\angle BAD$ and $\angle FBD$, whence $\angle BAD = \angle FBD$; further, $\angle DCB = $ st.$\angle - \angle BAD = $ st.$\angle - \angle FBD = \angle EBD$, Q.E.D.]

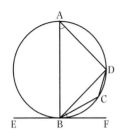

5.5.13. PT is the common tangent to the touching circles with radii r_1, r_2, centered at O, Q. Prove that $PT^2 = 4r_1r_2$.

[Extend PT and OQ to meet at S (Fig. 5.5.13). Let $PT = x$, $TS = y$, $QS = z$; OQ is $r_1 + r_2$. From the similar triangles OPS, QTS, we have (1) $(x + y) : r_1 = y : r_2$ and (2) $r : (r_1 + r_2 + z) = r_2 : z$. From (2), $z = r_2(r_1 + r_2)/ (r_1 - r_2)$. The Pythagorean theorem applied to $\triangle QTS$ gives $z^2 = y^2 + r_2^2$, whence $y^2 = r_2^2 (r_1 + r_2)^2/ (r_1 + r_2)^2 - r_2^2 = r_2^2 (4r_1r_2)/ (r_1 - r_2)^2$. From (1), $x^2 = y^2(r_1 - r_2)^2/ r_2^2$, which, with the preceding equation, gives $x^2 = 4r_1r_2$, Q.E.D.]

5.5.14. The preceding problem comes from the *Ladies diary* for 1800. The *Diary* for 1801 reports three solutions, all different from the demonstration just given. The nature of the first, and most elegant, of these solutions is indicated in Fig. 5.5.14, where QH is drawn parallel to PT. Show that $OH = r_1 - r_2$ and thence that $PT^2 = 4r_1r_2$. Another route to the same result is obtained by raising the perpendicular from the point of intersection of the circles M to the tangent PT at N (Fig. 5.5.15) and showing that $\angle ONQ = $ rt.$\angle$. [From Leybourn, *3*, 367.]

[The construction of Fig. 5.5.14 makes QHPT a rectangle; hence. $PT^2 = QH^2 = OQ^2 - OH^2 = (OQ + OH)(OQ - OH)$. But $OH = PO - TQ = r_1 - r_2$ and $OQ = r_1 + r_2$, whence $PT^2 = (r_1 + r_2 + r_1 - r_2) (r_1 + r_2 - r_1 + r_2) = 4r_1r_2$. In Fig. 5.5.15, we already know (APS 5.2.6) that N is the center of a circle through P, M, and T. In the quadrilateral OPTQ, $\angle POQ + \angle TQO = 2$st.$\angle - 2$rt.$\angle = 1$ st.$\angle$. But $\angle POQ = 2($rt.$\angle$

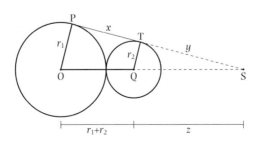

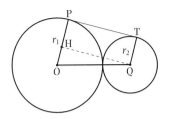

Fig. 5.5.14 Exercise 5.5.14.

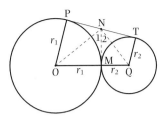

Fig. 5.5.15 Further to Exercise 5.5.14.

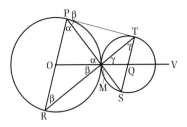

Fig. 5.5.16 Further to Exercise 5.5.14.

$- \angle 1)$ and $\angle TQO = 2(\text{rt.}\angle - \angle 2)$, from which $\angle 1 + \angle 2 = \text{rt.}\angle$. Therefore $(r_1 + r_2)^2 = ON^2 + QN^2 = r_1^2 + PN^2 + r_2^2 + NT^2$, or $2r_1 r_2 = 2(PT^2/4)$.

A third proof uses Fig. 5.5.16, in which the radii PO and TQ are extended to diameters PR, TS, and P, R, S, and T are each joined to M. Show that RMT and PMS are straight lines and that therefore $\triangle RPT \sim \triangle STP$. From this similarity, $PT : 2r_1 = 2r_2 : PT$, Q.E.D.

[Since $\angle RMP$ is inscribed in a semicircle, $\alpha + \beta = \text{rt.}\angle$. Therefore $\angle MPT = \text{rt.}\angle - \alpha = \beta$. All depends on seeing that $\angle PTM = \alpha$. Introduce $\gamma = \angle MTQ$. Then $\gamma + \angle PTM = \text{rt.}\angle$ and (from $\triangle PMT$) $\alpha + \gamma + \text{st.}\angle - (\beta + \angle PTM) = \text{st.}\angle$. Eliminating γ, you have $\angle PTM = \alpha$. Hence $\angle PMT = \text{st.}\angle - (\alpha + \beta) = \angle PMR = \text{st.}\angle$: RMT is a straight line. Similarly, PMS may be proved straight. All four of the angles around M are right. Now, since $\angle RPT = \angle STP = \text{rt.}\angle$ and $\angle PRT = \angle TPS = \beta$, $\triangle RPT \sim \triangle STP$. Q.E.D.]

5.5.15. In Fig. 5.5.17, B is an arbitrary point on the diameter AC and BD⊥AC. If semicircles are drawn tangent to BD and to the semicircle ADC as shown, prove that the shaded area equals the area of a circle whose diameter is BD. [*Ladies diary* (1808), in Leybourn, 4, 106.]

[Let r_1, r_2, $r = AC/2$ be the radii of the three semicircles; then $r = r_1 + r_2$. The shaded area A therefore is

$$A = (\pi/2)(r^2 - r_1^2 - r_2^2) = (\pi/2)(2r_1 r_2) = \pi r_1 r_2.$$

But we know DB is a mean proportional between $2r_1$ and $2r_2$, that is, $DB^2 = 4r_1 r_2$. The area of a circle of diameter DB is $\pi DB^2/4 = r_1 r_2 = A$, Q.E.D.]

5.5.16. In Fig. 5.5.18, B now is at the center of AC and the circle centered at D touches the three semicircles centered at B, E, and F, where $AB = BF = r/2$. Show that s, the radius of the circle centered at D, is $AB/3 = r/3$. [*Ladies diary* (1809), in Leybourn, 4, 117.]

[Raise the perpendicular at B; it evidently goes through D. Join D and E. Then $DE^2 = DB^2 + EB^2$. But $DE = r/2 + s$ (Euclid III.12); $DB = r - s$; $EB = r/2$. Therefore

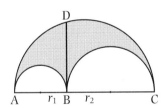

Fig. 5.5.17 Exercise 5.5.15.

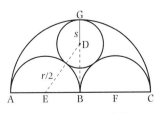

Fig. 5.5.18 Exercise 5.5.16.

$$\left(r/2+s\right)^2 = \left(r/2\right)^2 + (r-s)^2,$$

whence $s = r/3$. O.E.D.

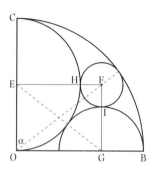

Fig. 5.5.19 Exercise 5.5.17.

5.5.17. In Fig. 5.5.19, arc CB is a quarter circle of radius r, OC is a semicircle of radius $r/2$, the semicircle centered at G touches the first semicircle and the quadrant, and the circle centered at F touches both semicircles and the quadrant. Prove that the area A_G of the semicircle centered at G is twice the area A_F of the circle centered on F, and that EFGO is a rectangle. [*Ladies diary* (1813), in Leybourn, 4, 184.]

[Draw EG, EF, FG and the radius from O through F; let GB = s and FH = FI = t. In $\triangle$EOG, EG = $r/2 + s$, EO = $r/2$, OG = $r - s$. Therefore

$$(r/2 + s)^2 = (r/2)^2 + (r - s)^2, \qquad \text{or } s = r/3.$$

To obtain t, apply the law of cosines to $\triangle$EOF and $\triangle$GOF, setting $\angle$EOF = α to simplify the writing. In $\triangle$EOF,

$$(r/2 + t)^2 = (r/2)^2 + (r - t)^2 - 2(r/2)(r - t) \cos \alpha,$$

whence

$$\cos \alpha = (r - 3t)/(r - t)$$

In $\triangle$GOF,

$$(r/3 + t)^2 = (r - t)^2 + (2r/3)^2 - 2(2r/3)(r - t) \sin \alpha,$$

whence

$$\sin \alpha = (r - 2t)/(r - t).$$

Since $\sin^2\alpha + \cos^2\alpha = 1$,

$$(r - 2t)^2 - (3t - r)^2 = (r - t)^2,$$

from which $t = r/6$. Therefore $A_F = \pi r^2/36$, $A_G = (\pi/2)(r^2/9)$, so $A_G = 2A_F$, Q.E.D. Since $s + t = r/2$, GF = OE; and, since $r/2 + t = 2r/3$, EF = OG. Therefore EFGO is a parallelogram, and, because $\angle$EOB = rt.$\angle$, it is a rectangle.]

5.5.18. In Fig. 5.5.20, the diameter of the semicircle is extended an arbitrary distance to P, and PC is drawn at any convenient angle intersecting the circle at B and C. At B and C erect perpendiculars to PC and extend them to cut the diameter at D, E. Prove that, if O is the circle's center, DO = OE. [*Ladies diary* (1805), in Leybourn, 4, 67]

[Draw OQ⊥BC; since BC is a chord, BQ = QC. It follows from the similar triangles, DBP, OQP, ECP that DO = OE.]

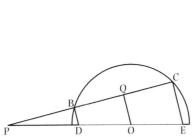
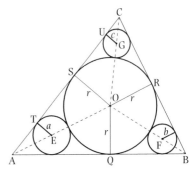

Fig. 5.5.20 Exercise 5.5.18.   **Fig. 5.5.21** Exercise 5.5.20.

5.5.19. If you are *h* feet tall and walk all the way around the Earth, keeping to the same circumference, how much farther has your head gone than your feet when you complete the journey? [*Ladies diary* (1715), in Leybourn, *1*, 44]

[Let the earth's radius be *R*. Then your head travels $2\pi(R + h)$ and your feet $2\pi R$; the difference, $2\pi h$, which would be 44 feet if you were 7 feet tall, is therefore independent of the size of the earth. Surprising, no?]

5.5.20. Circles are inscribed in $\triangle ABC$ as shown (Fig. 5.5.21). Derive the radius *r* of the largest circle and the sides of the triangle given the radii *a*, *b*, *c* of the smaller circles. [*Ladies diary* (1730), in Leybourn, *1*, 186.]

[First derive AB, BC, CA in terms of the unknown *r* and the givens *a*, *b*, *c*; then obtain *r* in terms of *a*, *b*, *c*. From $\triangle AET$ and $\triangle AOS$, which are similar since each contains a rt.$\angle$ and $\angle SAO$,

$$a : AE = r : (a + r + AE), \qquad \text{whence } AE = a(r + a)/(r - a).$$

Similarly, $CG = c(r + c)/(r - c)$, $BF = b(r + b)/(r - b)$. Now $AO = AE + a + r = r(r + a)/(r - a)$, $CO = r(r + c)/(r - c)$, $BO = r(r + b)/(r - b)$. Then, from $AS^2 = AO^2 - r^2$,

$$AS = \frac{2r}{r - a}\sqrt{ra} = AQ.$$

The equivalent quantities are $SC = 2r\sqrt{rc}/(r - c) = CR$, and $BR = 2r\sqrt{rb}/(r - b) = BQ$. Finally,

$$AC = AS + SC = 2r^{3/2}[\sqrt{a}/(r - a) + \sqrt{c}/(r - c)].$$

$$CB = CR + RB = 2r^{3/2}[\sqrt{c}/(r - c) + \sqrt{b}/(r - b)].$$

$$AB = AQ + QB = 2r^{3/2}[\sqrt{a}/(r - a) + \sqrt{b}/(r - b)].$$

Now you must find *r*. Use the property that OA, OB, OC are the bisectors of angles A, B, C, respectively. You have $\sin A/2 = r/OA$

= $(r - a)/(r + a)$; cos A/2 = AS/OA = $2\sqrt{ra}/(r + a)$. Similarly, sin B/2 = $(r - b)/(r + b)$, cos B/2 = $2\sqrt{rb}/(r + b)$, and sin C/2 = $(r - c)/(r + c)$. But also, $(A + B + C)/2 = $ rt.$\angle$, so

$$\sin C/2 = \cos(A/2 + B/2) = \cos A/2 \times \cos B/2 - \sin A/2 \times \sin B/2,$$

whence

$$\frac{r - c}{r + c} = \frac{4r\sqrt{ab}}{(r + a)(r + b)} - \frac{(r - a)(r - b)}{(r + a)(r + b)}.$$

We therefore have r as the solution of the equation

$$r^2 - 2r\sqrt{ab} + ab - ac - bc - 2c\sqrt{ab} = 0.]$$

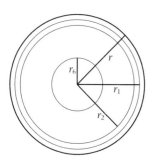

Fig. 5.5.22 Exercise 5.5.21.

5.5.21. Seven men brought equal shares in a grinding stone 5 feet in diameter and agreed that each should use it until he had ground his share. What part the diameter should each grind away? [*Ladies diary* (1709), in Leybourne, *1*, 5–8.]

[Algebraically, nothing could be easier. We need to divide a circle 30 inches in radius into six equal bands plus a central disk equal in area to a band. Call the radius of the stone r, that of the internal band of the first grinding r_1, that of the same for the second grinding r_2, etc. Then (Fig. 5.5.22).

$$\pi(r^2 - r_1^2) = (\pi/7)r^2, \qquad \text{or } r_1^2 = (6/7)r^2;$$

$$\pi(r_1^2 - r_2^2) = (\pi/7)r^2, \qquad \text{or } r_2^2 = (5/7)r^2, \text{ etc.,}$$

so that $r_6^2 = r^2/7$. Therefore, the first man to use the stone should grind $30 - 30\sqrt{6/7} = 2.25$ inches from the radius, the second $30(\sqrt{6/7} - \sqrt{5/7}) = $ an additional 2.42 inches, and so on.]

5.5.22. Show that the geometrical construction of Fig. 5.5.23 also satisfies the problem. There the radius AB is divided into seven equal parts at C, E, G, etc., arc AB is a semicircle of radius $r/2$ centered at O, and CD, EF, GH, etc., are perpendicular to AB. The circles drawn with A as center and radii AD, AF, AH, etc., give the limits of the bands, that is, $r_1 = $ AD, $r_2 = $ AF, etc.

[In Fig. 5.5.24, AB = r, OD = $r/2$, AC = $(6/7)r$, OC = OB − BC = $r/2 - (r - $ AC$) = $ AC − $r/2$. Hence CD^2 = OD^2 − OC^2 = $(r/2)^2 - ($AC − $r/2)^2$. But $r_1^2 = $ CD^2 + AC^2 = AC^2 + $(r/2)^2 - ($AC − $r/2)^2$ = $r \times$ AC. We have therefore

$$r_1^2 = r \times (6/7)r = (6/7)r^2,$$

as before, and, similarly, $r_2^2 = $ AE^2 = $(5/7)r^2$, etc.

In the first posing and answering of this problem, in 1709–10, only the algebraic form was given. A little later, a man named

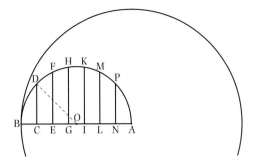

Fig. 5.5.23 Exercise 5.5.22.

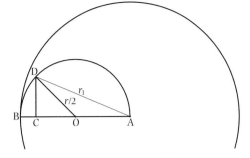

Fig. 5.5.24 Further to Exercise 5.5.22.

Hawney devised a complicated geometrical construction, which gave way to that just elucidated in a manner worth repeating as an example, if one is needed, of the truth that adults, and experts in mechanics and astronomy, can have difficulty with geometry.

One evening around 1770, James Ferguson, who made a living in England lecturing on applied mathematics, was entertaining the family of Charles Hutton, a mathematics professor at Britain's main military school and the editor of the *Ladies diary*. Ferguson showed a careful sketch of Hawney's construction. Hutton, who had not previously known the problem of the grindstone, which had appeared in the *Diary* long before his time, said he thought there must be an easier way. He spent the evening devising the construction of Fig. 5.5.23. Ferguson was pleased with it but could not understand the proof, whereupon Hutton asked where he had learned geometry. Ferguson, the lecturer in astronomy and other branches of applied mathematics, said that he had never been able to grasp a single demonstration of any of Euclid's propositions. His method of 'proof' was to draw a diagram exactly to scale. 'Accordingly the next morning, with a joyful countenance, he brought me the construction [of the grindstone problem], neatly drawn out on a large sheet of paste-board, saying he esteemed it a treasure, having found it quite right, as every point and line agreed to a hair's breadth.' Thus the power of geometrical proof: we can demonstrate the truth of Hutton's construction in a few quick lines, whereas even the most skillful draughtsman ignorant of geometry required many hours to make an exact drawing, which, however, at best could only illustrate, not demonstrate, the proposition.]

5.5.23. A round pond (Fig. 5.5.25) has its center at an inaccessible point in a rectangular garden. If you know the shortest distances from each corner of the rectangle to the pond, can you find its radius? (Compare exercise 4.4.19.) Such a pond and garden existed in eighteenth-century England. The distances from the corners to

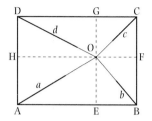

Fig. 5.5.25 Exercise 5.5.23.

the pond were, in order, 60, 52, 28, and 40 yards. What was the radius of the pond? [*Ladies diary* (1754), in Leybourn, *3*, 95–6.]

[Let the distances from the corners to the pond be *a, b, c, d*, and the pond's radius be *r*. Drop the perpendiculars from the pond's center, O, onto the sides at E, F, G, H. Then

$$(a + r)^2 + (c + r)^2 = AE^2 + OE^2 + OF^2 + CF^2$$
$$(b + r)^2 + (d + r)^2 = EB^2 + OE^2 + DG^2 + OG^2.$$

But AE = DG, OF = EB, CF = OG, so that

$$(a + r)^2 + (c + r)^2 = (b + r)^2 + (d + r)^2,$$

whence

$$2r = \frac{a^2 - b^2 + c^2 - d^2}{b - a + d - c},$$

Q.E.D. In the example given, *a* = 60, *b* = 52, *c* = 28, *d* = 40, and *r* = 10].

5.5.24. Two circles of radius 25 feet intersect so that the distance between their centers is 30 feet. What is the length of a side of the square inscribable within the space defined by the intersecting arcs? [*Ladies diary* (1755), in Leybourn, *3*, 107–8.]

[In Fig. 5.5.26, E and F are the centers of the intersecting circles, G the midpoint of the line EF, ABCD the desired square. Let *a* = EC = 25 feet, *b* = EG = 15 feet. Since CG is one half the diagonal of the square, CG = $s/\sqrt{2}$, where *s* is the side of the square, and ∠CGF =(rt.∠)/2. The law of cosines applied to ΔEGC yields

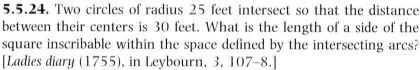

$$a^2 = b^2 + s^2/2 - (2bs/\sqrt{2})\cos \angle EGC.$$

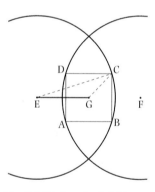

Fig. 5.5.26 Exercise 5.5.24.

But cos ∠EGC = cos [st.∠ – ∠CGF] = – cos∠CGF = – $1/\sqrt{2}$. Hence

$$s^2/2 + bs + (b^2 - a^2) = 0$$

which, when solved for the values *a* = 25, *b* = 15, gives *s* ≈ 17 feet.]

5.5.25. Draw a rectangle ABCD with sides *b* and *a* = $b/\sqrt{2}$, and raise a semicircle on one of the sides AB = *b* as in Fig. 5.5.27. Now pick any point E on the semicircle. Join E to the corners C and D of the rectangle. Call U and V, respectively, the points of intersection of ED and EC with AB. Can you prove that AV² + BU² = *b*²? (This problem was posed by Pierre Fermat, of Fermat's theorem, to John Wallis, an important English mathematician, during a geometrical duel in the later seventeenth century. The *Ladies diary* printed an answer in 1770, over a century after the publication of the problem in the *Commercium epistolicum* (Oxford, 1658), 188). [*Ladies diary* (1769), in Leybourn, *3*, 311.]

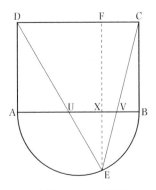

Fig. 5.5.27 Exercise 5.5.25.

[Drop the perpendicular from E onto AB at X and CD at F. You have AV = b – BV, BU = b – AU. From the similar triangles DAU, EXU, and XV = BX – BV, you get BV = aBX/$(a + x)s$ where x = EX. Likewise, from triangles VBC, VXE, you have AU = aAX/$(a + x)$. Therefore,

$$AV^2 + BU^2 = 2b^2 - 2ab(AX + BX)/(a + x) + a^2(AX^2 + BX^2)/(a + x)^2.$$

But AX + BX = b and, as we know from the geometry of the circle, AX × BX = x^2. Substituting for AX^2 + BX^2 and AX × BX in the previous equation and using the given $b^2 = 2a^2$, you will find

$$AV^2 + BU^2 = 2a^2(a + x)^2/(a + x)^2 = 2a^2 = b^2,$$

Q.E.D.

5.5.26. Three circles with radii a, b, c have their centers on a straight line. It is proposed to draw a circle tangent to all three. If the separation of the centers of the first and second circles is p, and that of the centers of the second and third circles is q, show that the diameter of the tangent circle d satisfies the equation

$$d = \frac{p(b^2 - c^2 + q^2) + q(b^2 - a^2 + p^2)}{q(a - b) + p(c - b)}.$$

[*Ladies diary* (1760), in Leybourn, 3, 185.]

[Let the circles be centered at A, B, C, O (Fig. 5.5.28) and set ∠OBA = α. Then the law of cosines applied to ΔOBC and ΔOBA gives, respectively,

$$(r + c)^2 = (r + b)^2 + q^2 + 2q(r + b)\cos \alpha,$$

$$(r + a)^2 = (r + b)^2 + p^2 - 2p(r + b)\cos \alpha.$$

(Recall that cos(st.∠ – α) = – cos α). Elimination of cos α between these equations gives

$$q(2r(a - b) + a^2 - b^2 - p^2) = -p(2r(c - b) + c^2 - b^2 - q^2),$$

which, when solved for $2r = d$, gives the expression desired.]

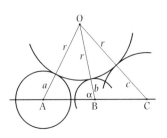

Fig. 5.5.28 Exercise 5.5.26.

5.5.27. Using trigonometric tables, show that the perimeters of polygons of 12, 24, 48, and 96 sides inscribed in a circle of diameter 10 l.u. are $\sqrt{965}$ l.u., $\sqrt{981}$ l.u., $\sqrt{986}$ l.u., and $\sqrt{987}$ l.u., respectively. From this sequence, it appears that the perimeter of a polygon with a large number of sides might approach $\sqrt{1000}$. Reasoning thus, perhaps, the Indian mathematician Brahmagupta, who lived in the seventh century, guessed that the circumference C of a circle of diameter 10 might be $\sqrt{1000}$, whence $10\sqrt{10} = \pi 10$, or $\pi = \sqrt{10} = 3.1623$. The value of π can then be exhibited as a straight line. Show how to obtain a line equal to $\sqrt{10}$ in a circle of diameter 10.

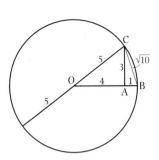

Fig. 5.5.29 Exercise 5.5.27.

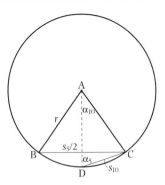

Fig. 5.5.30 Exercise 5.5.28.

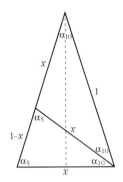

Fig. 5.5.31 Further to Exercise 5.5.28.

[In Fig. 5.5.29 the (3,4,5) basic Pythagorean triangle OAC is drawn with the radius of the circle of diameter 10 as hypothenuse, making AB = 1 and BC = $\sqrt{10}$, Q.E.D.]

5.5.28. Prove that $s_5^2 = s_{10}^2 + s_6^2$. Remember that s_6 equals the radius of the circle in which the hexagon is inscribed.

[In Fig. 5.5.30, $\angle BAC = \alpha_5$, the central angle subtended by one side of a regular pentagon; $\angle DAC = \alpha_5/2 = \alpha_{10}$, the central angle of a decagon; and $\angle ADC = (\text{st.}\angle - \alpha_{10})/2 = \alpha_5$. Hence the expressions for the sides in terms of the angles: $s_5 = 2r \sin \alpha_{10}$, $s_{10} = 2r \cos \alpha_5$. From Fig. 5.5.31, which reproduces the defining triangle for the central pentagonal angle α_5, $\sin \alpha_{10}/(1 - x) = \sin \alpha_5/x$, whence $\sin \alpha_{10} = x \sin \alpha_5$ (recall that, by definition of the pentagon, $(1 - x)/x = x$); also, from the same figure, $\cos \alpha_5 = x/2$. Therefore

$$s_5^2 = 4r^2\sin^2\alpha_{10} = 4r^2x^2\sin^2\alpha_5 = 4r^2x^2(1 - \cos^2\alpha_5) = r^2x^2(4 - x^2),$$

$$s_{10}^2 = 4r^2\cos^2\alpha_5 = r^2x^2.$$

Finally, $s_5^2 - s_{10}^2 = r^2x^2(4 - x^2 - 1) = r^2 = s_6^2$. Q.E.D.]

5.5.29. Deduce from the preceding demonstration the correctness of the following prescription for drawing s_5 and s_{10}: in a circle of radius r (Fig. 5.5.32), draw two diameters AB, CD, intersecting at right angles at O; bisect OB at E; lay off from E toward A a length EF = EC; then FO = s_{10} and FC = s_5.

[Since CE = $r\sqrt{5}/2$, FO = $(\sqrt{5}/2 - 1/2)r = (\sqrt{5} - 1)r/2 = s_{10}$. This last equality may be obtained from Fig. 5.5.31 and 5.5.32: $s_{10} = 2r \sin \alpha_{10}/2 = (2r)(x/2) = rx$; but, by definition, $x^2 = 1 - x$, or $x = (\sqrt{5} - 1)/2$. That AC = s_5 follows from the preceding problem.]

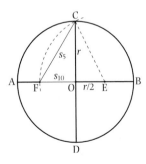

Fig. 5.5.32 Exercise 5.5.29.

5.5.30. Hero of Alexandria noticed that $s_7 \approx s_3/2$. Indeed, it is very close: draw a heptagon with compass and straight edge using this approximation. (An exact construction of a heptagon with these tools is not possible.) Show that the difference between s_7 and $s_3/2$ is

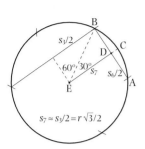

Fig. 5.5.33 Exercise 5.5.30.

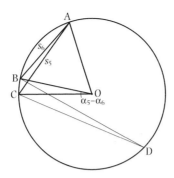

Fig. 5.5.34 Exercise 5.5.31.

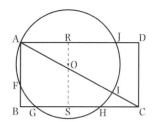

Fig. 5.5.35 Exercise 5.5.32.

under 0.25%. The medieval version takes the perpendicular from the center of a circle to the midpoint of the side of an inscribed hexagon as the approximation to s_7.[33] Evidently, it (ED in Fig. 5.5.33) has the same value as $s_3/2$.

5.5.31. Draw an angle of 6°. Hint: the difference between α_5 and α_6 is 12°.

[In Fig. 5.5.34, AB = s_6, AC = s_5, ∠BOC = $\alpha_5 - \alpha_6$ and any angle at the circumference, like ∠BDC, is ∠BOC/2 = 6°.]

5.5.32. Given the rectangle ABCD (Fig. 5.5.35), draw a circle centered at any point inside, passing through one vertex (A), and cutting the adjacent sides at F, J and the diameter through A at I. Prove that AB × AF + AD × AJ = AC × AI.

[From the theorem illustrated in Fig. 5.2.18, AC × IC = GC × HC and AB × BF = BH × BG. Substract the second equation from the first: AC × IC − AB × BF = GC × HC − BH × BG = HC (CG + BG) − aBG = a(HC − BG), where a is the length BC = AD of the rectangle. But BG = AR − GS = RJ − SH, where R, S are the feet of perpendiculars from the center of the circle O to the chords AJ, GH. Also, HC = RD − SH = a − RJ − SH, so HC − BG = a − 2RJ = JD. With this last result, AC × IC = AB × BF + AD × JD, or

$$AC(AC - AI) = AB(AB - AF) + AD(AD - AJ),$$

from which (since $AC^2 = AB^2 + AD^2$), AC × AI = AB × AF + AD × AJ, Q.E.D.

5.5.33. Two equal circles intersect at A and B so that the center of one is on the circumference of the other (Fig. 5.5.36). From A draw any chord intersecting the circles at C and D. Join D and C to B. Then CB = CD.

---

33. Steck, *Dürers Gestaltlehre* (1948), 47; Shelby, *Gothic design* (1977), 118–19.

Fig. 5.5.36 Exercise 5.5.33.

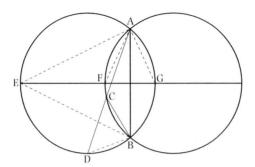

[Notice that ∠DAB enjoys the unusual status of a circumferential angle in two circles of equal radii; hence the arcs it subtends on the two circles are equal and so are the included chords: CB = BD. Now if, as the problem claims, CB also equals CD, ΔCDB must be equilateral, which would follow if one of the equal angles CDB or DCB were 60°. Consider ∠CDB = ∠ADB, a circumferential angle subtending chord AB. Connect A and B with E, one extremity of the diameter containing the centers of the two circles; ∠AEB also subtends chord AB and so must equal ∠ADB. Can you show that ∠AEB = 60°?

Connect A with F, G, the centers of the intersecting circles. Then ΔAFG is equilateral, since the radii of the circles are equal, and ∠FGA = 60°. But ∠EAG = rt.∠, since it is inscribed in a semi-circle; hence ∠AEG = 30° and ∠AEB = 2∠AEG = 60°, so that, as required, ∠ADB = ∠AEB = 60°. Remember that C can be any point on circle center G: that, nonetheless, ΔCDB should always be equilateral is remarkable.]

5.5.34. It is remarkable, and true, that if from any point P (Fig. 5.5.37) a secant is let fall across a given circle and the tangents drawn from the points S and T where the secant cuts the circle until they meet at Q, then the perpendicular dropped from Q onto the line joining P and the circle's center O is the locus of all points similarly defined by secants to the circle from P. Or, to speak more efficiently, X, the intersection of the perpendicular from Q with PO, is independent of the angle $\alpha = \angle OPT$.

[To prove it, connect Q, S, and T with O; let the intersection of OQ and ST be R. Then (OQ⊥ST) ΔXQO ~ ΔRPO and OR : OP = XO : QO. Also (OT⊥TQ) ∠OQT = ∠OTS = ∠OST, and, therefore, ΔOQT ~ ΔOTR; from which r : QO = OR : r. From

$$OR \times QO = r^2 \quad \text{and} \quad OR \times QO = XO \times OP,$$

XO = r^2/OP = constant for a given point P, Q.E.D. The line QO is called the *polar* of the 'pole' P. Can you see that, if Q is taken as pole, its polar is the line PR?]

Fig. 5.5.37 Exercise 5.5.34.

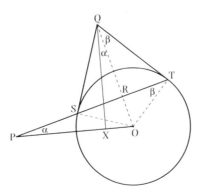

Fig. 5.5.38 Exercise 5.5.35 ('the butterfly').

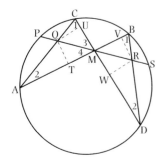

5.5.35. The butterfly. Draw any chord PS (Fig. 5.5.38) and bisect it at M. Through M draw any two chords AB, CD. Join A to C and B to D. You have the figure of the butterfly. Among its interesting properties is that QM = MR.

[Proofs of this relation tend to flutter just beyond reach. The obvious move to capture the unknowns is to use Euclid III.35. That gives you, with QM = x, MR = y, and PM = MS = a,

$$\text{PQ} \times \text{QS} = a^2 - x^2 = \text{CQ} \times \text{QA}, \qquad \text{PR} \times \text{RS} = a^2 - y^2 = \text{BR} \times \text{RD}.$$

The next move may not be so obvious. You need to create some similar triangles with the help of the pairs of equal angles marked 1 and 2; but what triangles? If you do not know, seek guidance and reassurance from the almost always useful law of sines. Applied to triangles QMC, AMQ and RMB, DMR, it gives, respectively,

$$\text{CQ} \times \text{QA} = x^2(\sin\angle 3 \sin\angle 4/\sin\angle 1 \sin\angle 2),$$
$$\text{BR} \times \text{RD} = y^2(\sin\angle 3 \sin\angle 4/\sin\angle 1 \sin\angle 2).$$

The result of the application therefore is $(a^2 - x^2)/x^2 = (a^2 - y^2)/y^2$, or $x = y$, as required.

With this clue, you can devise a proof free from trigonometry. Drop the perpendiculars QU, QT, RV, RW. Then, by equating the values of sin∠1 in the similar triangles PUC, RVB, you have QU : CQ = VR : BR. Similarly, you can write QT : QA = WR : RD, QU : QM = WR : MR, QT : QM = VR : MR. Therefore

$$\text{BR} \times \text{RD}/\text{CQ} \times \text{QA} = (\text{VR}/\text{QU})(\text{WR}/\text{QT}) = (\text{VR}/\text{QT})(\text{WR}/\text{QU})$$
$$= \text{QM}^2/\text{MR}^2 = x^2/y^2.$$

You return to $(a^2 - x^2)/(a^2 - y^2) = (x^2/y^2)$.

Note that the sine function served as a bookkeeper for the construction of the similar triangles needed to effect a Euclidean proof. There is no reason to eschew this use of the trig functions even if you are required to furnish a proof by geometric methods alone. Translation is usually easy and direct provided that you do not rely on the functions of sums or differences of angles. We know that the Greeks themselves did not always or for the most part devise their proofs as they delivered them. The lesson of the butterfly is to try whatever ways occur to you to find a proof and then, if necessary, reformulate it in Euclidean terms.

6 Tough knots

The geometrical principles developed in Chapters 2–5 support a range of applications well beyond those already discussed. A few of these applications, which require no more geometry than you know, but, perhaps, a little more attention, follow. They concern three topics previously introduced, Hero's formula (§1), the computation of π (§2), and Gothic tracery (§5), and two new ones, radian measure (§3) and burning mirrors (§4).

6.1. Hero's formula for the area of a triangle

A good example of the difference in approach between the geometer and the algebraist may be drawn from proofs of the famous formula of Hero for the area of a triangle. Hero refers not to the demanding discipline of the proofs but to an applied geometer of that name who worked in Alexandria probably in the second century AD. Hero begins the book in which he demonstrates his formula with an account of the origin of geometry in Egypt; and he presents the problems of the resurveying of fields after floods as the re-establishment of all boundary marks when only a few remain. His formula, which he appears to have taken from Archimedes, allows the computation of the area of a triangle without knowing, or measuring, the height.[1]

Hero's (or Archimedes') rewriting of the area Fig. 6.1.7 of a triangle ABC employs two unintuitive steps, illustrated in Fig. 6.1.1: draw the inscribed circle and its radii perpendicular to the sides (OD, OE, OF) and then-only an Archimedes would have thought of this—run a perpendicular to OC and extend it until it meets, at L, a perpendicular dropped from B. Join L and C. We are to prove that $\triangle$AOF $\sim$ $\triangle$CLB. Since both contain a right angle ($\angle$AFO, $\angle$CBL), we need to show only one other pair of angles equal, say $\angle$AOF, $\angle$CLB. Their equality follows from the fact, to be proved instantly, that both $\angle$AOF and $\angle$CLB are the supplement of $\angle$COB. Begin with $\angle$AOF. Since OA, OB, OC bisect the angles of $\triangle$ABC, $\angle$AOF = $\angle 3'$, $\angle$BOC = $\angle 1' + \angle 2'$, where the prime designates 'complement of'. Since $\angle 1 + \angle 2 + \angle 3 =$ rt.$\angle$, $\angle$AOF = $\angle 3' =$ st. $- (\angle 1' + \angle 2') =$ st.$\angle - \angle$BOC, as required.

---

1. Heath, *Greek math.*, 2, 307. 321–3.

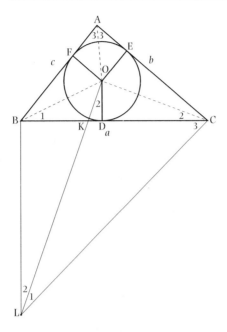

Note that L, B, O and C from a cyclic quadrilateral since the right
angles at B and O are on the same hypoteneuse, LC; therefore $\angle OLC$
= $\angle 1$ (they both subtend the chord OC). Further $\angle KOD = \angle 2$ (both
are complements of $\angle DOC$) = $\angle BLK$ (both are complements of
$\angle BKL$). Therefore $\angle BLC = \angle BLK + \angle OLC = \angle 1 + \angle 2 = $ st.$\angle$ $-$ (st.$\angle$
$- \angle 1 - \angle 2$) = st.$\angle - \angle BOC$, Q.E.D.

Since $\angle AOF$ and $\angle CLB$ are both the supplement of $\angle COB$, they are
equal; and since, in addition, triangles CLB and AFO each contain a
right angle, $\Delta CLB \sim \Delta AOF$. Therefore,

$$\frac{BC}{BL} = \frac{AF}{FO}, \quad \text{or} \quad \frac{a}{BL} = \frac{AF}{r}.$$

But from similar triangles LBK and ODK, BL: BK = r : KD, so that

$$a = \frac{AF}{r} \times BL = \frac{AF \times BK}{KD} = \frac{AF(BD - KD)}{KD} = AF\left(\frac{BD}{KD} - 1\right).$$

From the similar triangles KOD, DCO (remember, OL $\perp$ OC),

$$\frac{OD}{DC} = \frac{KD}{OD}, \quad \text{or} \quad KD = \frac{r^2}{DC},$$

whence

$$a = AF\left(\frac{BD \times DC}{r^2} - 1\right). \tag{6.1.1}$$

We have by definition

$$AF = s - BD - DC = s - a,$$
$$BD = s - AF - DC = s - AE - EC = s - b,$$
$$DC = s - AF - BD = s - c,$$

where s is the semi-perimeter. After substitution of these relations in equation (6.1.1), you will have

$$a = (s - a)\left(\frac{(s-b)(s-c)}{r^2} - 1\right),$$

$$r^2(a + s - a) = (s - a)(s - b)(s - c).$$

From equation (5.2.4),

$$\Gamma^2 = r^2 s^2 = s\,(s - a)\,(s - b)\,(s - c),\ \text{Q.E.D.}$$

The trigonometrician would proceed less directly and more simply as follows. Let the triangle whose area is sought be ABC in Fig. 6.1.2. He or she would want to express $\Gamma = ah/2$ in terms of s and the sides. Now $h = b \sin \angle ACB$, and, by the law of consines, $c^2 = a^2 + b^2 - 2ab \cos\angle ACB$, or $\cos\angle ACB = (a^2 + b^2 - c^2)/2ab$. Therefore

$$\Gamma^2 = a^2 h^2/4 = (a^2 b^2 \sin^2 \angle\,ACB)/4$$
$$= (a^2 b^2/4)\,[1 - (a^2 + b^2 - c^2)^2/4a^2 b^2],$$
$$16\Gamma^2 = 4a^2 b^2 - (a^2 + b^2 - c^2)^2 = 2\,(a^2 b^2 + a^2 c^2 + b^2 c^2) - a^4 - b^4 - c^4$$
$$= (a + b + c)[a^2(b + c - a) + b^2(a + c - b) + c^2\,(a + b - c) - 2abc]$$
$$= (a + b + c)(a + b - c)(a + c - b)(b + c - a). \tag{6.1.2}$$

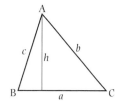

Fig. 6.1.2 A trigonometric route to the area of a triangle.

The algebra in the last steps is easily, if tediously, checked. The important point is that the trigonometrical derivation requires little attention, only the factoring of algebraic forms, whereas the geometrical proof demands a grasp of linear relationships at every step. Hence the claim, frequently made in geometry books of the last century, that geometry exercises, whereas algebra relaxes, the mind.

6.2. Huygens' improved method of computing π

Huygens set forth from a theorem of Archimedes on the summation of areas. Take a field of 400 acres; add to it one as fourth as large, that is, 100 acres; and another, one-quarter the second, or 25 acres; and still another, one-quarter the third, or $6\frac{1}{4}$ acres; and so on, as far as you please. What is the sum? To descend to algebra, call the area of the first field A, that of the second, $A/4$, of the third, $A/4^2 = A/16$, of the fourth, $A/4^{\,3}$, and so on; of which the total is $A(1 + \frac{1}{4} + (\frac{1}{4})^2 + (\frac{1}{4})^3 + ...)$. The sum of the power series in $\frac{1}{4}$ can easily be obtained by a trick. If x is a positive number less than 1, you can show immediately, by straightforward long division, that

$$\frac{1}{1-x} = 1 + x + x^2 + x^3 + \cdots$$

Therefore, if we were to continue to add smaller and smaller fields forever, their sum would be $S = A/(1 - \frac{1}{4}) = 4A/3$. But such a process would weary even an Archimedes. Let us then stop adding fields after some number, n say, where $n = 1$ designates the field A, $n = 2$ the field $A/4$, etc. We would then have a sum, S_n, less than S by an amount T_n that can be made as small as we please by enlarging n. In fine—and this is Archimedes' theorem— $S_n < 4A/3$.[2]

Huygens' application of Archimedes theorem makes use of two geometrical figures we have not yet defined. One, the circular segment, is the area of a circle cut off by any chord, such as seg ABC = AEBFCA in Fig. 6.2.1. The other, the circular sector, is a pie-shaped area created by two radii and the intercepted arc, such as sec AOC = OAEBFCO. Evidently, sec AOC = ΔAOC + seg ABC. Huygens filled up sectors with little triangles formed from regular polygons, as in Fig. 6.2.1, where sec AOC > ΔAOC + ΔABC + (ΔAEB + ΔBFC).

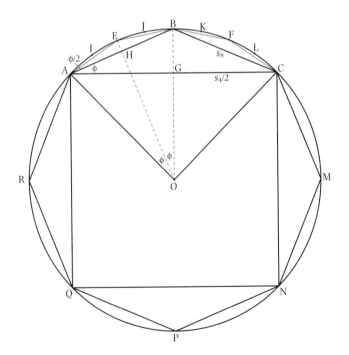

Fig. 6.2.1 Huygens' method for finding a lower limit to π.

2. Dijksterhuis, *Archimedes* (1987), 129–30, 344–5. The material that follows paraphrases Huygens, *Oeuvres, 12* (1910), 113–50.

The difference between the sector and the sum of the triangles can be made as small as we please by continuing the process of subdivision of the segments: if I, J, K, L designate the midpoints of the arcs AE, EB, BF, FC, respectively, we can write

$$\sec AOC = \Delta AOC + \Delta ABC + 2(\Delta AEB) + 4(\Delta AIE) + \cdots.$$

This equation makes use of the evident relations, $\Delta AEB = \Delta BFC$ and $\Delta AIE = \Delta EJB = \Delta BKF = \Delta FLC$.

The computation of the various areas is easy. We have $\Delta ABC = 2\Delta AGB = BG \times AG$, $\Delta AEB = 2\Delta AEH = AH \times EH$. Therefore

$$\frac{\Delta ABC}{\Delta AEB} = \frac{BG \times AG}{AH \times EH} = \frac{AB^2}{AE^2} 2\cos\phi.$$

The last step follows from Euclid III.20, which makes $\angle EAB = \angle EOB/2 = \phi/2$ and $\angle BAG = \angle BAC = \angle BOC/2 = (2\angle EOB)/2 = \phi$. We have therefore (recall that $\sin 2x = 2 \sin x \cos x$)

$$\begin{aligned}\Delta ABC &= (AB^2/AE^2) \cos\phi(2\Delta AEB) \\ &= (AB^2/AE^2)(\Delta AEB + \Delta BFC) \cos\phi < 4(\Delta AEB + \Delta BFC).\end{aligned}$$

The inequality is obtained by replacing $\cos\phi \leq 1$ by 1 and $AB < 2AE$ by $AB = 2AE$; $AB/2 < AE$ because $AB/2$ is the side of a right triangle with hypothenuse AE. In words: the sum of the areas of the largest triangles left in the segment ABC after removal of the base triangle ABC is larger than one-fourth of ABC.

Let seg_n be the magnitude of the segment defined by the side of an inscribed regular polygon of n sides, and let t_n be the area of the largest triangle inscribable in the segment. We have from Fig. 6.2.1, in which ACNQ is a square and ABCMNPQR an octagon,

$$\begin{aligned}\text{segABC} = \text{seg}_4 &= \Delta ABC + 2\text{seg } AEB = t_4 + 2\text{seg}_8, \\ \text{segAEB} = \text{seg}_8 &= \Delta AEB + 2\text{seg } AIE = t_8 + 2\text{seg}_{16},\end{aligned}$$

so that, in general,

$$\text{seg}_n = t_n + 2\text{seg}_{2n}, \tag{6.2.1}$$

and

$$\text{seg}_n = t_n + 2t_{2n} + 4t_{4n} + \dots \tag{6.2.2}$$

We know from the preceding paragraph that

$$t_n < 4 \, (2t_{2n}), \; t_{2n} < 4 \, (2t_{4n}).$$

Expressed otherwise,

$$t_n + r_n = 4 \, (2t_{2n}), \; t_{2n} + r_{2n} = 4 \, (2t_{4n}),$$

where the remainders r make up the difference between four times twice the smaller triangles and the larger ones. Introducing these last equivalents into equation (6.2.2), we have

$$\text{seg}_n = t_n + 2\left(\frac{t_n + r_n}{8}\right) + 4\left(\frac{t_{2n} + r_{2n}}{8}\right) + 8\left(\frac{t_{4n} + r_{4n}}{8}\right) + \cdots$$
$$= t_n(1 + 1/4 + 1/16 + \ldots) + r_n(1/4 + 1/16 + 1/64 + \ldots) + R_n,$$

where R_n signifies the sum of the various remainders beyond r_n. Now, according to Archimedes' theorem, $t_n(1 + 1/4 + 1/6 + \ldots) = (4/3)t_n$. We have, finally,

$$\text{seg}_n > (4/3)t_n + r_n/3. \tag{6.2.3}$$

This very nice relation leads immediately to a value for π—if we drop the term $r_n/3$. Referring again to Fig. 6.2.1, you see that

$$\text{sec}_4 = \Delta AOC + \text{seg}_4,$$
$$\text{sec}_8 = \Delta AOB + \text{seg}_8.$$

Now $4\text{sec}_4 = 4\Delta AOC + 4\text{seg}_4 = \square ACNQ + 4\text{seg}_4 = A_4 + 4\text{seg}_4$, where A_4 is the area of square ANCQ. Similarly $8\text{sec}_8 = 8\Delta AOB + 8\text{seg}_8 = A_8 + 8\text{seg}_8$. But $4\text{sec}_4 = 8\text{sec}_8 = A_c$, the area of the circle. In general, therefore,

$$A_c = A_n + n\text{seg}_n.$$

According to equation (6.2.3), $n\text{seg}_n > (4/3)nt_n$, or $n\text{seg}_n = (4/3)nt_n$ + remainder. Therefore

$$A_c = A_n + (4/3)nt_n + \text{remainder},$$
or
$$A_c > A_n + (4/3)nt_n.$$

Now nt_n is the difference in area between the $2n$-gon and the n-gon. (In Fig. 6.2.1, $4t_4 = 4\Delta ABC$, which is the difference between the octagon ABCMNPQR and the square ACNQ.) Finally, therefore,

$$A_c > A_n + 4/3(A_{2n} - A_n),$$
or
$$A_c > A_{2n} + (A_{2n} - A_n)/3. \tag{6.2.4}$$

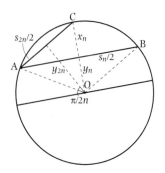

Fig. 6.2.2 Further to Fig. 6.2.1.

The lower bound of π that emerges from this inequality is not very good. That appears immediately from doing the calculation. In Fig. 6.2.2, $AB = s_n$, $AC = s_{2n}$, $A_n = ny_ns_n/2$, $A_{2n} = 2ny_{2n}s_{2n}/2$. But $y_n = r\cos 180°/n$, $s_n = 2r\sin 180°/n$, etc., so that

$$A_n = (nr^2\sin 360°/n)/2, \quad A_{2n} = nr^2\sin 180°/n,$$

and we have, for $n = 6$,

$$A_c = \pi r^2 > (4/3)(6r^2\sin 30°) - (1/3)(6r^2\sin 60°)/2,$$
or
$$\pi > 3.1339.$$

Huygens found a better way with the same equipment we have been developing. He wrote $A_n = ny_ns_n/2$ as $p_ny_n/2$, p_n being the perimeter of the n-gon, and substituted $r > y_n$ for y_n, transforming equation (6.2.4) into

$$2\pi r > p_{2n} + (p_{2n} - p_n)/3 \qquad (6.2.5)$$

It is true that this reworking improves the lower bound since the right-hand side of the equation has been increased in magnitude by substituting r for y. (It is easily checked that the substitution does not reverse the sign of inequality.) Again take $n = 6$, $2n = 12$; then $p_n = 6r$, $p_{2n} = 12s_{12}$. From $s_n = 2r \sin 180°/n$, $s_{12} = 2r(0.25882)$, $p_{12} = 12s_{12} > 6.21165r$, $(p_{12} - p_6)/3 > 0.07055r$. Hence, from equation (6.2.5),

$$\pi > (6.21165 + 0.07055)/2 = 3.1411.$$

But Archimedes' lower bound, 3 10/71, obtained from a polygon of 96 sides, was 3.1408. Hence Huygens came closer with less labor.

To check that the replacement of the y's by r does not raise the right side of equation (6.2.4) enough to reverse the inequality sign, reintroduce the differences, $r_n = 8t_{2n} - t_n$, between eight times the triangles of order $2n$ and one of order n. Then we have in place of equation (6.2.4)

$$A_c > (4/3) A_{2n} - (1/3) A_n + nr_n/3.$$

By definition,

$$nr_n/3 = n(8t_{2n} - t_n)/3.$$

Since r_n holds the answer to our question, we must calculate it. We have $A_n = ny_n s_n/2 = n(r - x_n)s_n/2$. Since (see Fig. 6.2.2) $x_n s_n/2 = t_n$,

$$A_n = rp_n/2 - nt_n, \quad A_{2n} = rp_{2n}/2 - 2nt_{2n}.$$

Hence we have

$$A_c > (4/3)(rp_{2n}/2 - 2nt_{2n}) - (1/3)(rp_n/2 - nt_n) + n(8t_{2n} - t_n)/3,$$
$$A_c > (2/3)rp_{2n} - rp_n/6,$$
$$2\pi r > p_{2n} + (p_{2n} - p_n)/3,$$

Q.E.D.

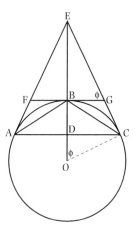

Fig. 6.2.3 Huygen's method for finding an upper limit for π.

It remains to find an upper bound. Huygens followed a strategy similar to his route to a lower bound. First, a basic inequality. Let AC (Fig. 6.2.3) be a chord, AE, CE, and FG tangents at A, C, and B, respectively. Huygens showed that the area of $\triangle$EFG is over half that of $\triangle$ABC by a method that can be paraphrased as follows. In Fig. 6.2.3, $\triangle$EFG = $2(EB \times BG)/2 = EB^2/\tan \phi = r^2 (1/\cos \phi - 1)^2/\tan \phi$, so

$$\triangle EFG = \frac{r^2(1 - \cos \phi)^2}{\sin \phi \cos \phi}.$$

Also, $\Delta ABC = 2(BD \times DC)/2 = r(1 - \cos \phi)\, r \sin \phi$, so $\Delta ABC = r^2 \sin \phi (1 - \cos \phi)$.

Therefore we have

$$\frac{\Delta EFG}{\Delta ABC} = \frac{(1 - \cos \phi)^2}{\sin^2 \phi \cos \phi (1 - \cos \phi)} = \frac{1}{\cos \phi (1 + \cos \phi)} > \frac{1}{2}.$$

Q.E.D. (The last inequality arises from the fact that the maximum value of $\cos \phi$ is 1.)

Figure 6.2.4 continues the construction begun in Fig. 6.2.3. Just as $\Delta EFG > \Delta ABC/2$, $\Delta JFK > \Delta AIB/2$ and $\Delta MGN > \Delta BLC/2$. The process can be repeated, the external triangles always being more than half of the internal ones. The sum of the internal ones will eventually equal the area of the segment ABC as nearly as desired, while the sum of the external ones becomes equal to the area between ΔAEC and the segment ABC. Hence

$$\Delta AEC - \mathrm{seg}ABC > \mathrm{seg}ABC/2,$$

or

$$\Delta AEC > (3/2)\mathrm{seg}ABC.$$

Let's change notation to agree with that used in the approximation by inscribed polygons. Greek letters will indicate circumscribed polygons and roman letters, as usual, internal ones. We have $\Delta ABC = t_n$, $\Delta AOC = A_n/n$, $AC = s_n$, $\mathrm{seg}ABC = \mathrm{seg}_n$, $\Delta AEC = \tau_n$, $FG = \sigma_{2n}$. The inequality $\Delta AEC > (3/2)\mathrm{seg}ABC$ becomes $\tau_n > (3/2)\mathrm{seg}_n$ so that

$$A_c = n\mathrm{seg}_n + A_n < 2n\tau_n/3 + A_n.$$

But $n\tau_n = \Gamma_n - A_n$, where Γ_n is the area of the circumscribed polygon with half-side $\sigma = EC$. In this notation,

$$A_c < 2\,\Gamma_n/3 + A_n/3 \qquad (6.2.6)$$

Fig. 6.2.4 Further to **Fig. 6.2.3**.

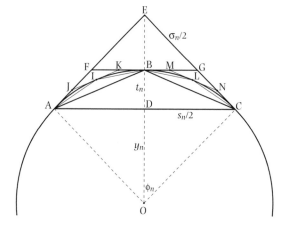

Now $\Gamma_n = nr\sigma_n/2 = \gamma_n r/2$, where $\gamma_n = n\sigma_n$ is the perimeter of the circumstribed n-gon; also, as before $A_n = ny_ns_n/2 = y_np_n/2$. Equation (6.2.6) transforms into

$$2\pi r < 2\gamma_n/3 + p_n/3. \qquad (6.2.7)$$

In the last step, y_n has been replaced by its greater, r, which makes the right side of equation (6.2.7) greater than that of equation (6.2.6), and so preserves the inequality. However, the maneuver raised the upper bound, and so worsened the approximation; for $n = 12$, equation (6.2.7) gives $\pi < 3.1788$ (note, from Fig. 6.2.4, $\sigma_n = 2r\tan\phi_n$, $s_n = 2r\sin\phi_n$, $\phi_n = 180°/n$).

Huygens lowered the upper bound by replacing y_n not by its greater r, as done above in moving from equation (6.2.6), but by its equal, $(p_n/\gamma_n)r$. The equality follows from the similarity of triangles OCE and ODC in Fig. 6.2.4, from which $s_n/\sigma_n = y_n/r$. We have in place of equation (6.2.7),

$$2\pi r < 2\gamma_n/3 + (p_n^2/\gamma_n)/3.$$

Since $\gamma_n = n\sigma_n = 2nr\tan\phi_n$, $p_n = ns_n = 2nr\sin\phi_n$, the inequality may be written

$$2\pi r < \gamma_n/3 + (p_n/3)\left(\frac{1+\cos^2\phi_n}{\cos\phi_n}\right).$$

Since the inequality holds for all values of n, it must hold for the minimum value of the factor involving $\cos\phi$, which occurs when $\phi = 0$. Therefore the best upper bound for π results from

$$2\pi r < \gamma_n/3 + 2p_n/3.$$

To calculate the upper bound on π from the dodecagon, take $s_{12} < 0.51764$, $\phi_{12} < 0.5359$. You find that

$$\pi < \frac{12}{2}\left(\frac{2(0.51764)}{3} + \frac{0.5359}{3}\right) = 3.1423.$$

From the 12-gon, therefore, Huygens worked out that

$$3.1411 < \pi < 3.1423,$$

as against Archimedes' limits from the 96-gon,

$$3.1408 < \pi < 3.1429.$$

If you care to go to 10 800 sides, Huygens can give you 14 secure decimals, Archimedes only 6. A lot of geometry, you might say, for a little gain. But Huygens' calculation offers, in addition to its own interest, an uplifting example of how even the greatest feats of geometry (in this case Archimedes' computation of π) can be bettered. If you know what you are doing.

6.3. Radian measure

From the relationship

$$C/n < 2s_n/3 + \sigma_n/3. \tag{6.3.1}$$

it follows that the length of any circular arc less than a quarter circle is less than two-thirds its sine plus one-third its tangent.

In Fig. 6.3.1, C/n = arc BYC, or $C/2n$ = arc BY; $BC = s_n$; $NP = \sigma_n$. Furthermore, BX $=$ BC/2 $=$ $r \sin \angle BOY$ and NY $=$ NP/2 $=$ $r \tan \angle BOY$. Hence, if we divide both sides of equation (6.3.1) by 2, we have

$$C/2n = \text{arc BY} < (2/3)r \sin \angle BOY + (1/3)r \tan \angle BOY.$$

This relationship holds for all arcs under a quarter circle since nothing in the derivation required that n be integral.

It follows, therefore, that, in a circle of radius r,

$$\text{arc BY}/r < (2 \sin \theta)/3 + (\tan \theta)/3, \tag{6.3.2}$$

where, as in Fig. 6.3.2, θ is any angle less than a rt.$\angle$. The preceding equation suggests the convenience of writing arc BY $= r\theta$, which amounts to measuring an angle by the circular arc it subtends. For example, half a right angle, or 45°, subtends an arc equal to one-eighth the circumference of a circle, or $(2\pi/8)r = (\pi/4)r$. Hence in our new measure, $r = 45° = (\pi/4)r$ or $45° = (\pi/4)$ angular unit. Over a century ago, two British professors christened this unit of angular measure the *radian*. It probably first occurred in print in an examination question presented to students at Queens College, Belfast, in 1873.[3]

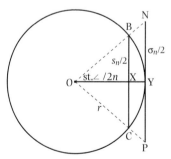

Fig. 6.3.1 Approach to radian
measure via Huygens' inequalities.

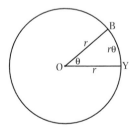

Fig. 6.3.2 Radian measure.

3. Thomson, *Nature*, *83* (1910), 217. Radian measure without the name was used as early as the mid-18th century (Jones, *MT*, *46* (1953), 424–5).

Table 6.3.1 Conversion of degrees and radians

Angle in degrees	Angle in radians
1	$0.017 = \pi/180$
30	$\pi/6$
45	$\pi/4$
57.30	1
60	$\pi/3$
90	$\pi/2$
180	π

Since $45° = \pi/4$ radian, 1 radian $= (180/\pi)° \approx 57.30°$, and $1° \approx 0.017$ radian. Note that a radian is much larger than a degree. Table 6.3.1 will fix ideas. Radian measure has many conveniences. In particular, it allows replacement of the cumbersome notations rt.$\angle$ and st.$\angle$ by $\pi/2$ and π, respectively.

Let us return to equation (6.3.2), which, in radian measure, runs

$$\theta < (2 \sin \theta)/3 + (\tan \theta)/3.$$

It is easy to guess that the right-hand side often exceeds θ considerably, since when θ is close to $\pi/2$, $\tan \theta$ becomes very large. Huygens gave a much closer approximation via the construction indicated in Fig. 6.3.3. Here $\theta = \angle BOY$, as in Fig. 6.3.1 and 6.3.2, and we know from equation (6.3.2) that

$$\text{arc } BY < (2r \sin \theta)/3 + (r \tan \theta)/3.$$

Huygens showed that

$$\text{arc } BY < GY,$$

where G marks the intersection of the line EB extended with the tangent NY. As for E, it is fixed by the conditions that it lie on the diameter YA extended to such a distance that ED $= r$, the radius of the circle.

Fig. 6.3.3 Huygens' upper limit for the rectification of an arc.

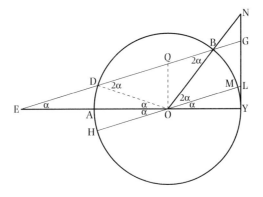

Draw a line through O parallel to EG cutting the circle at H and M
and the tangent at L. Call $\angle AOH$ α. Then $\angle DOH = 2\alpha$ ($\angle DEA = \alpha$,
by alternate interior angles, and $\angle DEA = \angle DOA$ since ΔEDO was
made isosceles). Also, $\angle BOM = \angle DBO$ (EB∥HM) $= \angle BDO$ (ΔDOB is
isosceles) $= \angle DOH$ (EB∥HM) $= 2\alpha$. Hence $\theta = \angle BOY = 3\alpha$ ($\angle MOY =$
$\angle AOH$ by vertical angles), or

$$\text{arc BY} = 3\text{arc MOY} < 3(2r \sin \alpha)/3 + 3(r \tan \alpha)/3$$
$$= 2r \sin \alpha + r \tan \alpha = GL + LY,$$

or arc FB < GY, Q.E.D. That $2r \sin \alpha = GL$ follows immediately from
raising OQ⊥AY. That makes OQ∥NY and QOLG a parallelgram,
whence $GL = OQ = 2r \sin \alpha$. Remember, this result has been
achieved using the powerful relation

$$\theta < 2 \sin \theta /3 + \tan \theta /3, \qquad\qquad 6.3.3$$

which holds only in radian measure.

To complete the story, we need a corresponding lower bound for θ.
Huygens achieved this goal by moving point E in Fig. 6.3.3 out
further along YA extended so as to make EA, rather ED, equal to r
(Fig. 6.3.4). The desired relation is

$$\text{arc BY} > G'Y.$$

Begin with equation (6.2.5)

$$C > p_{2n} + (p_{2n} - p_n)/3,$$

which gives

$$\text{arc BY} = \frac{C}{2n} > \frac{4}{3}\left(\frac{2ns_{2n}}{2n}\right) - \frac{1}{3}\left(\frac{ns_n}{2n}\right),$$

or

$$\text{arc BY} > 4s_{2n}/3 - (s_n/2)/3 \equiv Q.$$

We are to show that $Q \geqslant G'Y$.

Fig. 6.3.4 Huygens' lower limit
for the rectification of an arc.

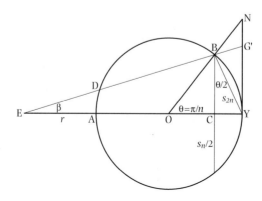

We have from $\Delta EG'Y$, $\tan \beta = G'Y/3r$, from ΔEBC, $\tan \beta = (s_n/2)/(2r + r \cos \theta)$. Hence

$$G'Y = 3(s_n/2)/(2 + \cos \theta).$$

Furthermore, as we read directly from Fig. 6.3.4, $\cos \theta/2 = (s_n/2)/s_{2n}$. Therefore $Q = (4/3)(s_n/2)/\cos \theta/2 - (1/3)(s_n/2)$ and

$$\frac{Q}{G'Y} = \frac{\dfrac{1}{3} \times \dfrac{s_n}{2}\left(\dfrac{4}{\cos \theta/2} - 1\right)}{\dfrac{3s_n/2}{2 + \cos \theta}} = \frac{1}{9}\left(\frac{4}{\cos \theta/2} - 1\right)(2 + \cos \theta).$$

When $\theta/2 = 0$, $Q/G'Y = (1/9)(4 - 1)(2 + 1) = 1$. When $\theta/2 > 0$, $\cos \theta/2 = 1 - x$, where x lies between 0 and 1, and

$$\frac{Q}{G'Y} = \frac{1}{9}(4(1 + x + x^2 + \cdots) - 1)(2 + \cos^2 \theta/2 - \sin^2 \theta/2)$$

$$= \frac{1}{9}(3 + 4x + 4x^2 + \cdots)(2 - 1 + 2(1 - x)^2)$$

$$> \frac{1}{9}(3 + 4x + 4x^2)(3 - 4x + 2x^2) = [9 + 2x^2(2x - 1)^2]/9 > 1.$$

We have therefore, as required, $Q \geqslant G'Y$ and arc $BY > G'Y$. The relationship holds even when θ is not π/n, since no step of the demonstration requires that BC be half the side of a regular polygon.

Expressed in radian measure, arc $BY > G'Y$ becomes $r\theta > G'Y$, or, since $G'Y = 3r \tan \beta$ and $\tan \beta = r \sin \theta/(2r + r \cos \theta)$,

$$\theta > 3 \sin \theta/(2 + \cos \theta). \tag{6.3.4}$$

From this relation, a good approximation to $\cos \theta$ for angles under $\pi/6$ can be obtained. In this region, take relation (6.3.4) to be an equality, whence

$$\theta^2(2 + \cos \theta)^2 = 9 \sin^2 \theta = 9(1 - \cos^2 \theta).$$

Applying the standard formula for solving a quadratic equation, we have

$$\cos \theta = \frac{-4\theta^2 + [16\theta^4 - 4(\theta^2 + 9)(4\theta^2 - 9)]^{1/2}}{2(\theta^2 + 9)} \approx 1 - \theta^2/2. \tag{6.3.5}$$

For $\theta = 30°$, the worst case under consideration, equation (6.3.5) gives

$$\cos 30° = 1 - 1/2(\pi/6)^2 = 0.8629,$$

whereas the correct value is $\sqrt{3}/2 = 0.8660$. For $\theta = 5°$, we have

$$\cos 5° = 1 - 1/2(\pi/36)^2 = 0.99619,$$

which is correct to five figures. In the same approximation, $\sin \theta = \theta$, since $\sin^2\theta = 1 - \cos^2\theta \approx 1 - (1 - \theta^2/2)^2 \approx \theta^2$. The approximate relations $\sin \theta = \theta$ and $\cos \theta = 1 - \theta^2/2$ are frequently used when only small angles enter the calculations. Remember that for the relations to hold, θ must be expressed in radian measure.

6.4. The burning mirror

In the year 214 BC the Romans besieged the city of Syracuse in Sicily. They came with a great fleet and many weapons. They proved no match for Archimedes. According to an ancient story, Archimedes, relying on mathematical principles, invented catapults so powerful that their missiles could crack open a ship in a single blow, and grappling machines so large and strong that they could lift a ship from the water, shake its men out into the sea, and crush it against the rocks. Two years after the siege began, the Romans took the city by land. A conquering soldier came upon Archimedes drawing a geometrical figure in the sand. The soldier bade the geometer follow him to his general. Archimedes insisted on finishing his demonstration; as you no doubt know from personal experience, it is annoying to be interrupted in the middle of a problem in geometry. Alas! the soldier was no geometrician: he grew angry at the delay and slew the greatest geometer of antiquity (Plate VIII).[4]

A late story credits Archimedes with destroying the Roman fleet with a subtler weapon than slings and grapples. This device reflected the sun's rays so intensely onto the ships that they burst into flames. We may doubt that even the Mediterranean sun could generate sufficient heat; but in theory the reflector might work, since it rests upon nothing more than the law of optical reflection and plane geometry.

Let us assume that the reflector had the shape of a piece of a spherical surface, as in Fig. 6.4.1, where OC is the radius of the sphere. The intersection of this surface with the plane of the paper is the circular arc ACB in Fig. 6.4.2. (A slightly different surface, whose intersection with the plane of the paper makes a parabola, gives a more intense focus.) Let a ray from the sun proceeding parallel to the mirror's axis OC strike the mirror at P and be reflected so as to cross the axis at Q. Since OP is a radius of the arc AB and, consequently, perpendicular to the surface at P, the law of reflection requires that $\angle SPO = \angle OPQ$. Call this angle ϕ. Since SP∥OQ, $\angle POQ = \angle SPO = \phi$, so $\triangle OQP$ is isosceles. Call the equal sides OQ and QP, s; the perpendicular dropped from P onto OC, x; and the radii OP and OC, r. Then

$$\sin \phi = PR/OP = x/r,$$
$$\cos \phi = OT/OQ = r/2s,$$

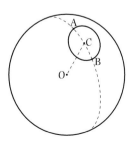

Fig. 6.4.1 Schematic of a spherical burning mirror.

4. Dijksterhuis, *Archimedes* (1987), 26–32.

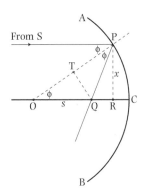

Fig. 6.4.2 Geometry of the focusing of a spherical mirror.

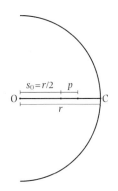

Fig. 6.4.3 Further to **Fig. 6.4.2**.

(T is the foot of the perpendicular dropped from Q on the radius OP; since ΔOQP is isosceles, QT bisects OP.)

From Pythagoras' theorem, $\sin^2 \phi + \cos^2 \phi = 1$,

$$\frac{r^2}{4s^2} = 1 - \frac{x^2}{r^2}. \tag{6.4.1}$$

The maximum value x can have occurs for the ray that strikes the mirror at A (or B). Let us call this value of x, t; it is half the diameter AB of the mirror. Typically, t is much smaller than r. We can therefore obtain a rough value for s, the distance from the center O at which any ray originally parallel to the axis crosses the axis after reflection, by ignoring x^2/r^2 in equation (6.4.1). This rough approximation to s, which we shall label s_0, is $r/2$: a spherical mirror whose diameter AB is small compared with its radius of curvature OC reflects rays originally parallel to its axis to a point half way between the center of curvature (O) and the apex (C) of the mirror.

In fact, of course, the reflected rays do not all cross the axis at the same point: those that strike the mirror closer to A cross further from O than those that strike the mirror closer to C. A practical case will indicate how widely spread the rays are when they reach the axis. Let $s = s_0 + p$, where p is small compared with s_0 (Fig. 6.4.3). Then, from equation (6.4.1).

$$\frac{r^2}{4} = (s_0^2 + 2ps_0 + p^2)(1 - x^2/r^2) \approx \frac{r^2}{4} + pr - \frac{x^2}{4},$$

whence $p = x^2/4r$. Between the first and second steps s_0 was replaced by its value $r/2$, and p^2 and x^2p/r ignored as quantities considerably small than those retained.

During the eighteenth century, all good collections of scientific instruments had burning mirrors. Typically, the diameter of these mirrors was less than half their radius of curvature; a good example, shown in Fig. 6.4.4, has $t/r \approx 1/4$. In such a mirror, the reflected rays cross the axis within an interval of $r/64$. Another example has $t = 23$ cm, $r = 100$ cm, and, consequently, $p = 1.3$ cm. Some of these machines could melt metals and stones.

A similar burner can be made by placing two spherical segments against one another as in Fig. 6.4.5. A spectacular version of this sort of machine, built for the French chemist Antoine Lavoisier in 1774, is shown in Fig. 6.4.6. (Lavoisier was a tax collector as well as a chemist, and died under the guillotine during the French Revolution.) The large lens A was composed of two spherical shells each 8 feet in radius, enclosing a volume filled with 140 pints of alcohol. (The builders used alcohol instead of water not to keep their spirits up but to prevent freezing and cracking of the glass.) The rays collected by the lens A and concentrated by the solid glass B roasted the specimens mounted on G: iron ran like water in a few seconds.

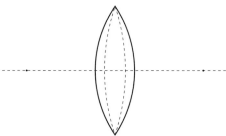

Fig. 6.4.5 Schematic of a
burning lens.

Cranks at E and F and the pivot C allowed assistants to maneuver the
machine to follow the sun. Note the sunglasses worn by the operator.

The geometry of the passage of the ray through the 'double
convex lens', as the glass of Fig. 6.4.5 is called, is too intricate for
presentation here. But the general idea may be illustrated by a
'plano-convex lens', made from one spherical shell, with its plane
face turned toward the sun. Figure 6.4.7. shows the intersection of
this lens with the plane of the paper. The center of curvature of the
arc ACB is O; the representative ray from the sun strikes the curved
surface at P; after refraction the ray crosses the axis at Q. Our task,
as in the case of the burning mirror, is to calculate the place of Q in
terms of the height $x = PD$ of P above (or below) the axis. We have
$\angle SPO = \phi$, the angle of incidence; $\angle TPQ = \theta$, the angle of refraction,

Fig. 6.4.6 Lavoisier's great burning machine. From Lavoisier, *Oeuvres* (1862), 2:3, plate IX.

Fig. 6.4.7 Geometry of the focusing of a planoconvex lens.

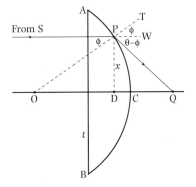

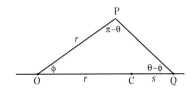

Fig. 6.4.8 Further to the same.

where $\sin\theta = n\sin\theta$. Since $\angle SPO = \angle TPW = \phi$ (vertical angles), $\angle WPQ = \theta - \phi$; and since the line SPW makes a straight angle, $\angle OPQ = \pi - \theta$ in radian measures.

Label the unknown quantity of interest, CQ, s, and apply the law of sines to $\triangle OPQ$, bearing in mind that $OP = OC = r$, the radius of curvature of the lens (Fig. 6.4.8). In redrawing the figure, $\angle POC$ has been labelled ϕ and $\angle PQC$, $\theta - \phi$, since $\angle POC = \angle SPO$, and $\angle PQC = \angle WPQ$, by alternate interior angles.

The law of sines gives

$$\frac{\sin(\theta - \phi)}{r} = \frac{\sin(\pi - \theta)}{r + s} = \frac{\sin\theta}{r + s},$$

$$\sin(\theta - \phi) = \frac{r\sin\theta}{r + s}.$$

The formula for the sine of the difference of two angles gives

$$\sin\theta\cos\phi - \cos\theta\sin\phi = (r/(r + s))\sin\theta,$$

or $\cos\phi - (\cos\theta/\sin\theta)\sin\phi = r/(r + s).$ (6.4.2)

Now $\sin\phi = x/r$, $\sin\theta = nx/r$, $\cos^2\phi = 1 - x^2/r^2$, $\cos^2\theta = 1 - n^2x^2/r^2$. For rays for which x is very small compared with r, we can neglect $(x/r)^2$ in these equations; then $\cos\theta = \cos 0 = 1$, and the preceding equation gives

$$1 - (x/r)/(nx/r) = r/(r + s_0)$$

(s_0 signifies the value of s when x differs negligibly from zero.) This equation yields for s_0:

$$1 - 1/n = r/(r + s_0), \quad s_0 = r/(n - 1).$$

To find how s changes with x to the next approximation, set $\cos\phi = 1 - x^2/2r^2$ and $\cos\theta = 1 - n^2x^2/2r^2$ and rewrite equation (6.4.2):

$$(1 - x^2/2r^2) - (1 - n^2x^2/2r^2)(x/r)/(nx/r) = r/(r + s).$$

When solved for s in the approximation $1 - n^2x^4/4r^4 = 1$, the preceding expression produces

$$s = s_0(1 - n^2x^2/2r^2),$$

where $s_0 = r/(n - 1)$.

We come to the bottom line:

$$p = s - s_0 = \frac{-n^2x^2}{2(n - 1)r}.$$

For one shell of Lavoisier's burning lens, the maximum value of x was $t = 2$ feet; $r = 8$ feet; and the index of refraction of the glass, n, probably fell between 1.5 and 1.6. Hence the rays refracted by this shell alone would have been spread over a width

$$p = (1.5)^2(2)^2/2(0.5)8 = 1.1 \text{ feet.}$$

The negative sign means that the heated area began at $s_0 = r/(n - 1)$ = 16 feet from point C and ran back toward the lens for a little more than 1 foot. It would not have been a good place to put your hand.

6.5. Gothic lights

Most Gothic tracery and all of its examples discussed in this book can be drawn by anyone who knows three theorems of Euclid: I.47 (the Pythagorean theorem) and III.11–12 (the collinearity of the centers of touching circles and their point of tangency). Late medieval architects and stone masons must have had rules equivalent to these theorems and their consequences; and it appears from a German treatise of the fifteenth century that they may have had an understanding of the underlying geometry. Of course, a mason could have placed a large rose window within a standard arch (Fig. 6.5.1) by cut and try; but the coincidence of the beginning of the high Gothic style with the diffusion of Euclid in the Latin West in the late twelfth century suggests that someone had found an easier way.

That someone needed only to identify the right triangle ADE in Fig. 6.5.1. The desired radius r then follows almost immediately:

$$(2a - r)^2 = r^2 + a^2, \quad r = 3a/4.$$

Note that $r^2 + a^2 = a^2(1 + 9/16) = 25\,a^2/16$, so that the hypotenuse AE = $5a/4$ and ΔADE is a (3, 4, 5) Pythagorean triangle. One of the pleasures and surprises of Gothic tracery is the recurrence of right triangles all of whose sides are expressible as rational numbers. No doubt the masons knew this property and made use of it. Of course, neither they nor the architects nor the mathematicians of the time possessed our efficient algebra and would have had to work hard to obtain their results.

Let's continue the play with circles a little further by putting small ones atop the big one and in the corners between it and the arch, as in Fig. 6.5.2. The radius, call it s, of the top little circle is easily obtained from the right triangle FDB:

$$(2a - s)^2 = a^2 + (r^2 + s)^2.$$

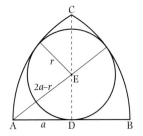

Fig. 6.5.1 Rose window in a Gothic arch.

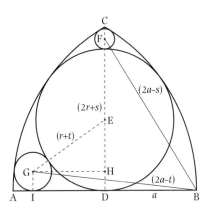

Fig. 6.5.2 The same, with roundels in the corners.

With the value previously found for r ($r = 3a/4$), this equation gives $s = 3a/28$. For the hypotenuse FB we then have $2a - s = 53a/28$ and, since $a = 28a/28$, the Pythagorean triplet (53, 45, 28) for the triangle FDB. The little circles in the lower corners brings something new. We have two unknowns: the radius t and the perpendicular distance GH of the circles' centers from the axis CD. They satisfy equations derived from the right triangles EHG, GIB:

$$(r + t)^2 = (r - t)^2 + GH^2,$$
$$(2a - t)^2 = t^2 + (a + GH)^2.$$

Subtract the first from the second and obtain GH $= (3a - 7t)/2$; substitute this value and $r = 3a/4$ into the first and obtain for t

$$9a^2 - 54at + 49t^2 = 0.$$

This equation does not have rational roots; t comes out a little larger than $a/5$; neither of the triangles EHG, GIB, is Pythagorean. Not all Gothic tracery is rational.

Instead of inscribing a complete circle in the parent Gothic windows, we might choose a half circle, as did the masons at Sens Cathedral (Fig. 6.5.3). Obviously its radius is a (Fig. 6.5.4). Now place a circle on top of the half circle and tangent to it and to the arch ACB. Let the radius be b; its value comes immediately from the right triangle ADP:

$$(2a - b)^2 = a^2 + (a + b)^2, \quad \text{whence } b = a/3, \ a + b = 4a/3.$$

$\triangle$ADP is a (3, 4, 5) Pythagorean triangle.

Continue. Draw (or imagine drawn) a circle centered on Q and tangent to the circles centered at A, D, and P. To find its radius, c, three right triangles must be involved: $\triangle$AUQ, $\triangle$PEQ, $\triangle$DUQ. They give the three equations

$$(2a - c)^2 = QU^2 + (a + DU)^2,$$
$$(b + c)^2 = PE^2 + DU^2 = (PD - QU)^2 + DU^2 = (a + b - QU)^2 + DU^2,$$
$$(a + c)^2 = QU^2 + DU^2.$$

The result of subtracting the third equation from the first is

$$DU = a - 3c.$$

The result of subtracting the second from the first and substituting $b = a/3$, DU $= a - 3c$, is

$$QU = a + c/2$$

Fig. 6.5.3 The rose window of the cathedral of Sens (ca. 1500). From Binding, *Masswerk* (1989), 112.

Fig. 6.5.4 Alison's window.

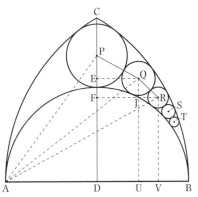

These values of DU, QU placed in the third equation give a quadratic for c:

$$8\tfrac{1}{4}c^2 - 7ac + a^2 = 0.$$

The solutions are $c = 2a/11$ and $2a/3$. Evidently the first alone applies.

All the triangles must be rational because their sides are linear combinations of rational fractions. Thus, in $\triangle$PEQ, EQ = DU = $a - 3c$ = $15a/33$; PE = PD − QU = $a + b - (a + c/2) = 8a/33$; PQ = $b + c$ = $17a/33$; which gives the triplet 8, 15, 17. $\triangle$AUQ is (3, 4, 5) and $\triangle$DUQ is (5, 12, 13).

Continue with the circle centered at R, with radius d. You have the equations

$$(2a - d)^2 = RV^2 + (a + DV)^2,$$
$$(c + d)^2 = (DV - DU)^2 + (QU - RV)^2,$$
$$(a + d)^2 = RV^2 + DV^2.$$

Proceed as before to obtain

$$DV = a - 3d,$$
$$RV = 2a/3 + 2d,$$
$$12d^2 - (16/3)ad + (4/9)a^2 = 0,$$

which gives the value $d = a/9$. Again, all the triangles are rational: $\triangle$AVR is (8, 15, 17), $\triangle$QJR is (20, 21, 29), $\triangle$DVR is (3, 4, 5). The triplet 20, 21, 29 suggests the useful approximation $\sqrt{2} = 29/20.5$.

This outcome leads to *Alison's conjecture*, so-called after the insightful geometer who made it, that if the series were continued, by drawing the tangent circles centered at S, T, ..., all the triangles involved would be rational. This conjecture amounts to guessing that the radii of the circles centered at S, T, etc., are rational fractions of a; for if they are, the sides of the triangles, being linear combinations of the radii, will be rational. We can test Alison's conjecture by obtaining a general formula for the radii, as follows.

In Fig. 6.5.5, let R and S be consecutive centers in the series and x_r, x_s and y_r, y_s, be their perpendicular distances from the lines DC, DB, respectively. The strategy is to obtain s (the radius of the circle centered on S) in terms of r (the radius of the circle centered on R) and y_r. The usual triangles yield the usual equations

$$(2a - s)^2 = (a + x_s)^2 + y_s^2,$$
$$(r + s)^2 = (x_s - x_r)^2 + (y_s - y_r)^2,$$
$$(a + s)^2 = x_s^2 + y_s^2. \tag{6.5.1}$$

Subtraction of the third equation from the first gives, as before,

$$x_s = a - 3s. \tag{6.5.2}$$

Fig. 6.5.5 Further to the same.

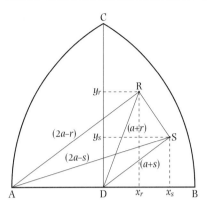

Subtraction of the second from the first gives

$$4as + 4ar - 10rs = y_r y_s. \qquad (6.5.3)$$

(To obtain this form you must use $x_r^2 + y_r^2 = (a + r)^2$ and $x_r = a - 3r$.)

Insert the values of x_s and y_s from equations (6.5.2) and (6.5.3) into the third of equations (6.5.1) to obtain a quadratic for s:

$$s^2 \left[\frac{(2a - 5r)^2}{y_r^2} + 2 \right] + as \left[\frac{4r(2a - 5r)}{y_r^2} - 2 \right] + \frac{4a^2 r^2}{y_r^2} = 0. \qquad (6.5.4)$$

You have from the third of equations (6.5.1), written for $s = r$,

$$y_r^2 = (a + r)^2 - (a - 3r)^2 = 8ar - 8r^2$$

Put this expression for y_r^2 into equation (6.5.4) and do some algebra. If you make no mistakes (Alison made several), you will obtain

$$s = \frac{ar(2a + r - y_r)}{2(a^2 - ar + 9r^2/4)}. \qquad (6.5.5)$$

The formula reproduces the results found earlier for c (= $2a/11$) and d (= $a/9 = 2a/18$). For $s = c$, $r = b = a/3$, $y_r = a + b = 4a/3$, $c = 2a/11$. For $s = d$, $r = c = 2a/11$, $y_r = (8r(a - r))^{1/2} = 12a/11$, and $d = a/9$, again as before. With this confirmation, there is good reason to regard equation (6.5.5) as correct.

We have found for the radii of the first three circles $b = a/3$, $c = 2a/11$, $d = a/9$, or, converted to fractions of a, 2/6, 2/11, 2/18. The numbers 6, 11, 18 satisfy the formula $n^2 + 2$, beginning with $n = 2$: $6 = 2^2 + 2$, $11 = 3^2 + 2$, $18 = 4^2 + 2$. A good guess is that the next next number in the series is 27 and that the fourth circle has a radius $e = 2a/27$. Try equation (6.5.5) with $r = d = a/9$, $y_r^2 = 8d(a - d) = 64a^2/81$, $y_r = 8a/9$:

$$e = \frac{1/9(19/9 - 8/9)a}{2(8/9 + 1/36)} = \frac{2a}{27}.$$

We have almost proved Alison's conjecture. We need only show that, assuming $r_n = 2a/(n^2 + 2)$, which is of course always a rational fraction, then equation (6.5.5) gives the proper expression for r_{n+1}. Otherwise stated, if, in equation (6.5.5), $2a/(n^2 + 2)$ is put for r, and the appropriate expression for y_r, then

$$s = \frac{2a}{(n+1)^2 + 2} = r_{n+1}.$$

First the appropriate expression for y_r:

$$y_r^2 = 8r(a - r) = \frac{16a^2}{n^2 + 2}\left(1 - \frac{2}{n^2 + 2}\right) = \frac{16a^2n^2}{(n^2 + 2)^2}.$$

Hence $y_r = 4an/(n^2 + 2)$. (Had you required $8r(a - r)$ to be a perfect square and set $r = 2a/x$, then $y_r^2 = (16a^2/x^2)(x - 2)$ and the only admissible values of x are $n^2 + 2$). Now for equation (6.5.5):

$$s = \frac{[4a^3/(n^2 + 2)^2][n^2 - 2n + 3]}{[2a^2/(n^2 + 2)^2][n^4 + 2n^2 + 9]} = \frac{2a}{(n+1)^2 + 2} = r_{n+1}.$$

The expression

$$r_n = \frac{2a}{(n^2 + 2)}$$

therefore holds in general; r_n is always a rational fraction; all the right triangles that come into play are rational; and the triangles with $a + r_m$ as hypotenuse are 'Platonic'.[5]

6.6. The Tantalus problem

In 1995 the *Washington Post* posed a geometrical puzzle that, for its apparent ease and real difficulty, might be called the Tantalus problem. Some readers who cracked their heads but not the problem complained that the puzzler had not supplied enough information for a solution. The United Press Syndicate, which had carried the quiz in which the puzzle first appeared, insisted it could be solved, but by a method too long to print. The man who set the puzzle, whose business is writing study guides for college aptitude tests, said that he had forgotten how to do it and could not repeat his lost performance. He had recourse to three dozen geometers, none of whom, he said, could find a solution. 'I contacted about 40 geniuses around the nation and they all gave me insights about the problem without being able to solve it.' With these insights and a weekend's labor he managed a solution, which requires a clever but non-intuitive expansion of the given figure.[6]

5. I.e., they express the triplet $(n^2 + 4, n^2 - 4, 4n)$, $n > 2$; cf. supra, Table 4.1.1.
6. Carolyn Hax. 'The genius solution', *Washington Post*, 13 September 1995, p. B.5.

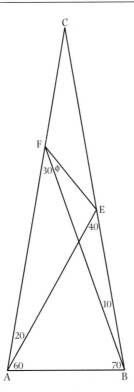

Fig. 6.6.1 The Tantalus
problem, version 1.

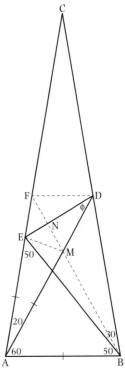

Fig. 6.6.2 The Tantalus
problem, version 2.

The problem is to find the value of the angle ϕ in Fig. 6.6.1 without recourse to trigonometry. As you see, $\triangle ABC$ is isosceles and ϕ is defined implicitly by the division of the base angles. The problem is frustrating not only because it appears to be elementary, but also because its solution is easily obtained by trigonometry (the law of sines) or from an accurate drawing. These non-Euclidean methods indicate that $\phi = 20°$. You will not be able to confirm the finding merely by addition and subtraction of the given angles because $\triangle ABC$ is not fully defined by them: it is but one of the class of isosceles triangles with base angles of 80°. To show that $\phi = 20°$ irrespective of the size of $\triangle ABC$, you will have to invoke lines as well as angles. The existence of a unique trigonometric solution indicates that ϕ is independent of the scale.

A similar but easier version of the problem divides the base angles into 50°, 30°, and 60°, 20° (Fig. 6.6.2). Now, as you will discover if you do the construction carefully, $\phi = 30°$. To prove it, you need to involve ϕ in some triangles other than the given ones. An obvious move is to draw FD parallel to AB, which makes $\angle FDA = 60°$; since we know by measurement that ϕ is probably 30°, DE must bisect $\angle FDA$. You need only to show it. Shut up the problematic angle in

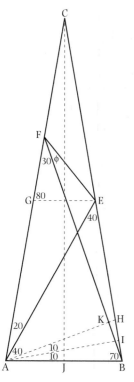

Fig. 6.6.3 Further to Fig. 6.6.1.

an equilateral triangle by drawing BF intersecting AD at M. The remaining challenge is to prove $\triangle EFD \cong \triangle EMD$, which comes down to showing that FE = EM.

It is really very easy once you think to connect E and M. You know that $\triangle EAB$ is isosceles with sides equal to the base of $\triangle ABC$. But $\triangle AMB$ is equilateral; therefore AE = AM and $\triangle AEM$ is isosceles with base angles of 80° (and thus is similar to the parent triangle ABC). Since $\angle BMN = 180°$, $\angle EMN = 180° - 80° - 60° = 40°$; and, since $\angle AFB = \angle ADB = 40°$, $\triangle FEM$ is also isosceles. Hence EF = EM and $\triangle EFD \cong \triangle EMD$: ED bisects the 60° angle FDM and $\phi = 30°$.[7]

Thus encouraged, let us return to the so-called genius problem. Reasoning as before, capture ϕ in $\triangle EFG$ by drawing GE parallel to AB (Fig. 6.6.3). Then $\angle CGE = 80°$ and $\angle AFB = 180° - 80° - 70° = 30°$. Since you know by measurement and trigonometry that ϕ is probably 20°, $\triangle GEF$ should be isosceles with base angles of 50°. To prove it, you must show that GE = GF. That can be done straightforwardly by calculating GE and GF separately.

The approach to GE is obvious. From the similar triangles GEC and ABC, GE : b = CE : a, where b is the base and a either of the sides of $\triangle ABC$. Evidently CE = a − BE. What is not so evident is that to find BE you should draw AH so that $\angle HAB = 20°$. Then $\triangle BAH$ is isosceles (AB = AH = b) and similar to $\triangle ABC$: hence HB : b = b : a or HB = b^2/a. But, since $\angle EAH = 60° - 20° = 40°$, $\triangle AEH$ is isosceles, whence EH = AH = b. Therefore BE = HB + EH = $b^2/a + b$ and

$$GE = bCE/a = (b/a)(a - BE) = (b/a)(a - b - b^2/a),$$
$$GE = b(1 - b/a - b^2/a^2).$$

That leaves GF = AF − AG = AF − BE. Since we know BE, we need only AF. Here we can exploit the right angle happily and unexpectedly created at K when we made the isosceles triangle ABH. Since $\angle AFB = 30°$, AF = 2AK (the side opposite the 30° angle in a 30°–60° right triangle is half the hypothenuse, as you can demonstrate in an impeccably Euclidean manner by drawing any altitude of an equilateral triangle). Now AK = AH − KH = b − KH. Drop the altitudes CJ and AI; you have two 10°–80° right triangles, from whose similarity you deduce KH : HB = b/2a or KH = $(b/2a)(b^2/a) = b^3/2a^2$. Therefore AK = $b(1 - b^2/2a^2)$ and

$$GF = 2AK - BE = 2b(1 - b^2/2a^2) - (b^2/2a + b),$$
$$GF = b(1 - b/a - b^2/a^2).$$

We have GE = GF, $\triangle GEF$ is isosceles, $\phi + 30° = 50°$, $\phi = 20°$, Q.E.D.

7. Cf. Coxeter and Greitzer, *Geometry revisited* (1967), 26.

When Alison (see § 6.5) read this proof, she observed that it demonstrated one thing very well: it does not take 40 geniuses, or even one, to do problems in plane geometry. All that is required is patience, resourcefulness, and the methods taught in this book.

Bibliography

AHES   Archive for history of exact sciences
HSPS   Historical studies in the physical sciences
MT   Mathematics teacher
NCTM   National Council of Teachers of Mathematics

AASHTO. American Association of State Highway and Transportation Officials. *A policy on geometric design of highways and streets.* AASHTO, Washington, 1984.

Alberti, Leon Battista. *The ten books of architecture.* Ed. James Leoni [1755]. Dover, New York, 1986.

Adams, William Howard. *Jefferson's Monticello.* Abbeville, New York, 1983.

Amma, T.A. Saraveti. *Geometry in ancient and medieval India.* Motilal Banersi dass, Delhi, 1979.

Archimedes. *The works.* Ed. T.L. Health. Dover, New York, 1952.

Arnauld, Antoine. *Nouveaux élémens de géometrie.* C. Savreux, Paris, 1667.

Arnauld, Antoine. *The Port-Royal logic.* Tr. and ed. T.S. Baynes. 2nd edn. Sutherland and Knox, Edinburgh, 1851.

Artmann, Benno. The cloisters of Hauterive. *The mathematical intelligencer,* 13:2 (1991), 44–9.

Bag, A.K. *Mathematics in ancient and medieval India,* Chaukhambha, Delhi, 1979.

Ball, W.W. Rouse. *A history of the study of mathematics at Cambridge.* Cambridge University Press, Cambridge, 1889.

Barrow, Isaac. *Euclide's elements. The whole fifteen books compendiously demonstrated.* R. Daniel, for Willson Nealand, London, 1660.

Bartol, William C. *The elements of solid geometry.* Leach, Shewell, and Sanborn, Boston, 1893.

Bartoli, Lando. *Il disegno della cupola del Brunelleschi.* Olschki, Florence, 1994.

Beatley, Ralph. Third report of the committee on geometry. *MT, 28* (Oct–Nov 1935), 329–79, 401–50.

Becher, Harvey W. William Whewell and Cambridge mathematics. *HSPS, 11* (1980), 1–48.

Beck, James. *Raphael. The stanza della segnatura, Rome.* Braziller, New York, 1993.

Beckmann, Petr. *A history of π.* Saint Martin's Press, New York, 1971.

Beman, Wooster Woodruff, and David Eugene Smith. *New plane and solid geometry.* Ginn, Boston, 1903 (1st edn. 1895).

Bennett, James A. *The divided circle. A history of instruments for astronomy, navigation and surveying.* Phaidon-Christie's, Oxford, 1987.

Bennett, James A., and Stephen Johnstone. *The geometry of war, 1500–1750.* Museum for History of Science, Oxford, 1996.

Bergery, C.L. *Géométrie appliquée à l'industri à l'usage des artistes et les ouvriers.* 2nd edn. Thiel, Metz. 1828.

Berggren, J.L. *Episodes in the mathematics of medieval Islam.* Springer, New York, 1986.

Beretta, Marco. *The starry messenger and the polar star. Scientific relations between Itay and Sweden from 1500 to 1800.* Giunti, Prato, 1995.

Berthaut, H.M.A. *Les ingénieurs géographes miliaires, 1624–1831.* 2 vols. Imprimerie du Service géographique, Paris, 1902.

Bettini, Mario. *Aerarium philosophiae mathematicae.* C.G. Ferroni, Bologna, 1648.

Betz, William. Five decades of mathematical reform. *MT, 43* (Dec 1950), 377–87.

Binding, Heinrich. *Masswerk.* Wissenschaftliche Buchgesellschaft, Darmstadt, 1989.

Blackhurst, J.H. *Humanized geometry.* Iowa University Press, Des Moines, 1935.

Bobynin, V. Elementare Geometrie. In Cantor, *Vorl., 4* (1907), 319–402.

Borkland, C.W. *The Department of Defense.* Praeger, New York, 1968.

Bony, Jean. *French gothic architecture of the 12th and 13th centuries.* University of California Press, Berkeley, 1983.

Bourgoin, J. *Arabic geometrical pattern and design.* Dover, New York, 1973. (Reprint of the plates from Bourgoin, *Les éléments de l'art arabe: le trait des entrelacs.* Firmin-Didot, Paris, 1879.)

Bouton, Charles, L. Provisional reports of the National Committee of Fifteen on geometry syllabus. *School science and mathematics, 11* (1911), 329–55, 434–60, 509–31.

Bowser, E.A. *The elements of plane and solid geometry.* 2nd edn. D.C. Heath, Boston, 1891.

Branner, Robert. Villard de Honnecourt, Archimedes, and Chartres. Society of Architectural Historians. *Journal, 19* (1960), 91–6.

Brugsch-bey, Heinrich. *Steininschrift und Bibelwort.* 2nd edn. Allgemeiner Verein für deutsche Litteratur, Berlin, 1891.

Bruins, E.M. Quelques textes mathématiques de la mission de Suse. Akademie der Wetenschappen, Amsterdam. *Proceedings, 53* (1950), 1025–33.

Busard, H.L.L. *The Latin translation of the Arabic version of Euclid's* Elements *commonly ascribed to Gerard of Cremona.* Brill, Leiden, 1984.

Byerly, W.E. *Chauvenet's treatise on elementary geometry (plane) revised and abridged.* J.B. Lippincott, Philadelphia, 1901.

Byrne, Oliver. *The first six books of the elements of Euclid in which coloured diagrams and symbols are used instead of letters for the greater ease of learners.* William Pickering, London, 1847.

Cajori, Florian. *Mathematics in liberal education.* Christopher, Boston, [1928].

Calinger, Ronald, ed. *Vita mathematica. Historical research and integration with teaching.* Mathematical Association of America, Washington, D.C., 1996.

Campbell, Colin. *Two Theban princes ..., a land steward, and their tombs.* Oliver and Boyd, Edinburgh, 1910.

Cantor, Moritz. *Vorlesungen über Geschichte der Mathematik.* 4 vols. Teubner, Leipzig, 1907–08.

Carli, Enzo. *La Piazza del Duomo di Pisa.* Gherardo Casini, Rome, 1956.

Carroll, Lewis. *Euclid and his modern rivals.* 2nd edn. Macmillan, London 1885. Reprint, ed. H.S.M. Coxeter, Dover, New York, 1973.

Chace, Arnold Buffum, L. Bull, H.P. Manning, and R.C. Archibald. *The Rhind mathematical papyrus.* 2 vols. Mathematical Association of America, Oberlin, Ohio, 1927–29.

Chasles, Michel. *Aperçu historique sur l'origine et le développement des méthodes en géométrie.* Hayez, Brussels, 1837. Reprinted, Babay, Paris, 1989.

Chippendale, Christopher. *Stonehenge complete.* Cornell University Press, Ithaca, 1983.

Christofferson, Halbert C. *Geometry professionalized for teachers.* Christofferson, Oxford, Ohio, 1933.

Clarke, Somers, and R. Engelbach. *Ancient Egyptian construction and architecture.* Dover, New York, 1990.

Clavius, Christopher. *Euclidis Elementorum libri XV.* 2 vols. Vincenzo Accoli, Rome, 1574.

Combette, E. *Cours abrégé de géométrie élémentaire.* F. Alcan, Paris, 1898.

Committee of Ten. *Report on secondary school studies with reports of the conferences arranged by the committees.* American Book Company, New York, 1894.

Coolidge, Julian Lowell. *A history of geometrical methods.* Oxford University Press, Oxford, 1940.

Coolidge, Julian Lowell. *A treatise on the circle and the sphere.* Oxford University Press, Oxford, 1916.

Cooper, Lane. *Aristotle, Galileo, and the tower of Pisa.* Cornell University Press, Ithaca, 1935.

Coxeter, H.S.M., and S.L. Greitzer. *Geometry revisited.* Random House, New York, 1967.

Crowe, Michael J. *The extraterrestrial life debate 1750–1900. The idea of a plurality of worlds from Kant to Lowell.* Cambridge University Press, Cambridge, 1986.

Curtze, Maximilian. Die inquisicione capacitatis figurarum. Anonyme Abhandlung aus dem funftzehnten Jahrhundert. *Abhandlungen zur Geschichte der Mathematik, 8* (1898), 29–68.

Dante Alighieri. *The Divine comedy.* Ed. C.S. Singleton. 3 vols. in 6. Princeton University Press, Princeton, 1970–75.

Davies, Norman de Garis. *The tombs of two officials of Tuthmosis the fourth.* Egypt Exploration Society, London, 1923.

Davis, William. *Elementa. The elements of plane geometry; or, the first six books of Euclid.* London: 1872, 1874.

Deakin, Rupert. *Euclid, Books I, II.* W.B. Clive, London, 1903.

Dee, John. The mathematical preface. In H. Billingsley, *The elements of geometry of the most auncient Philosopher Euclide of Megara.* John Daye, London, 1570.

Delgado, Pat, and Colin Andrews. *Circular evidence. A detailed investigation of the flattened swirled crops phenomenon.* Bloomsbury, London, 1989.

De Morgan, Augustus. *A budget of paradoxes* [1872]. Reprinted, 2 vols. Dover, New York, 1954

Descartes, René. *Correspondance avec Elisabeth et autres lettres.* Flammarion, Paris, 1989.

Descartes, René. *Oeuvres.* Ed. Charles Adam and Paul Tannery. Vol. 6. *Discours de la méthode et essais.* Vrin, Paris, 1982.

Dijksterhuis, E.J. *Archimedes.* Princeton University Press, Princeton, NJ, 1987.

Dodgson, Charles L. *[Euclid's Elements,] Books I, II.* 2nd edn. Macmillan, London, 1883.

Dörrie, Heinrich. *100 great problems of elementary mathematics. Their history and solution.* Dover, New York, 1965.

Duarte, Francisco José M. *Bibliografiá: Euclides, Arquímedes, Newton.* Academia de ciencias, Caracas, 1967.

Eperson, D.B. Lewis Carrol—mathematician. *Mathematical gazette, 17* (May 1933), 92–100.

Euclid, *Elementa.* E. Ratdolt, Venice, 1482.

Euclid. *Euclidis quae supersunt omnia.* Ed. David Gregory. The Sheldonian, Oxford, 1703.

Euclid. *Elements of geometry. Book I. With an introduction from an essay by Paul Valéry.* Random House, New York, 1944.

Euler, Leonhard. 'Solutio facilis problematum quorundam geometricorum difficillimorum' [1765]. In Euler, *Opera omnia,* ser. 1, vol. 26. Füssli, Lausanne, 1953. Pp. 139–57.

Fernández-Armesto, Felipe. *Columbus.* Oxford University Press, Oxford, 1991.

Fichtner, Richard. *Die verborgene Geometrie in Raffaels 'Schule von Athen'*. Oldenbourg, Munich, 1984.

Fine, Oronce. *Protomathesis*. Paris, n.p., 1532.

Foster, Richard. *Patterns of thought. The hidden meaning of the great pavement of Westminster Abbey*. Jonathan Cape, London, 1991.

Frängsmyr, Tore, J.L. Heilbron, and Robin Rider, eds. *The quantifying spirit in the Eighteenth Century*. University of California Press, Berkeley, 1990.

Franklin, Benjamin. *Autobiographical writings*. Ed. Carl van Doren. Viking, New York, 1945.

Gerdes, Paulus. On culture, geometrical thinking and mathematics education. *Educational studies in mathematics*, 19 (1988), 137–162.

Gerdes, Paulus. On possible uses of traditional Angolan sand drawings in the mathematics classroom. *Educational studies in mathematics*, 19 (1988), 3–22.

G.-M., F. *Exercises de géométrie. Comprenant l'exposé des méthodes géométriques et 2000 questions resolues*. A. Mame, Tours, J. de Gigord, Paris, 1920.

Ghyka, Matila. *The geometry of art and life*. Dover, New York, 1977.

Gille, Bertrand. *Engineers of the Renaissance*. MIT Press, Cambridge, MA, 1966.

Gillet, J.A. *Euclidean geometry*. Henry Holt, New York, 1896.

Gillings, Richard J. *Mathematics in the time of the pharaohs*. 2nd edn. Dover, New York, 1982.

Grünbaum, Branko, and G.C. Shephard. *Tilings and patterns. An introduction*. Freeman, New York, 1989.

Guillemin, Amadée. *The forces of nature*. Ed. J.N. Lockyer. Macmillan, London, 1877.

Guinness, Desmond, and Julius T. Sadler, Jr. *Mr. Jefferson. Architect*. Viking Press, New York, 1973.

Gupta, R.C. Paramesvara's rule for the circumradius of a cyclic quadrilateral. *Historia mathematica*, 4 (1977), 67–74.

Hadamard, Jacques. *Leçons de géométrie élémentaire. I. Géométrie plane*. 11th edn. A. Collin, Paris, 1931.

Hall, Asaph. The discovery of the satellites of Mars. Royal Astronomical Society. *Monthly notices*, 38 (1878), 205–8.

Halsted, George Bruce. *An elementary treatise on mensuration*. Ginn, Heath, Boston, 1881.

Hammer, E. Zur Geschichte der arabischen Geodäsie. *Zeitschrift für Vermessungswesen*, 38 (1909), 721–7.

Harrison, J. School geometry reform. *Nature*, 67 (1903), 577–8.

Heath, Thomas Little. *Apollonius of Perga*. Barnes and Noble, New York, 1961.

Heath, Thomas Little. *Greek astronomy*. J.M. Dent, London, 1932.

Heath, Thomas Little. *A history of Greek mathematics*. 2 vols. Clarendon Press, Oxford, 1921.

Heath, Thomas Little. *The thirteen books of Euclid's elements*. 2nd edn. 3 vols. Cambridge University Press, Cambridge, 1926.

Hedgecoe, John, and Salma Samar Damluji. *Zillig. The art of Moroccan ceramics*. Garnet, Reading, 1992.

Heilbron, J.L. The measure of Enlightenment. In Frängsmyr *et al.*, *The quantifying spirit* (1990), 207–42.

Heilbron, J.L. *Weighing imponderables and other quantitative science around 1800*. University of California Press, Berkeley, 1993. (*Historical studies in the history of the physical and biological sciences*, 24:1, suppl.)

Henrici, O.F.E. *Elementary geometry*. Longmans Green, London, 1879.

Herodotus. *The history*. Ed. G. Rawlinson et al. 2nd edn. 4 vols. John Murray, London, 1858.

Hilbert, David. *Foundations of geometry*. 2nd edn. Open Court, La Salle, IL, 1971.

Hill, Henry. *The six first, together with the eleventh and twelfth books of Euclid's elements, demonstrated after a new, plain, and easie method.* William Pearson, for R. and J. Bonwicke *et al.*, London, 1726.

Hobson, Ernest Wilson. *Squaring the circle: A history of the problem.* Cambridge University Press, Cambridge, 1913.

Hogben, Lancelot. *Mathematics for the million.* Norton, New York, 1937.

Hölscher, Uvo. *Das Grabdenkmal des Königs Chephren.* J.C. Hinrichs, Leipzig, 1912.

How, W.W., and J. Wells. *A commentary on Herodotus.* 2 vols. Oxford University Press, Oxford, 1912.

Hunrath, Karl. Des Rheticus Canon doctrinae triangulorum und Vieta's Canon mathematicus. *Abhandlungen zur Geschichte der Mathematik, 9* (1899), 211–40.

Hutton, Charles. *The Diarian Miscellany: Consisting of all the useful and entertaining parts both mathematical and poetical, extracted from the* Ladies' diary. 5 vols. Robinson and Baldwin, London, 1775.

Huygens, Christiaan. *De circuli magnitudine inventa* (1654). In Huygens, *Oeuvres complètes.* 22 vols. Nijhoff, The Hague, 1888–1950. Vol. 12, pp. 113–215.

Jha, Parmeshwar. *Aryabhata I and his contributions to mathematics.* Biher Research Society, Patna, 1988.

Jones, Phillips. Angular measure—enough of its history to improve its teaching. *MT, 46* (1953), 419–26.

Jones, Phillips. Early American geometry. *MT, 37* (Jan 1944), 3–11.

Jordan, H. Die Gradmessung der Araber, 827 nach Chr. *Zeitschrift zur Vermessungswesen, 18* (1889), 100–9.

Joseph, George Gheverghese. *The crest of the peacock. Non-European roots of mathematics.* Penguin, London, 1991.

Kästner, A.G. *Geometrische Abhandlungen.* 2 parts. Vandenhoek and Ruprecht, Göttingen, 1790–91.

Kilmer, Ann. Summerian and Akkadian names for designs and geometric shapes. In Ann C. Gunter, ed. *Investigating artistic environments in the ancient Near East.* Smithsonian Institution, Washington, D.C., 1990. Pp. 83–91.

Kilwardby, Robert. *De ortu scientiarum.* Ed. A.G. Judy. British Academy, London, 1976.

Kitao, Timothy K. *Circle and oval in the square of Saint Peters.* New York University Press, New York, 1974.

Kline, Morris. *Mathematical thought from ancient to modern times.* 3 vols. Oxford University Press, Oxford, 1990.

Kochansky, Adam Amandus. Observationes cyclometricae ad facilitandem praxim accommodatae. *Acta eruditorum,* 1685, 394–8.

Kulkarni, R.P. *Layout and construction of citis according to Baudhāyanā, Mānava, and Āpastamba Śulbasūtras.* Bhandarkar Oriental Research Institute, Poona, 1987.

Kusserow, Wilhelm. *Los von Euklid* [1928]! Reprint, F. Schöningh, Paderborn, 1985.

Land, Frank. *The language of mathematics.* Doubleday, New York, 1963.

Lam, Lay-Yong, and Shen Kangsheng. Right angled triangles in ancient China. *AHES, 30* (1984), 87–112.

Lamartine, A.M.L. de Prat de. Les destinées de la poésie [1834]. In *Oeuvres.* Maline, Cans, Brussels, 1838. Pp. 397–412.

Lardner, Dionysius. *The first six books of the Elements of Euclid with a commentary and geometrical exercises* [1828]. Bohn, London, 1855.

Lavoisier, A.L. *Oeuvres.* Parte II. *Mémoires de chimie et de physique,* Vol. III. Imprimerie royale, Paris, 1862.

Leeke, John, and George Serle, eds. *Euclid's elements of geometry.* Leybourn, London, 1661.

Legendre, A.M. *Elements of geometry*. Tr. John Farrar. Cambridge University Press, Cambridge, 1825.

Leybourn, Thomas. *The mathematical questions proposed in the* Ladies' diary, *and their original answers*. 4 vols. J. Mawman, London, 1817.

Libbrecht, Ulrich. *Chinese mathematics in the thirteenth century: The Shu-Shu Chiu-Chang of Ch'in Chiu-Shao*. MIT Press, Cambridge, MA, 1973.

Loewe, Michael. *Early Chinese texts: A bibliographical guide*. University of California, Institute of East Asian Studies, Berkeley, 1993.

Lorch, Richard. Abū Kāmil on the pentagon and decagon. In M. Folkerts and J.P. Hogendijk, eds. *Vestigia mathematica. Studies in medieval and early modern mathematics in honour of H.L.L. Busard*. Rodopi, Amsterdam, 1993. Pp. 215–52.

MacCurdy, Edward. *The notebooks of Leonardo da Vinci*. 2 vols. Reynal and Hitchcock, New York, 1938.

Malton, Thomas. *The new royal road to geometry, and familiar introduction to the mathematics. Part. I. Elements of geometry abridged. Containing the substance of Euclid's first six, the eleventh and twelfth books*. 2nd edn. for the author, London, 1793.

Martzloff, J.C. *Histoire des mathématiques chinoises*. Masson, Paris, 1988.

McLaughlin, Jack. *Jefferson of Monticello*. Henry Holt, New York, 1988.

Melnick, Mimi. *Manhole covers*. MIT Press, Cambridge, MA, 1994.

Millon, Henry A., and V.M. Lampugnani, eds. *The Renaissance from Brunelleschi to Michelangelo. The representation of architecture*. Bompiani, Milan, 1994.

Minchin, George M. Geometry *versus* Euclid. *Nature*, 59 (1899), 369–70.

Moxon, Joseph. *A tutor to astronomy and geography*. Moxon, London, 1686.

Naber, Henri Adrien. *Das Theorem des Pythagoras*. P. Visser, Haarlem, 1908.

Müller, Johann Heinrich. *Das regulierte Oval. Zu den Ovalkonstruktionen im Primo libro di architettura des Sebastiano Serlio*. Bremen, 1967.

Needham, Joseph. *The grand titration. Science and society in East and West*. George Allen & Unwin, London, 1979.

Needham, Joseph. *Science and civilization in China*. Vol. 3. *Mathematics and the sciences of the heavens and the earth*. Cambridge University Press, Cambridge, 1959.

Nelson, David, George Gheverghese Joseph, and Julian Williams. *Multicultural mathematics*. Oxford University Press, New York, 1993.

Neugebauer, Otto. *Vorlesungen über Geschichte der antiken mathematischen Wissenschaften*. Vol. 1. *Vorgriechische Mathematik*. Springer, Berlin, 1934.

Neugebauer, Otto. *The exact sciences in antiquity*, 2nd edn. Brown University Press, Providence, RI, 1957.

Neugebauer, Otto, and A. Sachs. *Mathematical cuneiform texts*. American Oriental Society, New Haven, CT, 1945.

Noyes, Ralph, ed. *The crop circle enigma*. Gateway, Bath, 1990.

Palmer, Claude Irwin, and Daniel Pomeroy Taylor. *Plane geometry*. Scott Forsman, Chicago, 1915.

Pennant, Thomas. *Journey from Chester to London*. B. White, London, 1782.

Perry, John. *England's neglect of science*. T.F. Unwin, London, 1900.

Pitiscus, Bartholomeus. *Trigonometriae, sive De dimensione triangulorum libri quinque*. 3rd edn. J. Rosa, Frankfurt, 1612.

Playfair, John. *Elements of geometry* [1813]. Lippincott, Philadelphia, 1864.

Pressland, Arthur John. *An introduction to the study of geometry*. Rivingtons, London, 1904.

Proclus. *A commentary on the first book of Euclid's Elements*. Trans. Glenn R. Morrow. Princeton University Press, Princeton, NJ, 1970.

Quintilian. *Institutio oratoria*. 4 vols. Ed. H.E. Butler. Harvard University Press, Cambridge, MA, 1920.

Rebière, Alphonse. *Mathématiques et mathématiciens*. 7th edn. Vuibert, Paris, 1926.

Reisch, Gregor. *Margarita philosophica, totius philosophiae rationalis, naturalis et moralis principia ... complectans.* N.p., n.p., 1512.

Riccardi, Pietro. *Saggio di una bibliografia euclidea.* 5 parts. Gamberini and Paermggiani, Bologna, 1887–93.

Rickey, V. Frederick. The necessity of history in teaching mathematics. In Calinger, *Vita mathematica* (1996), 251–6.

Robins, Gay, and Charles Shute. *The Rhind mathematical papyrus.* British Museum, London, 1987.

Rossiter, Margaret W. *Women scientists in America. Struggles and strategies to 1940.* Johns Hopkins University Press, Baltimore, MD, 1982.

Schiebinger, Londa. *The mind has no sex? Women in the origins of modern science.* Harvard University Press, Cambridge, MA, 1989.

Schubring, Paul. *Pisa.* Seeman, Leipzig, 1902.

Seherr-Thoss, Sonia P. *Design and color in Islamic architecture: Afghanistan, Iran, Turkey.* Smithsonian Institution, Washington, 1968.

Seidenberg, A. The ritual origin of geometry. *AHES, 1* (1962), 458–527; *2* (1963), 1–40.

Shea, William R., ed. *Storia delle scienze.* Vol. 2. *Le scienze fisiche e astronomiche.* Einaudi, Turin, 1992.

Shelby, L.R. *Gothic design techniques, the fifteenth century design booklets of M. Roriczer and H. Schmuttermayer.* Southern Illinois Press, Carbendale, IL, 1977.

Skyring Walters, R.C. Greek and Roman engineering instruments. Newcomen Society. *Transactions, 2* (1921), 45–60.

Slaught, H.E., and N.J. Lennes. *Plane geometry.* Allyn and Bacon, Boston, MA, 1918.

Smith, David Eugene. *History of mathematics.* 2 vols. Dover, New York, 1958.

Spencer, William George. *Inventional geometry.* Appleton, New York, 1877.

Steck, Max. *Dürers Gestaltlehre der Mathematik und der bildenden Künste.* M. Niemayer, Halle, 1948.

Steck, Max. *Bibliographia euclidiana.* Gerstenberg, Hildesheim, 1981.

Stone, E. *Euclid's Elements of geometry.* 2nd edn. T. Payne, London, 1763.

Swetz, Frank J. *The sea island mathematical manual: Surveying and mathematics in ancient China.* Pennsylvania State University Press, University Park, PA, 1992.

Swetz, Frank, and T.I. Kao. *Was Pythagoras Chinese? An examination of right triangle theory in ancient China.* Pennsylvania State University Press, University Park, PA, 1977.

Tacquet, André. *Elementa geometriae planae et solidae.* J. Mersius, Antwerp, 1654.

Tacquet, André. *Elementa Euclidea geometriae planae, solidae ... plurimis corallariis, notis, ac schematibus quadraginta illustrata a Guilielmo Whiston.* Typ. Premondiana, Venice, 1762.

Tannery, Paul. *Mémoires scientifiques.* Ed. J.L. Heiberg and H.G. Zeuthen. Vol. 5. E. Privat, Toulouse, 1922.

Tartaglia, Nicolò. *Euclide megarense.* G. Barilelto, Venice, 1569.

Thibaut, G. On the sulvasutras. Asiatic Society of Bengal. *Journal, 44* (1875), 227–75.

Thomas-Stafford, Charles. *Early editions of Euclid's Elements.* Bibliographical Society, London, 1926.

Thomson, James. The term 'radian' in trigonometry. *Nature, 83* (1910), 217.

Thureau-Dangin, F. *Textes mathématiques babyloniens.* Brill, Leyden, 1938.

Todhunter, Isaac. *The elements of Euclid for the use of schools and colleges; comprising the first six books and portions of the eleventh and twelfth books; with notes, an appendix, and exercises.* Macmillan, London, 1884.

Tompkins, Peter. *Secrets of the great pyramid.* Harper and Row, New York, 1971.

Tournes, Jean de. *Les six premiers livres des Elements géométriques d'Euclides. Avec les demonstratons de Jacques Peletier du Mans*. Tr. en français, et dédiés à la noblesse françoise. Jean de Tournes, Geneva, 1611.

Valéry, Paul. La crise de l'espirit. In Valéry, *Variété*. Gallimard, Paris, [1924]. Pp. 9–49.

Vitruvius. *The ten books of architecture*. Tr. M.H. Morgan. Harvard University Press, Cambridge, MA, 1914.

Wakeling, Edward. The centenary of *Euclid and his modern rivals*. *Jabberwocky, 8:3* (1979), 51–5.

Warren, S.E. *A primary geometry ... suited to all beginners*. J. Wiley, New York, 1887.

Watson, Foster. *The beginning of the teaching of modern subjects in England*. Pitman, London, 1909.

Weaver, Warren. Lewis Carroll as a geometrical paradox. *American mathematical monthly, 45* (1938), 234–6.

Wentworth, G.A. *A textbook of geometry*. Ginn, Boston, MA, 1888, 1895.

Wentworth, G.A. *Plane and solid geometry*. Revised edn. Ginn, Boston, 1899.

Wentworth, G.A., and David Eugene Smith. *Revised plane geometry*. Athenaeum Press, Boston, 1910.

Whewell, William. *History of scientific ideas*. 2 vols. Parker and Son, London, 1858.

Williamson, James. *The elements of Euclid, with dissertations*. 2 vols. Clarendon Press, Oxford, 1781; T. Spilsbury, London, 1788.

Wilson, Eva. *North American Indian designs for artists and craftspeople*. Dover, New York, 1984.

Wilson, Eva. *Islamic designs*. Dover, New York, 1988.

Wilson, Henry, ed. *Euclid's Elements*. J. and B. Sprint, London, 1722.

Winter, Franz. Der Tod des Archimedes. Archäologische Gesellschaft, Berlin. *Winkelmannsprogram, 82* (1924).

Index

Bold numbers denote references to illustrations; 't' following a number denotes reference to a table.